Phytochemicals in Medicinal Plants

Also of Interest

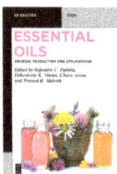

Essential Oils.
Sources, Production and Applications
Padalia, Verma, Arora, Mahish (Eds.), 2023
ISBN 978-3-11-079159-4, e-ISBN (PDF) 978-3-11-079160-0

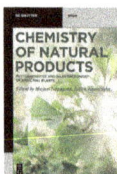

Chemistry of Natural Products.
Phytochemistry and Pharmacognosy of Medicinal Plants
Napagoda, Jayasinghe (Eds.), 2022
ISBN 978-3-11-059589-5, e-ISBN (PDF) 978-3-11-059594-9

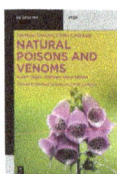

Natural Poisons and Venoms.
Plant Toxins: Terpenes and Steroids
Teuscher, Lindequist (Eds), 2023
ISBN 978-3-11-072472-1, e-ISBN (PDF) 978-3-11-072473-8

Natural Poisons and Venoms.
Animal Toxins
Teuscher, Lindequist (Eds), 2023
ISBN 978-3-11-072854-5, e-ISBN (PDF) 978-3-11-072855-2

Drug Delivery Technology.
Herbal Bioenhancers in Pharmaceuticals
Pingale (Ed.), 2022
ISBN 978-3-11-074679-2, e-ISBN (PDF) 978-3-11-074680-8

Phytochemicals in Medicinal Plants

Biodiversity, Bioactivity and Drug Discovery

Edited by
Charu Arora, Dakeshwar Kumar Verma, Jeenat Aslam
and Pramod Kumar Mahish

DE GRUYTER

Editors

Charu Arora
Department of Chemistry
Guru Ghasidas University
Koni, Bilaspur 495009
Chhattisgarh
India

Jeenat Aslam
Department of Chemistry
College of Science
Taibah University
Al-Madina
Yanbu 30799
Saudi Arabia

Dakeshwar Kumar Verma
Government Digvijay Autonomous
Postgraduate College
Rajnandgaon 491441
Chhattisgarh
India

Pramod Kumar Mahish
Government Digvijay Autonomous Postgraduate
College
Rajnandgaon 491441
Chhattisgarh
India

ISBN 978-3-11-079176-1
e-ISBN (PDF) 978-3-11-079189-1
e-ISBN (EPUB) 978-3-11-079194-5

Library of Congress Control Number: 2023932370

Bibliographic information published by the Deutsche Nationalbibliothek
The Deutsche Nationalbibliothek lists this publication in the Deutsche Nationalbibliografie;
detailed bibliographic data are available on the Internet at http://dnb.dnb.de.

© 2023 Walter de Gruyter GmbH, Berlin/Boston
Cover image: S847/iStock/Getty Images Plus
Typesetting: Integra Software Services Pvt. Ltd.
Printing and binding: CPI books GmbH, Leck

www.degruyter.com

Preface

It has been demonstrated that medicinally significant plants play a vital role in maintaining people's and societies' health. There are several chemical constituents responsible for their therapeutic value that causes a specific physiological change in the human body. Alkaloids, tannins, flavonoids, and phenolic compounds are the chemical components of plants with significant potential as therapeutic agent. The study of plants is the one of oldest science. Among various areas of plant biology, plant phytochemicals have been proven enormous utility, especially in the medical and nutritional aspects. Therefore, identification, characterization, extraction, and utilization of biomolecules from plant sources are the focused area of the scientific community. Further advances in this speedily growing field of knowledge will no doubt lead to even more remarkable improvements in our aptitude to understand nature. The book emphasizes several themes like the diversity of medicinal plants, phytochemical constituents of plants and their characterization, the bioactivity of phytochemicals, nanotechnology, and drug discovery based on plant phytochemicals. Based on these themes, various chapters have been received from authors around the globe. Under these themes, several aspects of plant phytochemicals like the distribution and biodiversity of plants worldwide; threats and conservational strategies of medicinal plants; and plant phytochemicals like alkaloids, flavonoids, terpenoids, steroids, phenolic compounds, etc. have been uncovered. Characterization techniques used for the analysis of phytochemicals such as photometric methods, microscopic methods, and chromatographic methods have also been discussed in the book. Additionally, the book covers the bioactivity of plant's phytochemicals such as anticancerous, anti-HIV, antimicrobial, anti-inflammatory, wound healing, antioxidant, anti-corona, and other viruses. Some recent advances in the field of plant phytochemicals like nanomaterial synthesis from medicinal plants, the discovery of drugs through plant phytochemicals, etc. have also been placed in the book.

All these chapters have been organized with recent aspects and development of the field with the aid of suitable tables and figures. The book has been conceived to provide fundamental applied knowledge of plant biomolecules, which targets a larger audience like students of undergraduate and postgraduate, researchers, academicians, environmentalists, policymakers, industries, R&D, etc. The text has been prepared by esteemed authors; therefore, we do not claim for our original writing. We thankfully acknowledge all the authors. We shall welcome suggestions for improvement in the next book/edition.

Editors,
Charu Arora
Jeenat Aslam
Dakeshwar Kumar Verma
Pramod Kumar Mahish

https://doi.org/10.1515/9783110791891-202

Contents

About the editors

Prof. Charu Arora is Professor of Physical Chemistry at Guru Ghasidas University (A Central University), Bilaspur, Chhattisgarh, India. She has 21 years of teaching and research experience. She has served as Head, Department of Chemistry at Guru Ghasidas University and as Division Chair, Division of Nanochemistry at Galgotias University, Greater Noida, and several other reputed institutes and universities. Her areas of interest and research are environmental chemistry, kinetics, and natural product chemistry. She has completed her Ph.D. and postgraduation in Chemistry from Gurukula Kangri University, Haridwar, and Indian Institute of Technology, Roorkee, in 2003 and 1995, respectively. She is the recipient of SERS Fellow Award 2017 for her contribution in the field of Chemistry during International Conference on Innovative Approaches in Applied Sciences and Technologies, organized at Nanyang Technological University, Singapore, in 2017. She is the recipient of Gold Medal Award of Hi-Tech Horticulture Society for outstanding contribution in Chemistry, during Global Agriculture and Innovation Conference (27–29 November, 2016) at Noida International University, Noida, and Science Initiator Award by Scientific Educational Research Society during International Conference on Innovative Approaches in Applied Sciences and Technologies, organized at Faculty of Science, Kasetsart University, Bangkok, during 1–5 February, 2016. She received SAARC Gold Medal Award 2011 conferred by Scientific and Applied Research Centre, CCS University, Meerut, for outstanding contribution in the field of Chemistry Research. She has published more than 68 research papers, 1 book, and 26 book chapters with reputed publishers, viz. Elsevier, Bentham Science, Wiley, Academic Excellence, Springer, etc.

Dr. Dakeshwar Kumar Verma is Assistant Professor of Chemistry at the Govt. Digvijay Autonomous Postgraduate College, Rajnandgaon, Chhattisgarh, India. He has completed his Ph.D. program at the National Institute of Technology, Raipur, Chhattisgarh, India. He has more than 45 peer-reviewed international publications in Scopus and Web of Science journals of ACS, RSC, Wiley, Elsevier, Springer, and Taylor & Francis. He and has delivered numerous oral and posters presentations in varoius international conferences. He has contributed over 20 book chapters to various international books. His H-index is 19, and his total citation is 1,040 in Google Scholar database. His research is mainly focused on the preparation and designing of various materials used in the field of nanomaterials, green chemistry, quantum chemistry, and organic synthesis for various industrial applications. Recently, he has served as an author and editor/coeditor of various books published by ACS, Elsevier, Wiley, Taylor & Francis, and De Gruyter indexed in Scopus. Dr. Verma received the CSIR JRF and SRF awards in 2013 and 2015. Currently, four students are doing Ph.D. under his guidance.

https://doi.org/10.1515/9783110791891-204

Dr. Jeenat Aslam is Associate Professor at the Department of Chemistry, College of Science, Taibah University, Yanbu, Al-Madina, Saudi Arabia. She earned her Ph.D. degree in Surface Science/Chemistry from the Aligarh Muslim University, Aligarh, India. Materials and corrosion, nanotechnology, and surface chemistry are the primary areas of her research. Dr. Jeenat has published a number of research and review articles in peer-reviewed international journals like ACS, Wiley, Elsevier, Springer, Taylor & Francis, Bentham Science, and others. She has authored over 30 book chapters and edited more than 20 books for the American Chemical Society, Elsevier, Springer, Wiley, De Gruyter, and Taylor & Francis.

Dr. Pramod Kumar Mahish is Assistant Professor of Biotechnology in Govt. Digvijay Autonomous Postgraduate College, Rajnandgaon, Chhattisgarh, India, and Chairman Board of Studies Biotechnology at Hemchand Yadav Vishwavidyalaya Durg, India. He has 10 years of teaching and 13 years of research experience. Dr. Mahish had published 21 research papers and 8 books. He had also written nine book chapters in Springer, Wiley, De-Gruyter, Elsevier, and IGI. He had completed three research projects and presently guiding two Ph.D. scholars. He is a fellow member of the World Researchers Association and a life member in the Aerobiology Association of India and Microbiologists Society India. He was awarded UGC National Fellowship for his Ph.D. research work. He is also the Associate Editor of *Indian Journal of Aerobiology*. Dr. Mahish is engaged in the editing of three books of De Gruyter, Taylor & Francis, and ACS.

List of contributors

Chapter 1
Nabanita Hazarika
Department of Social Work
The Assam Royal Global University
NH-37, Betkuchi
Guwahati 781035
Assam
India

Hiranjyoti Deka
Assam Down Town University
Sankar Madhab Path
Gandhi Nagar
Panikhaiti, Guwahati 781026
Assam
India

Amenuo Susan Kulnu
Department of Environmental Science
Nagaland University (Central)
Lumami 798627
Zunheboto
Nagaland
India

Puranjay Mipun
Department of Botany
B. N. College
Dhubri 783323
Assam
India

Petekhrienuo Rio
Department of Forestry
Nagaland University (Central)
Lumami 798627
Zunheboto
Nagaland
India

Mayur Mausoom Phukan
Department of Forestry
Nagaland University (Central)
Lumami 798627
Zunheboto
Nagaland
India

Debajit Kalita
Microbiology Laboratory
Poohar Essence Private Limited
H.No.-39, Beltola
Guwahati 781028
Assam
India
pooharessence@gmail.com

Chapter 2
Saeid Hazrati
Department of Agronomy
Faculty of Agriculture
Azarbaijan Shahid Madani University
Tabriz, Iran
E-mail: saeid.hazrati@azaruniv.ac.ir

Maryam Mohammadi-Cheraghabadi
Department of Agronomy
Faculty of Agriculture
Tarbiat Modares University
PO Box 14115-336
Tehran
Iran

Saeed Mollaei
Phytochemical Laboratory
Department of Chemistry
Faculty of Sciences
Azarbaijan Shahid Madani University
Tabriz
Iran

Chapter 3
Sonal Mishra
Department of Botany Government Digvijay
Autonomous Post Graduate College
Rajnandgaon
Chhattisgarh
India
sonalmishra2017@gmail.com

https://doi.org/10.1515/9783110791891-205

Trilok Kumar
Department of Botany Government Digvijay
Autonomous Post Graduate College
Rajnandgaon
Chhattisgarh
India
trilokdev111@gmail.com

Chapter 4
Tarun Kumar Patle
Department of Chemistry
Pandit Sundarlal Sharma (Open) University
Bilaspur 495009
Chhattisgarh
India

Pramod Kumar Mahish
Department of Biotechnology
Government Digvijay (Autonomous)
Postgraduate College
Rajnandgaon 491 441
Chhattisgarh
India

Ravishankar Chauhan
Department of Botany
Pandit Ravishankar Tripathi Government College
Bhaiyathan 497 231
Surajpur
Chhattisgarh
India
ravi18bt@gmail.com

Chapter 5
Maryam Mohammadi-Cheraghabadi
Department of Agronomy
Faculty of Agriculture
Tarbiat Modares University
PO Box 14115-336
Tehran
Iran

Saeid Hazrati
Department of Agronomy
Faculty of Agriculture
Azarbaijan Shahid Madani University
Tabriz
Iran

Chapter 6
Charu Arora
Department of Chemistry
Guru Ghasidas University
Bilaspur 495009
Chhattisgarh
India
charuarora77@gmail.com

Dipti Bharti
Department of Applied Science and Humanities
Darbhanga College of Engineering
Darbhanga 846005
Bihar
India

Brij Kishore Tiwari
Department of Applied Science and Humanities
G.L Bajaj Institute of Technology and
Management
Greater Noida 201306
Uttar Pradesh
India

Ashish Kumar
Department of Applied Science and Humanities
G.L Bajaj Institute of Technology and
Management
Greater Noida 201306
Uttar Pradesh
India

Dakeshwar Kumar Verma
Department of Chemistry
Government Digvijay (Autonomous)
Postgraduate College
Rajnandgaon
Chhattisgarh
India

Bhupender Singh
Institute Instrument Centre
Indian Institute of Technology
Delhi
India

Chapter 7

Suparna Paul
Surface Engineering and Tribology Group
CSIR – Central Mechanical Engineering Research
Institute (CMERI)
Mahatma Gandhi Avenue
Durgapur 713209
West Bengal
India
Webpage: www.cmeri.res.in
and
Academy of Scientific and Innovative Research
CSIR – Central Mechanical Engineering Research
Institute (CMERI)
Mahatma Gandhi Avenue
Durgapur 713209
West Bengal
India
and
Department of Chemistry
Seacom Skills University
Kendradangal
Bolpur 731236
Birbhum, West Bengal
India

Subhajit Mukherjee
Department of Chemistry
Seacom Skills University
Kendradangal
Bolpur 731236
Birbhum, West Bengal
India

Priyabrata Banerjee
Surface Engineering and Tribology Group
CSIR – Central Mechanical Engineering Research
Institute (CMERI)
Mahatma Gandhi Avenue
Durgapur 713209
West Bengal
India
Fax: +91-343- 2546 745
Tel: +91-343-6452220
E-mail: pr_banerjee@cmeri.res.in
Webpage: www.cmeri.res.in
and

Academy of Scientific and Innovative Research
CSIR – Central Mechanical Engineering Research
Institute (CMERI)
Mahatma Gandhi Avenue
Durgapur 713209
West Bengal
India

Chapter 8

Hilal Ahmed
Department of Zoology
University of Jammu
Jammu 180006
Jammu and Kashmir
India

Tousief Irshad Ahmed
Department of Clinical Biochemistry
SKIMS
Soura, Srinagar
Jammu and Kashmir
India

Roli Jain
Department of Chemistry
Dr. Hari Singh Gour University
Sagar
Madhya Pradesh
India

Jyoti Rathore
Post Graduate College
Govt. Engineer Vishwesarraiya
Korba 495677
India

Shanthi Natarajan
Department of Botany
Pachaiyappa's College
Affiliated to University of Madras
Chennai 600030
Tamil Nadu
India

Reena Rawat
Department of Chemistry
Echelon Institute of Technology
Faridabad 121101
Haryana
India

Bhawana Jain
Siddhachalam Laboratory
Raipur 493221
Chhattisgarh
India

Chapter 9
Mayur Mausoom Phukan
Department of Forestry
School of Science
Nagaland University
Lumami 798627
Nagaland
India

Pranay Punj Pankaj
Department of Zoology
School of Science
Nagaland University
Lumami 798627
Nagaland
India

Samson Rosly Sangma
Department of Forestry
School of Science
Nagaland University
Lumami 798627
Nagaland
India

Ramzan Ahmed
Department of Applied Biology
School of Biological Sciences
University of Science and Technology
Meghalaya, Baridua
Ribhoi 793101
Meghalaya
India

Kumar Manoj
Department of Botany
Marwari College
Tilka Manjhi Bhagalpur University
Bhagalpur 812007
Bihar
India

Jayabrata Saha
Department of Applied Biology
School of Biological Sciences
University of Science and Technology
Meghalaya, Baridua
Ribhoi 793101
Meghalaya
India

Manjit Kumar Ray
Department of Applied Biology
School of Biological Sciences
University of Science and Technology
Meghalaya, Baridua
Ribhoi 793101
Meghalaya
India

Amenuo Susan Kulnu
Department of Environmental Science
School of Science
Nagaland University
Lumami 798627
Nagaland
India

Rupesh Kumar
Department of Biotechnology
The Assam Royal Global University
Betkuchi 781035
Guwahati, Assam
India

Kalpana Sagar
Department of Botany and Microbiology
Faculty of Life Science
Gurukula Kangri University
Haridwar 249404
Uttarakhand
India

Pranjal Pratim Das
Department of Biotechnology
Guwahati University
Tezpur 784001
Assam
India

Plaban Bora
Department of Energy Engineering
Assam Science and Technology University
Guwahati 781013
Assam
India

Chapter 10
Walid Daoudi
Laboratory of Molecular Chemistry, Materials
and Environment (LCM2E)
Departement of Chemistry
Multidisciplinary Faculty of Nador
University Mohamed I
60700 Nador
Morocco
walid.daoudi@ump.ac.ma

Abdelmalik El Aatiaoui
Laboratory of Molecular Chemistry, Materials
and Environment (LCM2E)
Departement of Chemistry
Multidisciplinary Faculty of Nador
University Mohamed I
60700 Nador
Morocco

Selma Lamghafri
Laboratory of Applied Sciences
National School of Applied Sciences Al-Hoceima
Abdelmalek Essaâdi University
Tetouan
Morocco

Abdelouahad Oussaid
Laboratory of Molecular Chemistry, Materials
and Environment (LCM2E)
Departement of Chemistry
Multidisciplinary Faculty of Nador
University Mohamed I
60700 Nador
Morocco

Adyl Oussaid
Laboratory of Molecular Chemistry, Materials
and Environment (LCM2E)
Departement of Chemistry
Multidisciplinary Faculty of Nador
University Mohamed I
60700 Nador
Morocco

Chapter 11
Anuragh Singh
Department of Pharmacology
SRM College of Pharmacy
SRM Institute of Science and Technology
Kattankulathur 603 203
Chengalpattu
Tamil Nadu
India

Rushendran R
Department of Pharmacology
SRM College of Pharmacy
SRM Institute of Science and Technology
Kattankulathur 603 203
Chengalpattu
Tamil Nadu
India

Siva Kumar B
Department of Pharmaceutical Chemistry
SRM College of Pharmacy
SRM Institute of Science and Technology
Kattankulathur 603 203
Chengalpattu
Tamil Nadu
India

Ilango K
Department of Pharmaceutical Quality
Assurance
SRM College of Pharmacy
SRM Institute of Science and Technology
Kattankulathur 603 203
Chengalpattu
Tamil Nadu
India

Chapter 12
Shailendra Yadav
Green Chemistry Laboratory
Department of Chemistry
Faculty of Basic Science
AKS University
Satna 485001
Madhya Pradesh
India
syshailendra5@gmail.com

Dheeraj Singh Chauhan
Modern National Chemicals
Second Industrial City
Dammam 31421
Saudi Arabia

Vandana Srivastava
Department of Chemistry
Indian Institute of Technology (Banaras Hindu
University)
Varanasi 221005
India

Chapter 13
Ankush Kerketta
Department of Botany
Government Kalidas College
Pratappur 497223
Chhattisgarh
India

Balram Sahu
Department of Botany
Government Rani Durgawati College
Wadrafnagar 497225
Chhattisgarh
India
balramsahu@hotmail.com

Chapter 14
Shweta Singh Chauhan
Department of Biotechnology and Research
Centre
Government Digvijay (Autonomous)
Postgraduate College
Rajnandgaon 491 441
Chhattisgarh
India

Ravishankar Chauhan
Department of Botany
Pandit Ravishankar Tripathi Government College
Bhaiyathan 497 231
Surajpur
Chhattisgarh
India
and
School of Studies in Biotechnology
Pandit Ravishankar Shukla University
Raipur 492010
Chhattisgarh
India
ravi18bt@gmail.com

Nagendra Kumar Chandrawanshi
School of Studies in Biotechnology
Pandit Ravishankar Shukla University
Raipur 492010
Chhattisgarh
India

Pramod Kumar Mahish
Department of Biotechnology and Research
Centre
Government Digvijay (Autonomous)
Postgraduate College
Rajnandgaon 491 441
Chhattisgarh
India
drpramodkumarmahish@gmail.com

Chapter 15
Shushil Kumar Rai
Shri Rawatpura Sarkar Institute of Pharmacy
Kumhari
Durg 490042
Chhattisgarh
India
and
Center of Innovative and Applied Bioprocessing
(CIAB)
Sector – 81
Knowledge City
Mohali 140 306
Punjab
India
rainiper2411@gmail.com

Ravishankar Chauhan
Department of Botany
Pandit Ravishankar Tripathi Government College
Bhaiyathan 497 231
Surajpur
Chhattisgarh
India
and
School of Studies in Biotechnology
Pandit Ravishankar Shukla University
Raipur 492 010
Chhattisgarh
India
ravi18bt@gmail.com

Chapter 16

Anton Soria-Lopez
Nutrition and Bromatology Group
Department of Analytical Chemistry and Food
Science
Faculty of Science
Universidade de Vigo – Ourense Campus
E-32004 Ourense
Spain

Nuno Muñoz-Seijas
Faculty of Science
Universidade de Vigo – Ourense Campus
E-32004 Ourense
Spain

Rosa Perez-Gregorio
Associated Laboratory for Green Chemistry
of the Network of Chemistry and
Technology (LAQV-REQUIMTE)
Departamento de Química e Bioquímica
Facultade de Ciencias da Universidade de Porto
Porto
Portugal
and

Nutrition and Bromatology Group
Department of Analytical Chemistry and Food
Science
Faculty of Science
Universidade de Vigo – Ourense Campus
E-32004 Ourense
Spain

Jesus Simal-Gandara
Nutrition and Bromatology Group
Department of Analytical Chemistry and Food
Science
Faculty of Science
Universidade de Vigo – Ourense Campus
E-32004 Ourense
Spain

Paz Otero
Nutrition and Bromatology Group
Department of Analytical Chemistry and Food
Science
Faculty of Science
Universidade de Vigo – Ourense Campus
E-32004 Ourense
Spain
paz.otero@uvigo.es

Nabanita Hazarika, Hiranjyoti Deka, Amenuo Susan Kulnu,
Puranjay Mipun, Petekhrienuo Rio, Mayur Mausoom Phukan
and Debajit Kalita*

Chapter 1
Habitat and distribution of medicinal plants

Abstract: Medicinal plants are valuable sources of herbal medicine recognized globally. Medicinal plants are a source of new drug molecules and widely distributed throughout the world ranging from forest, desert, polar region, ocean, and fresh water ecosystems. Around 1,000 BC, the knowledge of habitat of medicinal plants and its application in various ailments of disease have been reported by the Indian saint Charak in Charak Samhita and Shusruta in Shusruta Samhita. Around 197 different plant species have been mentioned in Ayurveda along with its distribution and use. A total of 34 global "Biodiversity Hotspots" have been recognized on the basis of specific criteria which also includes medicinal plants. The article focuses on the habitat and distribution of medicinal plants with specific emphasis on Biodiversity Hotspots. Remote sensing and geographical information system-based approaches for distribution studies and additionally socioeconomic importance of traditional practices pertaining to medicinal plants are further discussed.

1.1 Introduction

Plants are main part of ecosystem and are grossly responsible for its proper functioning. Their main function in the ecosystem is to provide oxygen and food. Additionally, they have some other applications such as sources of fuel, food additives, pesticides, pigments, resins, perfumes, and other important industrial and agricultural raw materials. Other than these applications, one of the most important aspects is their use as herbal medicine.

*Corresponding author: Debajit Kalita, Microbiology Laboratory, Poohar Essence Private Limited, H. No. 39, Beltola, Guwahati - 781028, Assam, India, e-mail: pooharessence@gmail.com
Nabanita Hazarika, Department of Social Work, The Assam Royal Global University, Betkuchi, Guwahati 781035, Assam, India
Hiranjyoti Deka, Assam Down Town University, Sankar Madhab Path, Gandhi Nagar, Panikhaiti, Guwahati 781026, Assam, India
Amenuo Susan Kulnu, Department of Environmental Science, Nagaland University (Central), Lumami, Zunheboto 798627, Nagaland, India
Puranjay Mipun, Department of Botany, B. N. College, Dhubri 783323, Assam, India
Petekhrienuo Rio, Mayur Mausoom Phukan, Department of Forestry, Nagaland University (Central), Lumami, Zunheboto 798627, Nagaland, India

https://doi.org/10.1515/9783110791891-001

Since time immemorial India has been known for its traditional medicinal systems – Ayurveda, Siddha, and Unani. Medical systems are found mentioned even in the ancient Vedas and other scriptures. The Ayurvedic concept appeared and developed between 2,500 and 500 BC in India [1]. The Indian subcontinent is a vast repository of medicinal plants that are used in traditional medical treatments.

Around 1,000 BC, the knowledge of habitat of medicinal plants and its application in various ailments of disease has been reported by Indian saint Charak in Charak Samhita and Shusruta in Shusruta Samhita. Around 197 different plant species have been mentioned in Ayurveda along with its distribution and use. Some of the examples are

काश्मर्यत्रिफलाद्राक्षाकासमर्दपरूषकैः ||५२||
पुनर्नवाद्विरजनीकाकनासासहचरैः |
शतावर्या गुडूच्याश्च प्रस्थमक्षसमैर्घृतात् ||५३||
साधितं योनिवातघ्नं गर्भदं परमं पिबेत् |

Meaning: One prastha (approx. 640 g) ghrita (ghee) cooked with the paste of the medicinal plants of kashmarya, triphala, draksha, kasamarda, parushaka, punarnava, haridra, daruharidra, kakanasa, sahachara, shatavari, and guduchi. Use for vaginal problem (vataja yoni roga) and is best for conception.

पिप्पलीकुञ्चिकाजाजिवृषकं सैन्धवं वचाम् ||५४||
यवक्षाराजमोदे च शर्करां चित्रकं तथा |
पिष्ट्वा सर्पिषि भृष्टानि पाययेत् प्रसन्नया ||५५||
योनिपार्श्वार्तिहृद्रोगगुल्माशिविनिवृत्तये |

Meaning: The equal amount crushed of these medicinal plants of pippali, upakunchika, jiraka, vrishaka, saindhava, vacha, yavakshara, ajmoda, sharkara, and chitrakamula and fried in ghrita (Ghee) when consumed will help to reduce the pain in vagina, flanks, heart diseases, gulma, and the piles.

Medicinal plants are distributed in different ecosystem in worldwide. These include

1.1.1 Forest ecosystem

Forests play a cardinal role for life on Earth to exist. There are about 750 million people worldwide living in forests who depend directly on them for their livelihoods [2]. Forests are also home to a diverse range of medicinal plants. Forests are of paramount importance for their intricate association with the ecosystem services that they provide which also includes various medicines from plants.

According to the biomes, forests have been classified based on the variations present in the elements. One such variation is the plant's elements that change its natural habitat tremendously. On the basis of biomes forest can be generally classified as tropical forests, temperate forests, and coniferous forests [3].

Some of the medicinal plants are reported from tropical forest such as *Achyranthes aspera* used for treating snake bite, *Andrographis paniculata* used for all kind of fevers, *Boerhaavia diffusa* for jaundice, *Elaeocarpus serratus* leaves for rheumatoid arthritis and antidote to poisons, *Mesua ferrea* bark for treating gastritis and bronchitis, *Ficus retusa* bark for liver diseases, *Piper nigrum* fruit for fever, vertigo, coma, and as stomachic in dyspepsia and flatulence, and *Vitex leucoxylon* for anemia [4, 5].

Similarly from temperate forest medicinal plants such as *Taxus wallichiana* twigs, barks, and leaves are used to treat cancer; *Oenothera rosea* leaves are used against obesity; *Indigofera heterantha* leaves are used as antimicrobial; *Rosa webbiana* flower is used against cold and cough; *Euphorbia wallichii* fruits and leaves are used against skin infections; and *Bergenia ciliata* root's bark is used to treat stomach ulcers and intestinal ailments [6, 7]. Medicinal plants from coniferous forests such as *Cinnamomum impressinervium* bark is used for its antifungal, antimicrobial, stimulant and carminative properties, *Paris polyphylla* rhizome is used to treat helminths and cancer, *Rumex nepalensis* roots are used to treat hair loss, and *Zanthoxylum alatum* fruits are used to treat blood vessel dilator and various skin infections [8].

1.1.2 Desert and arid ecosystem

Desert and arid habitats are ecosystems that do not receive enough rainfall. Most deserts receive less than 10 in of rain per year, evaporation usually exceeds rainfall. Desert plants, animals, and other species have evolved to cope with harsh circumstances, limited water, and barren terrain. While evolving these plant species have changed in their secondary metabolites, leading to a different forms of the drug in curing many different ailments. Many native species from Gobi Desert region have been reported to have medicinal properties such as *Allium altaicum, Euphorbia mongolica, Thermopsis mongolica, Arnebia guttata, Olgaea leucophylla, Potaninia mongolica,* and *Zygophyllum xanthoxylum* [9]. Similarly people in Thar desert (Sindh) of Pakistan have been using 87 plant species belonging to 32 families as traditional medical; among the families Amaranthaceae was the highest, followed by Cucurbitaceae and Euphorbiaceae [10].

1.1.3 Polar regions and mountains ecosystem

Surviving in the planet's polar region can be difficult due to their harsh environmental conditions. In the winter, temperatures can drop below zero and the darkness can last for months. However, despite its condition these seemingly barren landscapes are home to diverse range of plants both on land and beneath the water surface that has evolved to withstand these harsh conditions. In a study by Uprety et al. [11], 546 medicinal plant species were reportedly used by aboriginal people from the Canadian boreal forest in their traditional health care systems. Among them, the most commonly used plants were *Abies balsamea, Achillea millefolium, Acorus calamus, Aralia nudicaulis, Betula papyrifera, Cornus sericea, Heracleum maximum, Juniperus communis, Larix laricina, Mentha arvensis, Nuphar lutea, Picea glauca, Picea mariana*, etc. for treating many commonly prevailing ailments.

On mountains, weather changes rapidly as elevation increases and similarly both plants and animals too change their adaptational characteristic. But even these bleak landscapes are home to a diverse array of plants ranging from angiosperm, gymnosperm to many pteridophytes species such as *Pteris semipinnata, Schefflera octophylla, Zanthoxylum avicennae, Lygodium japonicum, Dicranopteris dichotoma*, and *Rhodomyrtus tomentosa* [12]. Many endemic, rare, or vulnerable medicinal taxa from the mountain range of Europe and Romania are reported such as *Gentiana lutea, Angelica archangelica, Arnica montana, Rhododendron myrtifolium, Leontopodium alpinum, Hepatica transsilvanica*, and *Streptopus amplexifolius* for treatment against various local ailments [13]. Such plant species surviving in harsh environment can easily get extinct due to various prevailing anthropogenic activities and as such warrant conservation efforts.

1.1.4 Ocean ecosystem

Oceans comprise the highest diversity of life on the planet right from polar regions to tropics and deep sea to shallow seagrass beds [2]. The vast medicinal repertoire from the ocean body includes vascular plant – *Zostera marina, Costaria costata*; Diatom species – *Phaeodactylum tricornutum*; Bacteria – *Penicillium chrysogenum*; Marine fungus – *Cephalosporium acremonium*; Algae – *Laurencia obtuse*, Cyanobacteria – *Lyngbya majuscule*; Sponge – *Spongia officinalis, Dysidea etheria, Bugula neritina*; Seaweed – *Ulva reticulata, Polysiphonia* sp., *Barleria Prionitis lyallii*; Snail – *Conus magus*; phytoplankton – Dinoflagellates and many other aquatic lives [14, 15].

1.1.5 Freshwater ecosystem

Freshwater ecosystems account for less than 0.01% of the planet's total surface area but they support more than 100,000 species [2]. Freshwater medicinal plants have been known as cure to many local aliments for thousands of years. *Azolla* sp. is used in rheumatic pain, difficulty in micturition, edema, cnidosis, skin itching, myocutaneous disease, erysipelas, scalding, and burns; *Nelumbo nucifera* is used in small pox, throat complications, pigmentation problems in skin, and diarrhea; *Portulaca oleracea* is used in insect or snake bites on the skin, boils, sores, pain from bee stings, bacillary dysentery, diarrhea, hemorrhoids, postpartum bleeding, and intestinal bleeding; *Centella asiatica* is used in wounds, burns, and ulcerous skin ailments and prevention of keloid and hypertrophic scars; *Commelina benghalensis* is used as a diuretic, febrifuge, and for treating inflammations; *Scirpoides holoschoenus* is used in atherosclerosis, cardiovascular diseases, uterus contraction, and wound healing; *Trapanatans* is used in the treatment of dyspepsia, hemorrhages, diarrhea, dysentery, strangury, intermittent fever, leprosy, fatigue, inflammation, erysipelas, bronchitis, and general debility [16].

1.1.6 Worldwide distribution of medicinal plant and biodiversity hotspot

In 1988, Norman Myers, a British ecologist identified 10 tropical forest hotspots. In certain areas, there was significant habitat degradation as well as outstanding plant endemism. Conservation international (CI), one of the CEPF's global donor groups, agreed to implement Myers' hotspot model in 1989. In 1996, CI decided to reassess the hotspot's idea, examining whether any critical regions had gone undetected. A broad worldwide study was conducted in 1999, resulting in the introduction of 25 biodiversity indicators. The overall number of biodiversity hotspots was increased to 34 in 2005 based on further research. In 2011, the Forests of East Australia were named the 35th hotspot by a study team from the Commonwealth Scientific and Industrial Research Organization and CI. The North American Coastal Plain was named the Earth's 36th hotspot in 2016 (https://www.cepf.net/our-work/biodiversity-hotspots/hotspots-defined).

There are now 36 designated biodiversity hotspots on the planet. These hotspots are the most ecologically diverse and also most endangered terrestrial locations on the planet (https://www.cepf.net/our-work/biodiversity-hotspots/hotspots-defined).

1.1.7 Criteria for being designated as a biodiversity hotspot

a) At least 1,500 indigenous vascular plant species (discovered nowhere else) have to be present.
b) At least 70% of number one plant life has to have perished or be endangered (https://www.cepf.net/our-work/biodiversity-hotspots/hotspots-defined).

Biodiversity hotspots are areas where unusually high concentrations of common species, such as flora and wildlife, are vanishing (https://www.cepf.net/our-work/biodiversity-hotspots/hotspots-defined).

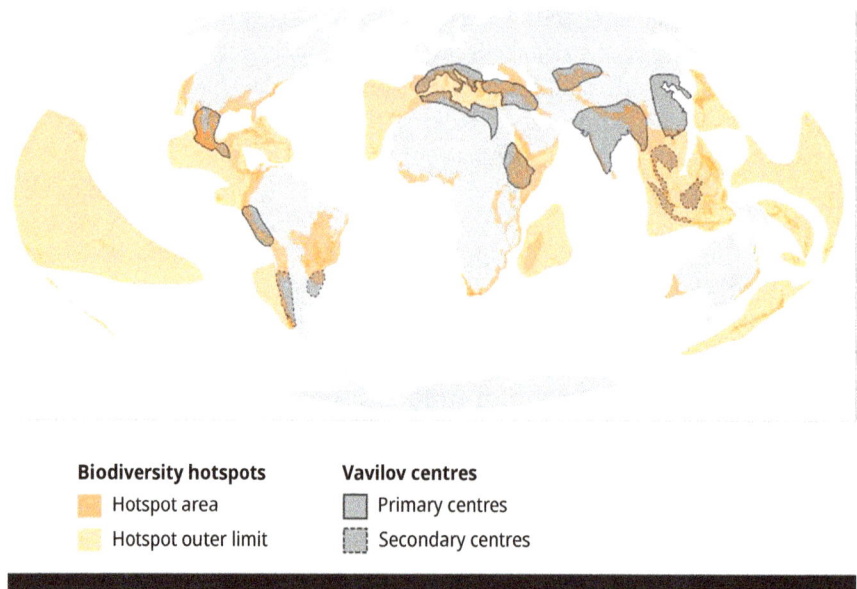

Biodiversity hotspots
▪ Hotspot area
▪ Hotspot outer limit

Vavilov centres
▪ Primary centres
▪ Secondary centres

Fig. 1.1: Distribution of biodiversity hotspots and Vavilov centers throughout the world. Myers initially described biodiversity hotspots in 1988 (Myers, N. Threatened biotas: "Hot spots" in tropical forests). According to Mittermeier and colleagues (Mittermeier, R.A.; Turner, W.R.; Larsen, F.W.; Brooks, T.M.; Gascon, C. Global biodiversity conservation: The critical role of hotspots, Environmentalist 1988, 8, 187–208), there are now 35 regions of high species richness, endemism, and threat, as last updated in 2011 (Mittermeier, R.A.; Turner, W.R.; Larsen Springer: Berlin/Heidelberg, Germany, 2011; pp. 3–22. ISBN 978-3-642-20,991-8). In Biodiversity Hotspots; Zachos, F.E., and Habel, J.C., Eds. Outer hotspot bounds draw attention to islands that make up biodiversity hotspots. In 1924, Vavilov established the centers of origin of farmed species and wild relatives; in 1935, he updated the list to include eight major and three minor centers (Vavilov, N.I.). Origin and Geography of Cultivated Plants; Cambridge University Press: Cambridge, UK, 1992; ISBN 978-0-521-40,427-3); Vavylov, M.I.; Vavlov, N.; Dorofeev, V.F.

1.1.8 Worldwide distribution of medicinal plant

Plant products are used for nourishment, spices, cosmetics, and adornment, and medicinal plants constitute a key category of nontimber forest products (NTFP). Medicinal plants are the most numerous of these types. Approximately 50,000–70,000 medicinal plant species are used globally [17]. Around 3,000 species were traded globally in the 1990s, but no current data are available. According to the World Health Organization, the worldwide market for medicinal plant products is substantial [18].

Medicinal plants are a particularly important type of NTFP because of its direct impact on local, national, and global health. According to the World Health Organization, 80% of the population in underdeveloped countries relies on traditional medical care;

alternative or supplementary medicine is used by 50% of the population in the developed world [19]. These health care systems are built on medicinal herbs [20]. Similarly, the majority of active mechanisms in current pharmaceutic medications were obtained either directly or indirectly from natural sources, which include botanical, microbiological, and other origins. Despite the development of synthetic and combinatorial chemistry, this finding still holds true. Only 29% of the 1,355 medications authorized between 1981 and 2010 were completely synthetic, 46% were semisynthetic but had an active mechanism drawn from or inspired by nature, and 19% were natural compounds without any synthetic alteration [21].

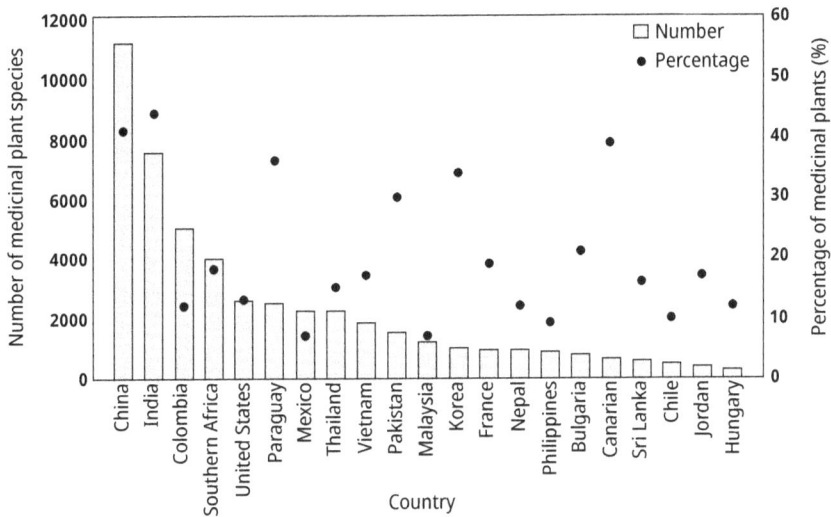

Fig. 1.2: Number and percentage of medicinal plant species in different countries. The light bar indicates the number of medicinal plant species and dark dots indicate the percentage of medicinal plants compared with number of medicinal plant species.

1.1.9 GIS model for ecologically suitable distributions of medicinal plants

For many years, determining optimal places for medicinal plant growth has solely relied on traditional knowledge, which is susceptible to the observation of one or few climatic parameters. However, this procedure is inefficient and produces erroneous findings [22]. As a result, there is an unmet need to establish an effective strategy based on the geographical distribution of key producers and wild populations, a thorough examination of climatic and soil conditions, and variables influencing medicinal plant growth. The goal of such system would be to identify ecologically acceptable places across the world, which provides efficient recommendations for introducing medicinal plants, and build a coherent plan for geographically dispersing plant production [23].

The Institute of Chinese Materia Medica, China Academy of Chinese Medical Sciences, created the Global Medicinal Plant Geographic Information System (GMPGIS). GMPGIS has the benefit of relying on or requiring data on confirmed absences from certain locations because it uses presence-only data. The device consists of a database of ecological settings which might be conducive to medicinal plant cultivation. The information is basically derived from databases: World Climate and the Harmonized World Soil Database, and it affords customers with uniform accuracy and coordinates. In addition, the GMPGIS is sensible than different geographical distribution fashions in forecasting potential medicinal plant ideal locations because of the subsequent features. To begin with, the system incorporates several sample locations and more than 240 medicinal plant species from geographical sampling areas, which cover both large producers' and wild populations' geographical distribution zones. Second, the system may utilize its built-in databases to extract values for ecological parameters, soil classifications, and the extent of prospective growth zones in order to provide tables for its users. Third, while developing an algorithm, the system eliminates outlier points; as a result, the regions in which sampling points are included should have a high level of ecological similarity. Fourth, the approach overcomes issues caused by a small number of sample points, which are particularly problematic for uncommon or limited-range species. Training algorithms with inadequate sampling points usually exhibit poor performance and errors. To efficiently handle these challenges, our system uses an unsupervised learning analysis technique. To ensure the correctness of the analytical results, a supervised verification approach is also used. Finally, the system has already completed certain fundamental tasks, making the analysis process easier and faster for consumers. Principal component analysis, for example, reduces the number of ecological components. In addition, customers may customize 16 soil parameters to meet their specific needs. Overall, the GMPGIS is an awesome distribution evaluation technique for global medicinal plant manufacturing because of these characteristics.

The Global Biodiversity Information Facility [24] and the Royal Botanic Gardens [25] provided the point locations of the medicinal plants with the widest global coverage for the proposed system. The selection criteria were the geographical distribution sites of both significant producers and wild populations and 158 occurrence records for this medicinal plant were found mostly in Iran, France, Turkey, Greece, Germany, Italy, Sweden, Austria, Bulgaria, Syria, and the United Kingdom [26].

With its unique agro-climatic conditions and regional geography, India has long been regarded as a botanical garden or treasure house of plant genetic resources. As a result, India is considered one of the world's top 12 mega diversity countries. More than 8,000 species make up India's herbal treasure, accounting for around half of all higher flowering plant species; over 70% of the country's medicinal plants are found in the Western Ghats' tropical forests. However, current data indicate approximately

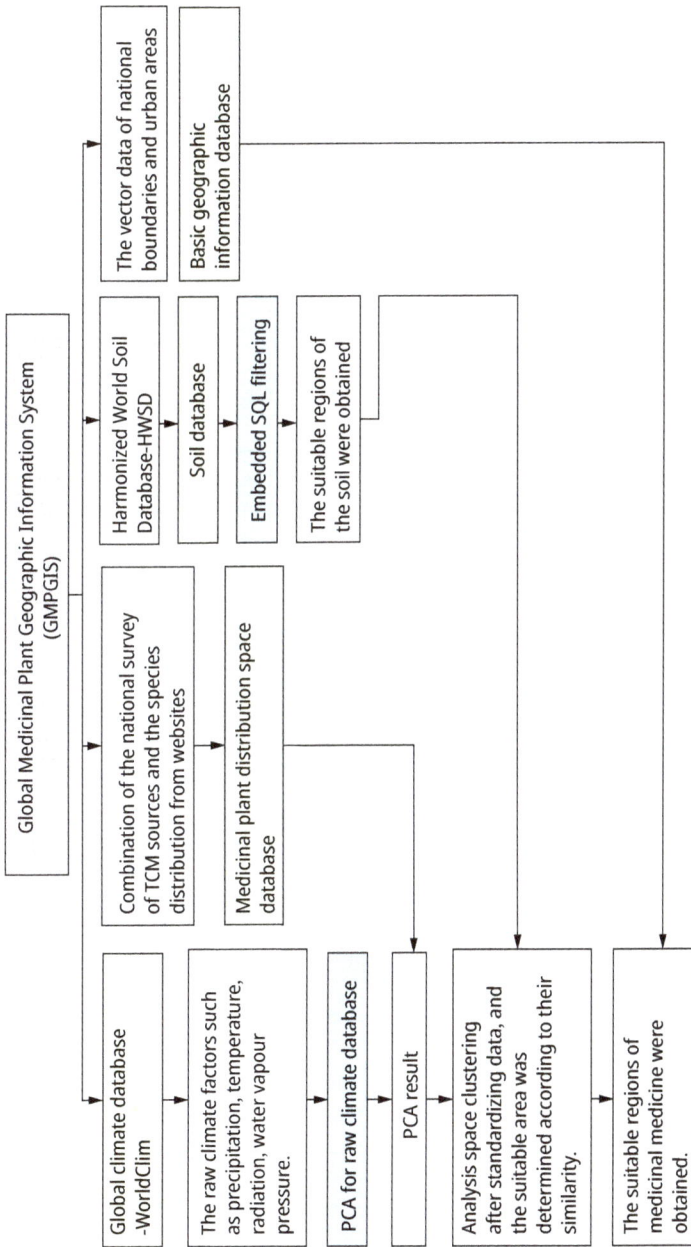

Fig. 1.3: Working principle of the Global Medicinal Plant Geographic Information System (GMPGIS) [26].

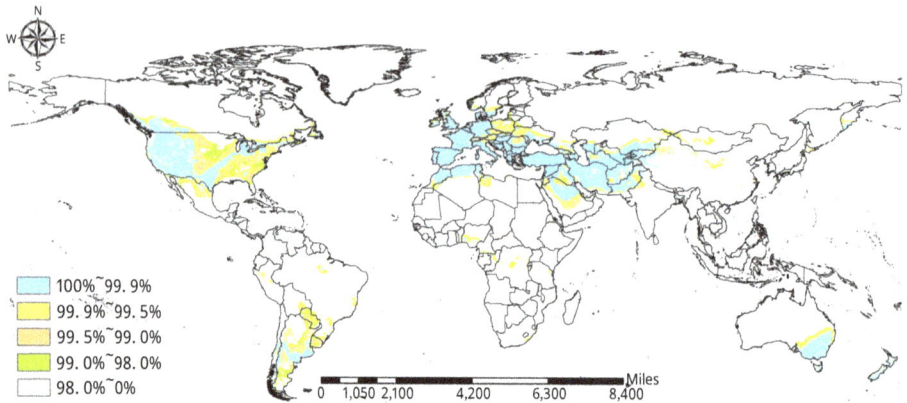

Fig. 1.4: Potential distribution region analysis for *Crocus sativus* L. based on the Global Medicinal Plant Geographic Information System (GMPGIS) results. (Ecologically suitable distribution regions originated in ArcGIS 10.4, The GADM database of Global Administrative Areas 2.0 can be downloaded from http://www.gadm.org.)

1,800 species are employed in classical Indian medicinal systems, 1,200 is used in Ayurveda, 900 in Siddha, 700 in Unani, 600 in Amchi, and 450 in Tibetan. The expanding sector of herbal goods business has a lot of promise for the Indian region's economic growth. Herbs are used as a source of nutrition. It is believed that 95% of therapeutic plants utilized in the Indian herbal business come from the wild. Every year, around half a million tons of dry debris is gathered indiscriminately by destructive ways and 1.65 lakh hectares of forest are cleared and destroyed. With rising population, fast expansion of land under food and commercial crops, deforestation, urbanization, the formation of businesses in rural areas, and other factors, plant genetic resources are rapidly depleting, with many species on the verge of extinction [27, 28].

There are 427 Indian medicinal plant entries in the red data book on endangered species, with 28 deemed extinct, 124 endangered, 81 uncommon, and 34 unknown (Tab. 1.1).

Tab. 1.1: Botanical representation of traded medicinal plants (family wise).

Family	Species
Fabaceae	67
Asteraceae	54
Euphorbiaceae	48
Caesalpiniaceae	41
Apiaceae	37
Lamiaceae	37
Solanaceae	35
Cucurbitaceae	32
Rubiaceae	29
Malvaceae	28

1.2 Biodiversity hotspot in the world with special reference to India

The term "BIODIVERSITY HOTSPOTS" has been coined as those areas serving the richest spot for biodiversity location. This concept was proposed by Sir Norman Myers in the year 1988 on the basis of high significance levels of habitat loss and the presence of an extraordinary number of plant endemism [29]. After two years, in 1990, he divided eight hotspots including four areas of Mediterranean type ecosystems [30].

Biodiversity hotspots have been considered as a tool to set conservation priorities which play a significant role in decision-making for cost-effective strategies to preserve biodiversity in terrestrial as well as marine ecosystems. This type of approach can be applied to any geographical scale and it has been considered to be one of the best ways for maintaining a large proportion of the world's biological diversity [31]. Maintaining biodiversity is very essential to the supply of ecosystem services and also important to support their health and resilience [32]. The number of hotspots increased to 25 covering 1.4% of the Earth's land area and maintaining 44% of the world's plant species and 35% of terrestrial vertebrate species and then again to 34. This number of hotspots lasted until 2011 comprising 2.3% of the land surface and supporting more than 50% of endemic plant species and 42% of the world's endemic terrestrial vertebrate species [33].

Till date, there are 36 hotspots, covering 2.4% of the land surface. Recently, forests of East Australia and North American Coastal Plain were identified in 2011 and 2016, respectively. Particularly, hotspots maintain 77% of all endemic plant species, 43% of vertebrates (including 60% of threatened mammals and birds), and 80% of all threatened amphibians [29, 34, 35].

India belongs to the 18 Mega Biodiversity countries of the world, which holds 3 of the 25 identified hot spots [36]. It covers a geographical area of ca. 329 million hectors and coastal line of over 6,000 km, which is 11% of the world's flora in approximately 2.4% of global land mass. Approximately 28% of the total Indian flora and 33% of angiosperms found in India are listed as endemic [37]. India possesses abundant biodiversity owing to its larger climatic and topographic gradient. Indian forests area cover 22.5% of total country's geographical area and harbor more than 17,000 angiosperms species [38, 39]. In 2006, CI demarcated 34 global "Biodiversity Hotspots;" out of these 34, 4 are partly within India: (1) Himalaya, (2) Western Ghats, (3) Indo-Burma, and (4) Sundaland [40] (Fig. 1.5).

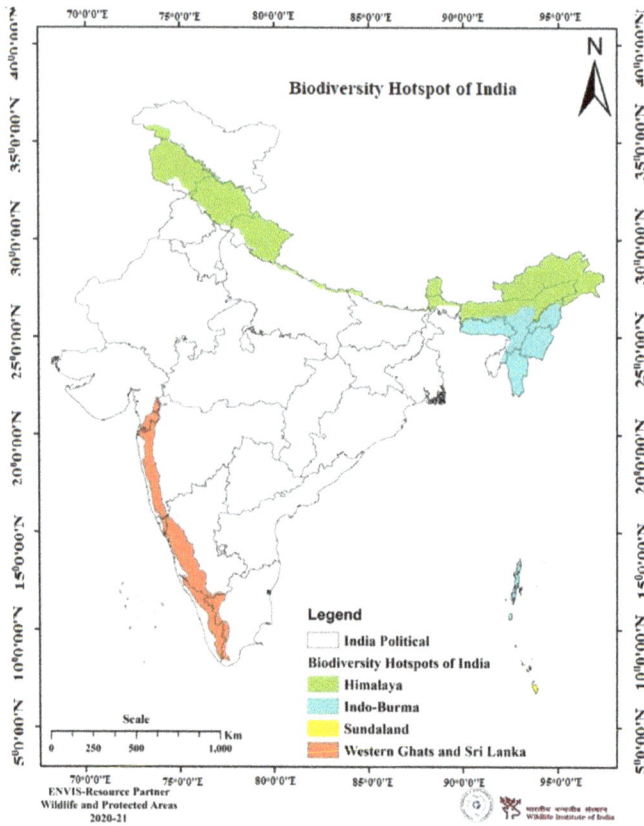

Fig. 1.5: Courtesy: [33].

1) Himalaya biodiversity hotspot: With a large geographical area of approximately 600,000 km², the Himalaya holds a series of mountain ranges that includes the boundaries of five nations: China, India, Bhutan, Nepal, and Pakistan [40]. From the geographical perspective, the Himalayas comprises four distinct tectonic units: the Outer Himalaya (or Siwaliks), Lesser Himalaya, Greater Himalaya, and Trans-Himalaya [41]. It is estimated that the Himalaya range has more than 10,000 species of vascular plants, with almost a third (31.6%) being endemic [42], although the number remains controversial, and many new plant species are also reported from this region [e.g., 43, 44].

Plant diversity of the Himalayas holds over 8,000 angiosperms, 44 gymnosperms, 600 pteridophytes, 1,737 bryophytes, 1,159 lichens, etc. [45]. Of these, 1,748 are reported to be used for various therapeutic purposes [46]. Different plants including orchids, wild edible plants, NTFPs, and medicinal plants of high value are recorded from the Himalayan region. More than 40 species of Rhododendron species are native to the Eastern

Himalayas [47]. Some examples are *Juniperu sindica* Bertol, *Uvaria lurida* var. *sikkimensis*, *Bhutan therahimalayana* Renz, *Allium sikkimense* Baker, *Viola bhutanica* Hara, *Microgynoecium tibeticum* Hook.f, *Bryocarpum himalaicum* Hook.f. and Thomson, *Rhdododendron sikkimense* Pradhan, and *Lachungpa* [47].

The state of Jammu and Kashmir (located at the northwestern boundary of the Himalayan biodiversity hotspot) has huge biodiversity including valuable medicinal plants [48]. The Himalaya, which harbors one of the world's richest floras, additionally supports the livelihoods of more than 1 billion people [49].

Tab. 1.2: Representative table of medicinal plant biodiversity of Himalayan biodiversity.

Plant species (family)	Region/area using	Medicinal property	References
Abies pindrow	Sewa River area of Jammu and Kashmir, India	Bronchitis and asthma, constipation, diuretic, and purgative.	[93]
Achillea millefolium L.	Sewa River area of Jammu and Kashmir, India	Diuretic, colds and fever	[93]
Acorus calamus L.	Sewa river area of Jammu and Kashmir, Baitadi and Darchula districts of far-western Nepal, Rasuwa District of central Nepal, Seti River area of western Nepal, Jutpani Village, Chitwan district of central Nepal	Anthelmintic, stomach ache, coughs, colds, and sore throat, wounds and swelling, and anti-inflammatory	[11, 50–51]
Aegle marmelos L.	In every part of India, Kumaun of Uttarakhand	Treat abscesses, cuts, wounds, and ulcers, gastrointestinal problems (vomiting, dysentery, and diarrhea)	[52]
Ajuga parviflora Benth	Mornaula Reserve Forest of Kumoun, west Himalaya, India	Anthelmintic	[53]
Artemisia dracunculus L.	Nubra Valley (Kashmir), Kibber Wildlife Sanctuary (Himachal Pradesh), and the Lahual Valley (Himachal Pradesh)	Flavoring food, toothache, and gastrointestinal problems	[54]
Artemisia japonica Thunb	Garhwal Himalaya (Uttarakhand)	Incense and insecticide	[55]

Tab. 1.2 (continued)

Plant species (family)	Region/area using	Medicinal property	References
Artemisia nilagirica (C.B. Clarke] Pamp	Darjeeling (West Bengal) India	Oral ulcers	[56]
Nepeta ciliaris Benth	Kedarnath Wildlife Sanctuary of Uttarakhand	Reduce fever	[47]

2) Western Ghat biodiversity hotspot: Western Ghats area is considered as one of the most significant biographic zones of India [57] as it is one of the richest center of endemism which comprises 56 genera and 2,000 species. Biogeographically, the hill chain of the Western Ghats covering the Malabar province of the Oriental realm, parallel to the west coast of India from 8° N to 21° N latitudes, 73° E to 77° E longitudes for around 1,600 km. From a relatively narrow strip of coast at its western border, the hills rising up to a height of 2,800 m before they merge to the east with the Deccan plateau at an altitude of 500–600 m. The average width of this mountain range is about 100 km. This bioregion is highly dense with a variety of species and under constant threat from human it is considered as one of the 18 biodiversity hot spots of the world. Tremendous environmental heterogeneity occurs across the Ghats; topographically, soils, rainfall, number of dry months per year, and temperature prepares this biogeographic area extremely environmentally heterogeneous, with a remarkable amount of diversity in both the cases of plants and animals [58]. In developing countries, WHO considered that medicinal plants play an important role in the health care of about 80% of world population. Herbal medicines constitute the most prominent part [59]. The rest of the 20% also depend significantly on the plant-based medicines.

Tab. 1.3: Representative table of medicinal plant biodiversity of Western Ghat biodiversity.

Plant species (family)	Region/area using	Medicinal property	References
Abutilon indicum	Paliyars tribes of Madurai, Tamil Nadu	Treat piles, body heat, and skin diseases	[60]
Adhatoda zeylanica	Mudhuvars tribes of Dindigul, Tamil Nadu	Cold, cough, breathing problems, and throat pain	[60]
Aerva lanata	Theni of Tamil Nadu	Kidney stone inflammation	[60]
Albizia amara	Tirunelveli of Tamil Nadu	To get rid of dandruff	[60]

Tab. 1.3 (continued)

Plant species (family)	Region/area using	Medicinal property	References
Dioscorea pentaphylla	Virudhunagar of Tamil Nadu	To treat piles	[60]
Trigonella foenum	Idukki of Kerala	To treat dysentery	[60]
Emilia sonchifolia	Kurichyastibes of Kannur district, Kerala	To treat worm infection	[61]
Aristolochia indica	Kani tribes of Pechiparai Hills, Western Ghats, India	Leaves are used to treat fever	[62]

3) The Indo-Burma region and the Sundaland biodiversity hotspot: Indo-Burma region is a newly designated biodiversity hotspot by CI, which ranges from eastern India and southern China across Southeast Asia. This Indo-Burma region and hotspot encircle the eastern India (including the Andaman and Nicobar Islands), southern-most China, most of Myanmar (excluding the northern tip), most of Thailand (excluding the southern tip), and all of Cambodia, Laos, and Vietnam, along with 2 million km^2 of tropical Asia east of the Ganges-Brahmaputra lowlands.

The Northeastern (NE) State of India contains eight states that harbor more than 180 major tribal communities of the total 427 tribal communities of India [63]. Chhetri et al. [64] reported that 37 species of plants belonging to 28 families are used as antidiabetic agents in the folk medicinal practices and 81% of these plants are unreported as hypoglycemic agents.

In Sikkim, a North East state of India, six important species of medicinal plants (*Aconitum heterophyllum* Wall. ex Royle, *Nardostachys jatamansi* (D. Don) DC., *Podophyllum hexandrum* Royle, *Picrorhiza kurrooa* Royle ex Benth., and *Swertia chirayita* (Roxb.) are used by tribal people for various ailments.

As per report of Forest Survey of India [65], India has a total area of 4,871 km^2 under mangroves region. Out of total mangroves, about 59% are found in east coast (Bay of Bengal), 23% on the west coast (Arabian sea), and the remaining 18% on the Bay Islands (Andaman and Nicobar Islands in Bay of Bengal) [66]. Sunderbans of West Bengal is the largest well-known coastal area of India.

Tab. 1.4: Representative table of medicinal plant biodiversity of the Indo-Burma region.

Plant species (family)	Region/area using	Medicinal property	References
Leucas linifolia (Roth) Spreng	Assam	Snake bite treatment	[67]
Oxalis corniculata	Assam	Food and antidiarrheal	[68]
Paederia foetida	Boghora hill (Morigaon District of Aassam)		[69]
Centella asiatica	Tiwa Tribes of Middle Assam, India	Gastrointestinal troubles and skin diseases	[70]
Chamaecostus cuspidatus	Mizo tribe of Mizoram, India	Blood sugar reducer	[71]
Myrica esculenta	Khasi, Jaintia, and Garo tribes of Meghalaya, NE India	Pickle and diarrheal trouble	[72]

1.2.1 Socioeconomic importance of traditional practices of medicinal plants

The history of traditional cultural practices among South Asian countries plays an important role in protecting and maintaining medicinally significant plant varieties. Majority of the global population are still dependent on various plant products for their primary requirements. Nearly, 21,000 plant species are used for healthcare purpose worldwide. An approximate of 2,500 plant species is applied in various alternative and complementary medicines in India [73, 74]. The explanation of various medicines, its formulations, and applications is found in ancient medicinal literature in terms of Rig Veda and others [75]. Medicinal plants have been an essential source in treating various diseases and an ingredient used in medical ground for the well-being of individuals. These include garlic and ginger [76]. The application of medicinal plants is considered vital either in treatment, cure, or prevention of various issues and are often guided by the body of knowledge arising between the relationship of individuals and environment [77]. Therefore, ethnobotany and medicinal plants being a responsible study illustrate its relationship between individuals and men from the ancient period till today.

1.2.2 Economic perspective

Effective practices for medicinal plants have been taken up to regulate the production, ensure quality, and facilitate standards of herbal medicines [78]. It ensures safe and pollution-free medicines or drugs by application of knowledge-based research to explore several problems. It takes into consideration the ecological environment of production sites, cultivation, and microscopic authentication for various elements [79]. Studies have found that traditional practices of medicinal plants influence the economic factor as the products are free of side effects and are not harmful to health and easily acquired [80]. Several countries are promoting and opting for good effective practices in traditionally cultivated region. Organic farming has achieved attention in the ability to develop economically sustainable production for medicinal plants [81]. The main objective of traditional organic farming practices includes better quality material with high productivity that ensures sustainable utilization of medicinal plants. Studies have reported that medicine being an important commodity in one's life, 90% of it comes from plant-based resources [82]. Ayurveda, which propagate the science of Ayurved, derives strong synergy between ancient wisdom of medicinal plants and modern technology that has created a captive market with their significant efforts since the previous decades. Indian Origin Patanjali that believes in the idea of Ayurved focuses on the philosophy of Swadeshi or home grown that considers ideal for the potential resources that every traditional medicinal plant carry [83]. There are several case studies which reported that people across the globe pay out of pockets for traditional drugs or complementary medicine (often derived from Ayurved). Several traditional and Ayurvedic institutes across the country are now promoting traditional medicines on scientific lines which may prove to be boon for its global acceptance [84]. Additionally, in many areas of the country, the traditional medicinal practices are mainly used by the rural setting as livelihood and often a major source of income for them. People living in rural areas often visit Kaviraj (traditional healer), who has in-depth knowledge on traditional medicinal plants for various treatments. It has been observed that very few traditional medicinal plants have been researched or cultivated technically for the purpose of medicinal extraction [85]. Moreover, the expansion of industries and urbanization has risked the biological diversity to a huge extent.

1.2.3 Market size and economic growth of Indian Ayurveda industry

In recent years, as a holistic healing system, Ayurveda has witnessed evolution in the form of Ayurvedic products and services. Rising awareness about the importance of a healthy lifestyle, increasing preference in favor of chemical-free natural products as well as favorable government initiatives have led to the expansion of the Ayurveda market in India. In 2018, around 75% of Indian households used Ayurvedic products as against only 67% in 2015. Of late, manufacturers have been using herbal ingredients in the production of personal care products like lotions, oils, and shampoos.

Players in the food processing industry are also making use of herbal ingredients in manufacturing products like packaged juices and nutritional supplements.

Ayurvedic market in India was valued at INR 300 billion in 2018 and is expected to reach INR 710.87 billion by 2024, expanding at a compounded annual growth rate of ~16.06%, during the forecast period (2019–2024). Some of the key Ayurvedic medicine and cosmetics companies of India are Patanjali Ayurved Limited, Dabur India Limited, Emami Limited, Sandu Pharmaceuticals Limited, Charak Pharma Private Limited, Himalaya Drug Company Private Limited, Shree Baidyanath Ayurved Bhawan Private Limited, etc. According to industry estimates, top 50 companies (across both food and Ayurveda nonfood products) reported revenue of around INR 22,500 crores for the financial year April 2017–March 2018 from sale of Ayurveda products only. India's major trading partners for export and import of Ayurveda products are varied. India's total trade in Ayurveda products (including medicants and medicaments) in the year 2017–2018 was USD150.96 million. Promotion and export of more Ayurved-based industry helps in a high level of output from a country's factories and industrial facilities as well as a greater number of people that are being employed in order to keep these factories in operation.

1.2.4 Social and cultural perspective

Since years, traditional medicinal plants have been used as a major source of therapeutic agents. Availability of resources and existing socioeconomic and cultural condition plays a significant role in people's lives. Most of these medicinal plants are collected from nature and used for not only treatment of various ailments but also connected with ancestral beliefs and protection from evil spirits [86]. Studies have found that people living in rural areas are believed to be infected with various kinds of evil spirits and witches and people showed evil plants that were used for witchery. The evil or the witch practices are often controlled by traditional medicinal plants and when burned it is used to communicate with unseen elements or power [87]. Socially and culturally traditional medicinal plants are considered very significant for the rural people as they care about their resources and agree together to stop the commercial gathering which is harmful for their own living. Traditional medicinal plants are considered vital for the rural people for primary health delivery as there are limited mobile clinics available and majority of the people believes in traditional medicine for treatment and healing. Moreover, apart from medical importance, traditional medicinal plants are a holistic approach that treats not only the physical ailment but also the unexplainable.

The meaning of traditional medicinal plants is significant as it is a part of rural people's understanding, their livelihood, health, and culture. The rural people's belief on traditional medical plants gives its power that it possesses potent significant sociocultural importance. Their divine traditional knowledge is highly integrated into the social and ethnic fabric of the people. Within a rural setting, such knowledge is a means of survival and education about livelihoods, family, conflict resolution, and relationships among each other [88].

1.2.5 Distribution and evaluation of medicinal plants through GIS and GPS mapping

Medicinal plants are vital for the development of health sector as well as socioeconomic sector to provide livelihood to millions of people in the world. However, the environmental and anthropological pressure has put medicinal plants on threatened conditions. To conserve and monitor the biodiversity, employing advance Geospatial technology is highly approved in today's time. Remote sensing (RS) and GIS techniques received an explosive interest in monitoring and conserving the medicinal plant biodiversity. RS along with GIS is a suitable tool for cataloging medicinal plant data through mapping and analysis. The knowledge of identification and distribution of plant species are prerequisite for safeguarding and restoration of medicinal plant species in an ecosystem [89]. RS offers a concise collection of multispectral information and multitemporal cover age of a particular or a large surface of the Earth and GIS is an integration of hardware and software for capturing, managing, analyzing, and displaying spatial data. Integrated RS and GIS in the field of Medicinal Plant Biodiversity provide data on analysis of plant species.

Distribution analysis of medicinal plants by geospatial technology can be carried out by multiple software (QGIS, ERDAS IMAGINE, ArcGIS, etc.) and models (MaxEnt, CaNaSTA, DSSAT). GIS software analyses, visualizes, and represents the data for distribution of medicinal plant species. The basic processes of handling GIS software include downloading and importing satellite imageries data in GIS software for the preparation of base map. GPS points collected during the vegetation survey along with the ethno-botanical information of the species are tabulated in Microsoft Excel and converted to attribute table in GIS platform which are then overlaid to the base map. The map generated by the induction of RS, GIS software, and GPS highlights the spatial patterns focusing on the species distribution [90].

ArcGIS software is diversely used in mapping the distribution of medicinal plant species in various research studies. Biswas et al. [91] have identified and mapped medicinal plants diversity using ArcGIS 10.2 which resulted in interactive and informative maps. Similarly, Dwivedi et al. [90] have also utilized ArcGIS 10.3 software to represent the multilayered database containing information on spatial distribution of antimalarial plants (Fig. 1.6). Another Geospatial technology is DIVA-GIS, a diversity analysis GIS tool that comprehends the diversity on a terrestrial extent and simplifies spaces in germplasm collection. Geospatial abilities of providing data on different time period for the same location enhance the detection of changes in a particular area or species. DIVA-GIS applies the same and analyzes the data for different time period and represents the collection gaps, diversity depletion, and extent of distribution in the plant species [91].

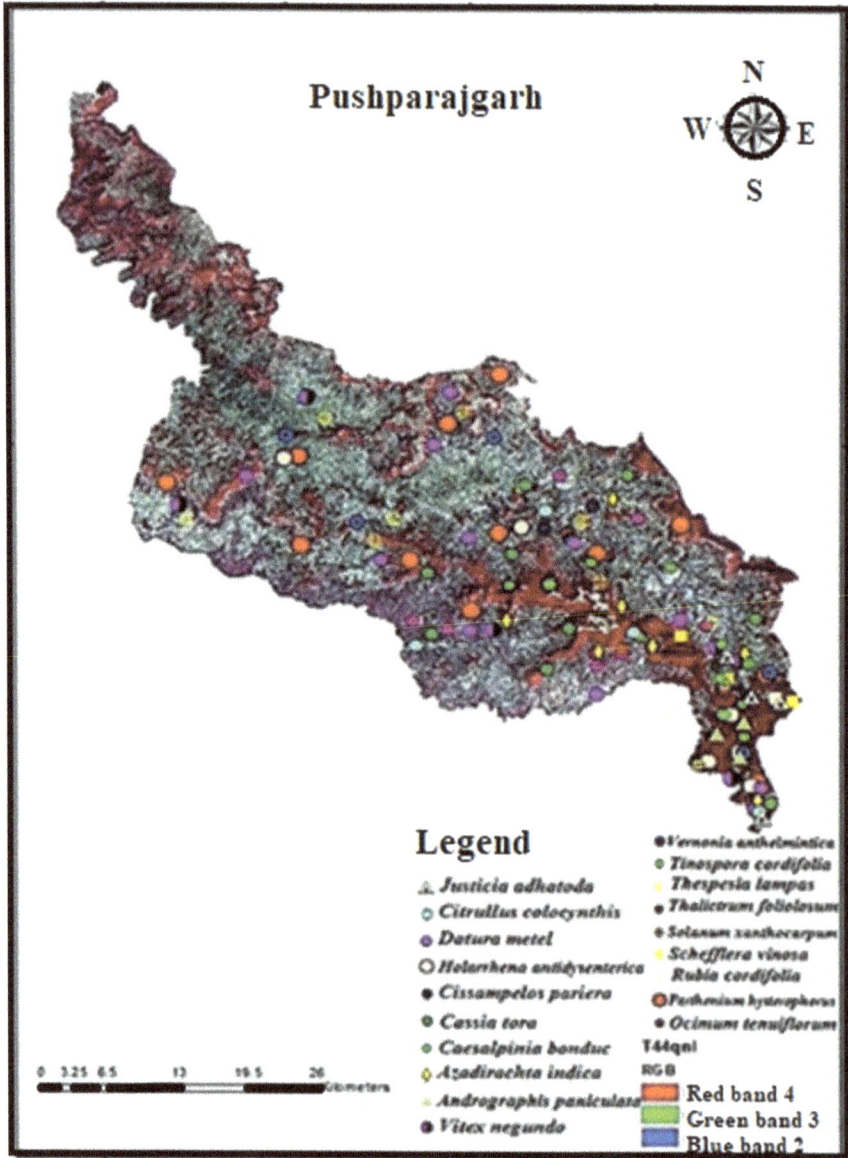

Fig. 1.6: Spatial distribution mapping of 19 antimalarial plants by using ArcGIS [91].

GIS models contribute widely in assessing the distribution of plant species for biodiversity conservation. Several models are effective in evaluating the association between plant taxa and environmental dynamics, identification of suitable area for adaptability of critically endangered endemic species, distribution of invasive species, and determining plant taxa response to climate change. GIS prediction model comprises Maximum Entropy (MaxEnt), Decision Support System for Agrotechnology Transfer (DSSAT), Crop Niche Selection in Tropical Agriculture (CaNaSTA), etc. The most accurate and relevant model among others for species distribution mapping is Maximum Entropy (MaxEnt). Since it is a niche modeling method involving the information of species distribution based on identified occurrences along with variables such as topography, biogeography, soil, and climate. Also, the model efficiently represents current and future potential of distribution pattern of plant species [91]. For instance, Yang et al. [92] used MaxEnt model to predict the potential distribution of medicinal plant in the lesser Himalayan foothills resulting in highly accurate statistical data for species restoration and conservation. MaxEnt models provide valuable data in conserving the endemic and critically endangered wild medicinal plants by determining suitable areas that has the potential for the growth of plant species under threat (Fig. 1.7).

RS with high resolution imageries can thoroughly investigate stocks of wild medicinal plants in a small area providing data on the height of the medicinal plants and the number of individuals per square meter and significantly reduces the physical barriers of inaccessible areas. Aerial photographs, multispectral, and hyperspectral near infrared cameras in a drone, aerial photography combined with unmanned aerial vehicles incorporated in geospatial technologies provides complete biometric information from key areas. These are high-resolution photographs containing qualitative and quantitative database of plants. Analysis through visualization can be time-consuming for a large area; however, it gives maximum precision in monitoring the stocks of medicinal plants in small areas. An example of this application is represented by Fadeev et al., 2017 for monitoring stocks of medicinal plants and mapping them using high resolution images over a small area.

The maps created by GIS are dynamic as they can contribute significantly in understanding and identifying the spatial pattern of the medicinal plants to formulate more targeted and conservative strategies. The geospatial technological maps and models can provide us with information on existing conservation status of the plant population and suitable sites for future plant species collection or restoration. Their enormous ability to monitor and analyze the wild medicinal plant species in large as well as small areas has greatly reduced the burden of inaccessibility of surface data for researchers.

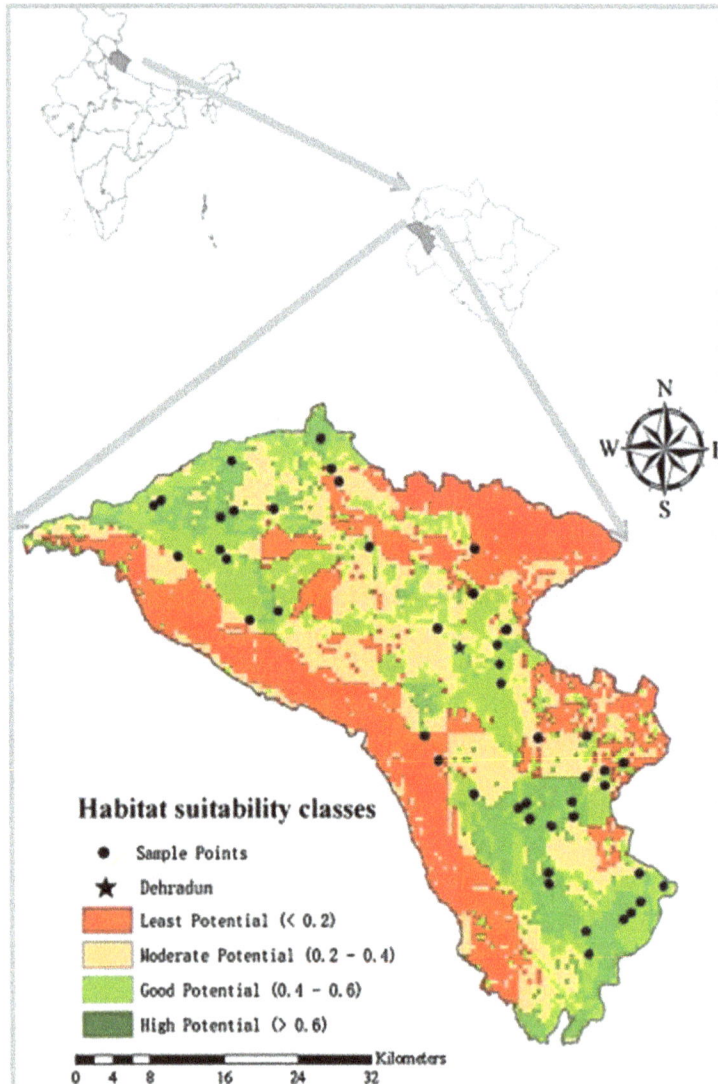

Fig. 1.7: Spatial distribution mapping of J. *adhatoda* within Dun valley using MaxEnt model [92].

1.3 Conclusion

The importance of medicinal plants in healthcare and therapeutics is irrefutable. Currently, the habitat of medicinal plants is depleting due to the rapid industrialization, unlimited population growth, and negligence of conservation. Consequently, there is an urgent need to identify their habitats and enforce strict conservation measures for judicious human utilization. GIS tools may be promising tools to analyze the present and future status of medicinal plants and may play a cardinal role in conservation and planning protection measures for medicinal plants diversity in the ensuing future. In addition to studying the habitat and distribution of medicinal plants there is also a burgeoning need to preserve the indigenous customs and knowledge related to medicinal plants for the future human generations.

References

[1]   V. Subhose, P. Srinivas, A. Narayana. Bull Indian Inst Hist Med. 2005, 35, 83–92.
[2]   WWF, 2022; https://www.un.org/development/desa/en/news/forest/forests-a-lifeline-for-people-and-planet.html.
[3]   C. B. Schmitt, A. Belokurov, C. Besançon, L. Boisrobert, N. D. Burgess, A. Campbell, L. Coad, L. Fish, D. Gliddon, K. Humphries, V. Kapos, et al. Analyses and Recommendations in View of the 10% Target for Forest Protection under the Convention on Biological Diversity (CBD). 2nd ed, Freiburg University Press, Freiburg, Germany, 2009.
[4]   G. Jegan, P. Kamalraj, K. Muthuchelian. Ethnobot Leafl. 2008, 12, 254–260.
[5]   S. K. Prajapati, K. Sharma, P. K. Singh. Trop Ecol. 2018, 59(3), 505–514.
[6]   M. Adnan, S. Begum, A. L. Khan, A. M. Tareen, I. J. Lee. J Med Plant Res. 2012, 6(24), 4113–4127.
[7]   S. Khana, T. H. Masoodia, M. A. Islama, A. A. Wania, A. A. Gattooa. Phytomed Plus. 2022, 2, 2.
[8]   N. Jamba, B. M. Kumar. Front Environ Sci. 2018, 5, 96.
[9]   U. Magsar, E. Baasansuren, M. E. Tovuudorj, et al. J Ecology Environ. 2018, 42, 4.
[10] G. Yaseen, M. Ahmad, S. Sultana, A. S. Alharrasi, J. Hussain, M. Zafar, S. U. Rehman. J Ethnopharmacol. 2015, 163(2), 43–59.
[11] R. M. Kunwar, Y. Uprety, C. Burlakoti, C. L. Chowdhary, R. W. Bussmann. Ethnobot Res Appl. 2009, 7, 5–28.
[12] H. Liang, Y. Huang, S. Lin, K. Zhao, Y. Zou, X. Yu, L. Yang, W. Xu, Z. Ming, Z. Zhou. Am J Plant Sci. 2016, 7, 2527–2552.
[13] M. Neblea, M. Marian, M. Duţă. ActaHortic. 2012, 955, 41–49.
[14] H. Malve. J Pharm Bioallied Sci. 2016, 8(2), 83–91.
[15] R. Kumar, A. Tewari. Synthesis of Medicinal Agents from Plants. 1st ed, 2018, pp. 257–282.
[16] W. Muhammad Amiruddin, S. A. M. Sukri, S. M. Al-Amsyar, N. D. Rusli, K. B. Mat, et al. Earth Environ Sci. 2021, 756, 012022.
[17] D. Lange, U. Schippmann. Bundesamt fur Naturschutz, Bonn, 1997.
[18] WHO traditional medicine strategy: 2014–2023; ISBN 978 92 4 150609 0, NLM classification: WB 55.
[19] C. Bodeker, C. K. Ong, C. Grundy, G. Burford, K. Shein. Vol 2 of WHO Global Atlas of Traditional, Complementary and Alternative Medicine. WHO Centre for Health Development, Geneva, 2015.
[20] A. M. Barata, F. Rocha, V. Lopes, A. M. Carvalho. Ind Crops Prod. 2016, 88, 8–11.
[21] D. J. Newman, G. Cragg2012. Natural products as sources of new drugs over the 30 years from 1981 to 2010. J Nat Prod. 2012, 75(3), 311–335.
[22] S. L. Chen, Y. Q. Zhou, C. X. Xie, R. H. Zhao, C. Z. Sun, J. H. Wei, et al. China J Chin Mater Med. 2008, 33(7), 741–745.
[23] F. M. Suo, S. L. Chen, D. Q. Ren. Chin J Chin Mater Med. 2005, 30(19), 1485–1488.
[24] http://www.gbif.org (retrieved on 25.03.2022)
[25] http://www.kew.org (retrieved on 29.03.2022)
[26] J. Wu, X. Li, L. Huang, et al. Chin Med. 2019, 4, 14.

[27] S. J. Vijayalatha. Indian J Arecanut Spices Med Plants. 2004, 6(3), 98–107.

[28] S. Singh. Agric Today. 2005, 3(3), 58–60.

[29] R. A. Mittermeier, W. R. Turner, F. W. Larsen, T. M. Brooks, C. Gascon. Global biodiversity conservation: The critical role of hotspots. In: Zachos, F. E., Habel, J. C., (Eds.), Biodiversity Hotspots. Springer, London, 2011, 3–22.

[30] N. Myers. Environmentalist. 1990, 10, 243–256.

[31] C. Marchese. Glob Ecol Conserv. 2015, 3, 297–309.

[32] H. M. Pereira, S. Ferrier, M. Walters, G. N. Geller, R. H. Jongman, R. J. Scholes, M. W. Bruford, N. Brummitt, S. H. Butchart, C. Cardoso, N. C. Coops, E. Dulloo, et al. Science. 2013, 339, 277–278.

[33] CEPF (Critical Ecosystem Partnership Fund) – The Biodiversity hotspots. Retrieved on November 7, 2014, http://www.cepf.net/resources/hotspots/Pages/default.aspx

[34] K. J. Williams, A. Ford, D. F. Rosauer, N. De Silva, R. Mittermeier, C. Bruce, F. W. Larsen, C. Margules. Forests of east Australia: The 35th biodiversity hotspot. In: Zachos, F. E., Habel, J. C., (Eds.), Biodiversity Hotspots. Springer, London, 2011, 295–310.

[35] R. F. Noss, W. J. Platt, B. A. Sorrie, A. S. Weakley, D. B. Means, J. Costanza, R. K. Peet. Divers Distrib. 2015, 21, 236–244.

[36] N. Myers, R. A. Mittermeier, C. G. Mittermeier, G. A. Da Fonseca, J. Kent. Nature. 2000, 403, 853–858.

[37] V. S. Chitale, M. D. Behera, P. S. Roy. PLoS ONE. 2014, 9(12), e115264.

[38] S. J. Irwin, D. Narasimhan. Rheedea. 2011, 21, 87–105.

[39] Forest Survey of India (FSI) Website, India State of Forest Report, 2011. http://www.fsiorgin/sfr_2011htm

[40] M. C. Wambulwa, R. I. Milne, Z.-Y. Wu, R. A. Spicer, J. Provan, Y.-H. Luo, G.-F. Zhu, W. T. Wang, H. Wang, L.-M. Gao, D.-Z. Li, J. Liu. Ecol Evol, 2021, https://doi.org/10.1002/ece3.7906.

[41] K. S. Valdiya. Prog Phys Geogr. 2002, 26, 360–399.

[42] R. A. Mittermeier, P. R. Gil, M. Hoffmann, J. Pilgrim, T. Brooks, C. G. Mittermeier, G. A. B. da Fonseca, CEMEX. 2004.

[43] D. Borah, N. Gap, R. K. Singh. Phytotaxa. 2020, 430, 287–293.

[44] D. Maity. Edinb J Bot. 2014, 71, 289–296.

[45] U. Dhar, R. S. Rawal, J. Upreti. Biol Conserv. 2000, 95, 57–65.

[46] C. P. Kala, A. Nehal, A. Farooquee, U. Dhar. Biodivers Conserv. 2004, 13, 453–469.

[47] W. C. D. ICIMOD, R. E. C. A. S. T. GBPNIHESD. Kathmandu: ICIMOD. Vol. 9, 2017.

[48] G. H. Dar, A. A. Khuroo. Sains Malays. 2013, 42(10), 1377–1386.

[49] J. Xu, R. E. Grumbine, A. Shrestha, M. Eriksson, X. Yang, Y. Wang, A. Wilkes. Conserv Biol. 2009, 23, 520–530.

[50] Y. Uprety, H. Asselin, E. K. Boon, S. Yadav, K. K. Shrestha. J Ethnobiol Ethnomed. 2010, 6.

[51] Y. Uprety, R. C. Poudel, H. Asselin, E. Boon. Environ Dev Sustain. 2011, 13, 463–492.

[52] A. Mehra, O. Bajpai, H. Joshi. Trop Plant Res. 2014, 1, 80–86.

[53] S. Pant, S. S. Samant. Ethnobot Leafl. 2010, 14, 193–217.

[54] R. S. Chauhan, S. Kitchlu, G. Ram, M. K. Kaul, A. Tava. Ind Crop Prod. 2010, 31, 546–549.

[55] J. A. Bhat, M. Kumar, R. W. Bussmann. J Ethnobiol Ethnomed. 2013, 9.

[56] P. Bantawa, R. Rai. Nat Prod Radiance. 2009, 8, 537–541.

[57] M. P. Nayar. Bull. Bot. Sarv. India, Vol. 22, 1982, pp. 12–33.

[58] M. Gadgil. J Indian Inst Sci. 1996, 76, 495–504.

[59] Farnsworth. Global Importance of Medicinal Plants, Conservation of Medicinal Plants – Proc. International Consultation. 21-27 March, Chian Mai University, Thailand, 1988, Akerale et al, 1991, Cambridge University Press, Cambridge, UK.

[60] K. Jeyaprakash, M. Ayyanar, K. N. Geetha, T. Sekar. Asian Pac J Trop Biomed. 2011, S20–S25.

[61] N. P. Rajith, V. S. Ramachandran. Indian J Nat Prod Resour. 2010, 1(2), 249–253.

[62] S. Sukumaran, R. M. Sujin, V. S. Geetha, S. Jeeva. Acta Ecol Sin. 2021, 41(5), 365–376.

[63] A. L. Sajem, J. Rout, M. Nath. Ethnobot Leafl. 2008, 12, 261–275.

[64] D. R. Chhetri, P. Parajuli, G. C. Subba. J Ethnopharmacol. 2005, 99, 199–202.

[65] Forest Survey of India, State of forest report. Dehradun, India, 1999

[66] K. Kathiresan. Envis Forestry Bulletin, 2004, 4.

[67] D. Kalita, J. Saikia, A. K. Mukherjee, R. Doley. Int J Med Arom Plants. 2014, 4(2), 97–106.

[68] D. Kalita. J Saha Asian J Pharm Biol Res. 2012, 2(4), 234–239.

[69] D. Kalita, J. Saikia, A. S. Sindagi, G. K. Anmol. Bioscan 2012. 7(2), 271–274.

[70] D. Kalita, J. Saikia. Int J of Phytomed. 2012, 4(3), 380–385.

[71] Tlau, Lalawmpuii. Sci Vision. 2020, 20(4), 156–161.

[72] Kayang H. (2007) Tribal knowledge on wild edible plants of Meghalaya, Northeast India. Indian Journal of Traditional Knowledge, 6(1), 177–181.

[73] J. P. Yadav, S. Kumar, P. Siwach. Indian J Tradit Knowl. 2006, 5(3), 323–326.

[74] Sujatha V. 2020. Globalization of south Asian medicines: knowledge, power, structure and sustainability. Society and culture in south Asia. 6(1):7–30.

[75] S. Sen, R. Chakraborty. J Tradit Complement Med. 2017, 7, 234–244.

[76] N. H. Rakotoarivelo, A. V. Ramarosandratana, V. H. Jeannoda, A. R. Kuhlman. J Ethnobiol Ethnomed. 2015, 11, 68.

[77] T. Ceolin, M. R. Heck, R. L. Barbieri, E. Swartz, R. M. Muniz, C. N. Pilon. Rev Esc Enferm USP. 2011, 45, 47–54.

[78] K. Chan, D. Shaw, M. S. J. Simmonds, C. J. Leon, Q. Xu, A. Lu, I. Sutherland, S. Ignatova, Y. P. Zhu, R. Verpoorte, E. M. Williamson, P. Duez. J Ethnopharma. 2012, 140, 469–475.

[79] N. P. Makunga, L. E. Philander, M. Smith. J Ethnopharma. 2008, 119, 365–375.

[80] A. R. A. Pereira, A. P. M. Velho, D. A. G. Cortez, L. L. D. Szerwieski, L. E. R. Cortez. Rev Rene. 2016, 17, 427–430.

[81] C. Macilwain. Nature. 2004, 428, 792–793.

[82] M. Rahman. Science World: VeshogUdviderRokomfer (in Bangla). Professor's Prokashon. 2003, 46.

[83] B. Singh, R. K. Gopal. PES Bus Rev, 2016, 10.

[84] A. Chaudhary, N. Singh. J Ayurveda Integr Med. 2011, 2, 179–186.

[85] P. G. Xiao. The Chinese approach to medicinal plants. In: Akerele, O., Heywood, V., Synge, H., (Eds.), Conservation of Medicinal Plants. 1991, pp. 305–313.

[86] A. Yadav, P. K. Verma, H. R. Verma. J Med Plants Stud. 2019, 7, 14–17.

[87] F. Liu, J. Vind, P. Promchote, P. Le. Tradit Healers Organ. 2007, 41.

[88] J. Mwitwa, Consultancy report prepared for WWF Southern Africa Regional Programme Office. 2009.

[89] B. Biswas, S. Walker, M. Varun. Plant Arch. 2017, 17(1), 8–20.

[90] M. K. Dwivedi, B. S. Shyam, R. Shukla, N. K. Sharma, P. K. Singh. J Herbs Spices Med Plants. 2020, 26(4), 356–378.

[91] N. Sivaraj, K. Venkateswaran, S. R. Pandravada, N. Dikshit, M. Thirupathi Reddy, P. Rajasekharan, et al. Conservation and Utilization of Threatened Medicinal Plants. Springer, Cham, 2020, pp. 229–274.

[92] X. Q. Yang, S. P. S. Kushwaha, S. Saran, J. Xu, P. S. Roy. Ecol Eng. 2013, 51, 83–87.

[93] M. Khan, S. Kumar, I. A. Hamal. Ethnobotanical Leaflets. 2009, 13, 1113–39.

Saeid Hazrati*, Maryam Mohammadi-Cheraghabadi and Saeed Mollaei

Chapter 2
Threats and conservation of the medicinal plants

Abstract: Medicinal plants (MPs) are universally worth sources of herbal products, and they are vanishing at a high rate. The MP resources are under constant threat of extinction because of population growth, overexploitation, environmental destruction, illegal trade, and unsound harvesting techniques. This chapter discusses about the developments, global trends, and prospects for the strategies and sustainable application of MP resources and methodologies about the conservation to supply a trusty referral for the sustainable application and MP conservation. We accented that both resource management (e.g., good sustainable use solutions and agricultural practices) and conservation strategies (e.g., in situ and ex situ cultivation) should be sufficiently taken to calculate for the sustainable application of MP resources. We recommend that biotechnical approaches (e.g., micropropagation, tissue culture, molecular marker, and synthetic seed technology-based approaches) should be applied to ameliorate yield and shift the power of MPs.

2.1 Introduction

Today, over half of the population in developing nations lacks access to adequate healthcare. This may be due to the fact that poor people neither have access to nor could afford the modern healthcare services. In addition to providing access and affordable medicine, MPs also provide them with an alternative remedy with remarkable employment and income opportunities. Natural products are becoming progressively popular due to their cheaper prices and lack of side effects, and both developing and developed countries are increasing the demand for herbal medicines. In addition to traditional healthcare, plant products are also used in the formulation of modern medicine [1]. MPs are universally worth sources of new drugs. There have been an increasing number of scientific and commercial studies done on MPs found in natural areas (IUCN Species Survival Commission, 2007). Globally, some 50,000–80,000 flowering plants are

*Corresponding author: Saeid Hazrati, Department of Agronomy, Faculty of Agriculture, Azarbaijan Shahid Madani University, Tabriz, Iran, e-mail: saeid.hazrati@azaruniv.ac.ir
Maryam Mohammadi-Cheraghabadi, Department of Agronomy, Faculty of Agriculture, Tarbiat Modares University, PO Box 14115-336, Tehran, Iran, e-mail: mohammadi.maryam@modares.ac.ir
Saeed Mollaei, Phytochemical Laboratory, Department of Chemistry, Faculty of Sciences, Azarbaijan Shahid Madani University, Tabriz, Iran, e-mail: s.mollaei@azaruniv.ac.ir

https://doi.org/10.1515/9783110791891-002

used as medicines. For example, Balunas and Kinghorn [2] stated that there are over 90% of the 1,300 medicinal plants (MPs) applied in Europe are harvested from wild resources; about 118 of the top 150 approval medicines are based on natural sources in the United States. The use of MPs worldwide is increasing rapidly due to the increasing demand for herbal medicines, natural health products, as well as secondary metabolites (SMs) of MPs [1]. In addition, 80% of all people in developing countries use herbal medicines for their primary healthcare, and over 25% of all prescribed medicines in developed countries are derived from wild plants [3].

Due to the growing human population and increasing demand for MPs, an overharvest of wild species is causing a continuous strain on existing resources that results in continuous extinctions of wild species, and, on the other hand, natural wild flora is disappearing at an alarming rate. A highly conservative apprise by Pimm et al. [4] states that the prevalent loss of plant species is among 100 and 1,000 times higher than the envisaged natural extinction speed and that the Earth is losing at minimum one potential main medicine every 2 years. The International Union for Conservation of Nature and the World Wildlife Fund reports that there are between 50,000 and 80,000 flowering plants used for medicinal purposes around the world. In addition, between 15% and 20% of these species are threatened with extinction through habitat destruction and overharvesting [5], and they are rapidly becoming more and more resource constrained due to the increasing population and plant utilization. Though this threat has been known for decades, the accelerated loss of species and habitat destruction worldwide has increased the risk of extinction of MPs, especially in China [1], India [6], Kenya [6], Nepal [6], Tanzania, and Uganda [7]. Larsen and Olsen [8] reported that the sustainable and conservation application of MPs has been studied widely. In order to conserve them, various recommendations have been developed, including the establishment of monitoring systems to monitor and inventory species status, as well as the need for adapted conservation practices based upon both in situ and ex situ strategies. For MPs with incriminatingly low supplies, sustainable use of wild resources could provide an effective conservation alternative. In South Africa and China, the condition is solely serious because of the high requests of large crowds. This chapter reviews universal trends, prospects of the strategies, developments, and methodologies concerning the conservation and sustainable application of MP resources [3].

2.2 Habitat destruction

Global MPs can lose their habitats as a result of habitat destruction, which is an extinction threat. Natural habitats of animals and plants are impacted by humans everywhere today (including high alpine regions, coastlines, rainforests, and deserts). Among the most significant concerns of habitat destruction is the rapid extinction of valuable plant species as a result of the destruction of ecosystems. In South America, Africa, and Southeast Asia tropical rainforests, the number of slash and burn clearings has increased dramatically. Furthermore, soils in these places that have lost their natural vegetation are prone to erosion if they are not allowed to regenerate back to their natural state. In many developing countries, mangrove forests are destroyed for the purpose of aquaculture, cultivation, and other similar activities, leading to the loss of wetlands and coastal plant species. The modification of habitat structures due to various human activities and climate change over the past 100 years has inevitably affected the plant species' ability to cope with climate change (UNEP 2007). More than 70% of the climate-dependent species in the Himalayas have lost their original habitat. Because of an increasing population in these regions, more forested, wetlands, and grasslands are being converted into agricultural lands and human settlements. In addition to fragmenting habitats, building roads through natural environments enhances the spread of invasive species, insects, and diseases, resulting in the loss of valuable medicinal habitats (WWF and IUCN 1997). Therefore, it is imperative that a better understanding of and threat assessment of several plant species be conducted, especially with respect to their ranges of adaptability.

2.3 Distribution of MPs by habitats

Of the 386 families and 2,200 genera that contain MPs, the families Asteraceae, Euphorbiaceae, Lamiaceae, Fabaceae, Rubiaceae, Poaceae, Acanthaceae, Rosaceae, and Apiaceae each hold the majority with the greatest number of MP species (419) occurring in the Asteraceae family. Approximately 90% of MPs are collected from the wild for use in industry. Despite the fact that the industry uses over 800 species of plants, fewer than 20 species of plants are cultivated. In some instances, over 70% of the plants in a collection are destroyed by harvesting the parts like roots, bark, wood, and stems, or even the whole plant in the case of herbs. This poses a threat to the genetic stock of MPs and the diversity of MPs. Medical plants have been a primary source of health for mankind for thousands of years. There is an increase in interest and use of traditional remedies (the so-called botanicals) in urban settings, especially those that are undergoing rapid growth. Research on the topic of conservation and sustainable use of MPs has lagged far behind the demand for this globally important resource. A strong market presence at all levels has driven an increased demand for MPs, which are generally collected from depleted wild populations in shrinking habitats. Over 20,000 species of plants are used as medicines somewhere on the planet, but we know very little about their conservation status, their harvesting within the limits of sustainability, and their cost-effective production alternatives, as a result. Nearly half of these species are potentially endangered by overharvest or habitat loss. MPs are important to local and regional health systems, regional markets, and global security and trade – a reason to conserve tropical forest ecosystems. Already knowledge and tools have overwhelmed the ability to effectively conduct conservation activities. Nevertheless, there are many other ecosystems around the world, which support a medicinal flora of widely varying importance to local health and economics, as well as to regional and global supplies of plant-based medicines. A variety of habitats, taxonomic groups, and social, economic, and cultural conditions influence how these resources are used and conserved, making their management difficult. While meeting these challenges for managing MP resources, the expertise, experience, and capacity developed for dealing with them will also enhance biodiversity resources management in any natural or social environment where plants are used [9].

2.4 Distribution of threatened MPs

MPs are found across a variety of habitats, and their distribution can be viewed formally as a distribution of species across diverse habitats and landscape elements. About 70% of the Indian MPs are found in tropical regions, almost entirely in different types of forests, while only about 30% are found in temperate and alpine regions, where many species have high medicinal value. Based on the studies, a greater proportion of MPs are found in dry and moist deciduous vegetation than in evergreen habitats, or even in temperate habitats. Statistics show that MPs are spread across diverse habitats, with up to one third of them being trees and shrubs, and the remaining one third being plants such as herbs, grasses, and climbers. However, most of the MPs are higher flowering plants. About 90% of the MPs used in industry are collected from the wild and are distributed across 386 families and 2,200 genera, with the largest number of species located in the Asteraceae, Euphorbiaceae, Lamiaceae, Fabaceae, Rubiaceae, Poaceae, Acanthaceae, Rosaceae, and Apiaceae families. Only about 20 species of plants are commercially cultivated, despite the fact that over 800 species are used for manufacturing purposes by the industry. Over 70% of plant collections involve destructive harvesting because parts such as root, bark, wood, and stem, as well as the entire plant for herbs, are used.

2.5 Overexploitation

The MP diversity is seriously threatened by anthropogenic overexploitation, as most people are unaware of the enormous bankruptcy of Earth's resource capital due to overexploitation of MP resources [10]. As a result of overharvesting of MPs, populations of some of the most valuable wild species and their habitats are being threatened. Human activities penetrating into the inner strata are a totally unsustainable practice. The World Conservation Union (International Union for Conservation of

Nature and Natural Resources (IUCN)) estimates that between 50,000 and 80,000 MPs are used worldwide in traditional medicine (TM), among which approximately 15,000 MP species are endangered [5]. Many species of wild plants are at risk of extinction because of overharvesting. Some MPs are insensitive to harvesting pressure, but others can be driven to extinction with the smallest amount of harvesting. The human race is therefore depleting the planet's valuable resources, thereby causing future generations to suffer [10].

2.6 Threats to MP

As conservationists as well as resource users have been concerned about the continuous decline of MP stocks, this is not surprising. Given that humans tend to abuse things that are good for them, many species that provide very real health benefits are also being abused. The global stock of MPs has been at risk for quite some time, a cause for concern among conservationists and resource users alike. There has been an increased demand for MPs by pharmaceutical industries in recent years due to herbal remedies and other natural products. Therefore, plants are being collected from the wild in an unsustainable manner, as they are exploited for their valuable medicinal properties. In addition to habitat specificity and overexploitation, other factors could lead to this loss, such as land use disturbances, the introduction of non-native species, the alteration of habitat, climatic change, heavy livestock grazing, population explosions, fragmentation and degradation of populations, population bottlenecks, and genetic drift. MPs are also traded for economic purposes in an unsustainable manner, resulting in the decline of several high-value MP species due to unsustainable exploitation. This has been clearly shown by studies that show the decline of several high-value MP species.

There are a number of factors that may contribute to the development of MPs, but the nature and extent of these factors as well as their impact may vary between different countries. Therefore, it is important to examine the geographical distribution and biological characteristics of MPs so that they can be used in a sustainable manner [11].

2.7 Categories of threatened plants

In 1995, the IUCN began to recognize updated categories of threatened plants based on geographic range, population size, and fragmentation of populations. "Extinct" (EX) means that there is no reasonable doubt that the last individual of a species has died, "extinct in the wild" (EW) means that a species may only survive in cultivation (80% decline in the last 10 years), and "critically endangered" (CR) means that a species may face extinction in the near future. Vulnerable (VU): A species is VU when it is not CR or endangered but faces a high risk of EW in the medium-term future (50% decline in the last 20 years). A taxon is considered conservation dependent (CD) if its range is targeted for continuing taxonomy-specific conservation programs, and the discontinuation of which would penalize the taxon under one of the threatened categories above within 5 years. Data deficient (DD): When there is not enough data to directly or indirectly estimate a taxon's risk of extinction. Low risk (LR): A taxon is LR when it has been evaluated and does not satisfy the criteria for any of the categories: CR, endangered, VU, CD, or DD. No evaluation: A taxon is not evaluated when it has not been evaluated against the criteria. It was noted throughout the world that several species are endangered, and several are critically rare. Several countries around the world have taken inventory of their rare plants and developed provisional or fairly accurate lists of their threatened plants. About 10% of vascular plants fall into one or the other category of threatened species. Several hundred rare species have been identified from different parts of India by the Botanical Survey of India. This was followed by the publication of a *Red Data Book of India*, which has gathered nearly 620 threatened species from different categories of ICUN, and later on several volumes of *Rare and Endangered* listed out the report of rare and endangered species in India.

2.8 Conservation technique to preserve them

MP resources are being harvested in incrementing masses, greatly from wild crowds. Lately, Bentley [5] enounced that the request for wild resources has added about 8–15% per year in North America, Europe, and Asia. There is a sill under which species reproductive valence becomes irreversibly decreased [12]. There are a number of approaches to ex situ and in situ conservation of MPs, including ensuring both in situ and ex situ conservation [13]. Natural reserves and wild nurseries represent typical examples of preserving medicinal efficacy of plants in their natural habitats, while seed banks and botanic gardens represent best practices for ex situ conservation and future replanting [14]. The biological attributes and geographic dispensation of MPs must be well known to head conservation activities, e.g., to recognize whether species conservation may take place in nursery or in a nature.

A conservation is defined as "the management of human activities that have the potential to enhance biodiversity in order to produce the greatest sustainable benefit for the present generation and ensure that future generations may benefit as well." The above explanation uses two complementary terms "conservation" and "sustainability." Sustainable development can be supported by conservation, which seeks to protect and use biological resources that do not annihilate global species diversity or destroy important habitats and ecosystems. There are many aspects to conservation of plant resources. These include collecting, propagating, describing, identifying, evaluating, eliminating diseases, storing, and sharing them. Plant resources have long been recognized as integral to biodiversity conservation.

Conservation of biodiversity aims at (1) sustaining ecological processes and life support systems upon which human survival and economic activities depend, (2) conserving species and genetic diversity, and (3) utilizing organisms and ecosystems in a way that is sustainable without causing damage to them. To conserve and sustainably utilize MPs, it is essential to adopt a holistic and systematic approach that takes into account the relevant aspects of protection, preservation, maintenance, exploitation, conservation, and sustainable utilization. An integrated approach relying on the interaction between social, economic, and ecological systems will be more advantageous [9].

There are two methods for the conservation of plant resources, namely, in situ and ex situ conservation (Fig. 2.1).

2.8.1 In situ conservation

2.8.1.1 In situ conservation by establishment of natural reserves or biosphere resources

Figueiredo and Grelle [15] reported that maximum MPs are endemic species, and their medicinal attributes are mostly because of the presence of SMs that respond to stimuli in natural environments, and that may not be represented under culture situations. In situ conservation of whole communities lets us to support native plants and retain natural communities, along with their complex network of communications [15].

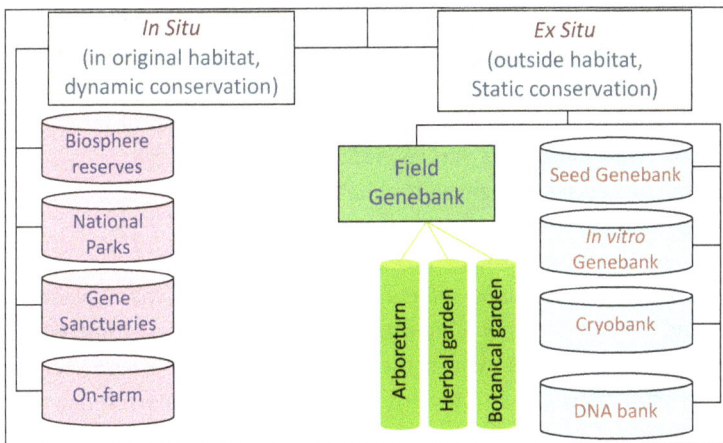

Fig. 2.1: Conservation strategies for threatened MPs.

In addition, in situ conservation increments the content of diversity that can be conserved and strengthens the link among sustainable application and resource conservation [16]. Ma et al. [17] stated that in situ conservation attempts worldwide have focused on organizing supported areas and taking an approximate that is ecosystem-oriented. Prosperous in situ conservation pertains on regulations, rules, and potential compliance of MPs within growth habitats [18]. In natural reserves, the depreciation and demolition of habitats is the main cause of the loss of MP resources. There are more than 12,700 protected areas around the world, which are enough to cover 13.2 million km^2 or 8.81% of the Earth's land surface. They were created in order to protect natural resources and restore biodiversity [13]. Protecting MPs by conserving key natural habitats needs recognizing the portions and ecosystem yields of individual habitats [19]. In wild nurseries, it is not possible to determine every natural wild plant habitat as a protected area, owing to expense competing and in consideration of land applications [20]. Specifically, a wild nursery is designed to cultivate and domesticate the VU plants in a natural habitat, a conserved area, or a place close to where the plants naturally grow [3]. Nonetheless, some wild species are facing hard times due to overexploitation, habitat destruction, and invasive species. However, nurseries offer an effective approach to protecting endangered, endemic, and threatened species that are in request [21]. There are some major benefits of this system:

(1) It is generally more likely to save a wide range of interesting alleles.
(2) It is especially suited to species, which cannot be established or reproduced outside of their natural habitats, including (a) those that are part of complex climax ecosystems, (b) those that have seeds nongerminating or dormant and cannot be manipulated by known artificial methods, and (c) those that have highly specialized breeding systems and depend on a single insect, bird, or bat species for pollination, dependent on other ecosystem components.
(3) In addition to allowing natural evolution to continue, it provides breeders with a source of native resistance for disease and pest-resistant species, which can co-evolve with their parasites.
(4) It can serve several sectors simultaneously and provide gene pools of interest to different sectors (e.g., crop breeding, forage production, and wildlife).
(5) It enables research on species in their natural habitats.
(6) It assures the protection of associated species.

Several studies have shown that the most effective and cost-effective way to preserve existing biological and genetic diversity is by conducting on-site conservation, which protects and preserves the wild species within their natural habitats. A species-centric approach is prone to extinctions and also toward increasing human interference with nature, as in the setting up of biosphere reserves, national parks, wildlife sanctuaries, sacred groves, and other areas protected. A total of more than 4.5% of India's land area is the protected area network, including 8 designated biospheres, 87 national parks, and 447 wild life sanctuaries. Today, the idea of setting up protected area networks is a central part of the biodiversity conservation policy decisions at national, international, and global levels. Biodiversity and medicinal and aromatic plants (MAPs) are present in a number of biogeographic zones and biomes of this network. In addition, there are a number of sacred groves scattered across the country, especially in the southwest and east, which are also active in situ conservation projects. Conservation areas contribute significantly to the conservation of biological diversity of our country, but experience has showed that declaring protected areas in a densely populated country like India is not enough to conserve the rapidly disappearing biological diversity. The success of any conservation program depends solely on the efficiency of management of protected areas, and the importance of local community involvement in conservation has now been realized. As a result, it is crucial to create an environment that emphasizes people–nature interaction. This will cultivate a sense of responsibility among the local people for the value and importance of biodiversity and the need to utilize it effectively for their own survival and those of the ecosystem as well. MPs can be effectively conserved in situ by actively engaging the people in or near the protected forest areas. Working with the local community during every phase of a conservation program, such as planning, policy-decision processes, and implementation, will contribute to sustainable MP management and utilization.

2.8.2 Ex situ conservation

The concept of ex situ conservation encompasses the practice of restoring a species outside its native habitat, usually to safeguard populations where their destruction, replacement, or degradation are imminent. In other words, off-site conservation, or ex situ conservation, is the conservation of a plant's genetic resources outside its natural habitat. Ex situ conservation is often achieved by establishing MP gardens, artificial plant regeneration in botanical gardens, and creating gene banks for MPs. There are several ways to conserve ex situ species, including seed storage, DNA storage, pollen storage, in vitro conservation, seed and field gene banks (FGBs), MP conservation parks (MPCPs), and botanical gardens.

In controlled conditions, ex situ conservation helps species recover, and it makes reintroductions into the wild possible.

Ex situ conservation is not always sharply a part from in situ protection, but it is an affective complement to it, mostly for those overexploited and imperiled MPs with low abundance, gradual growth, and great susceptibility to replanting illness [3]. Ex situ protection helps to cultivate and naturalize threatened species to make sure their included survival and rarely to generate large quantities of the planting material applied in the expansion of medicines, and it is mostly an instant action taken to retain MP resources. The number of many species of formerly wild MPs can not only maintain high power when grown in bowers far away from the habitats where they naturally happen, but also can have their reproductive materials elected and saved in seed banks for subsequent replanting [3]. In addition to their contribution to ex situ conservation, botanic gardens can enhance the local ecologies so that rare and endangered plant species can continue to thrive [8, 22]. Abiotic collection typically includes only a few individuals of a given species and so are of limited use in terms of genetic protection [23], but botanic gardens have several unique advantages. They involve a wide diversity of plant species grown together under common situations, and mostly contain ecologically and taxonomically various flora. Using botanical gardens for MP protection, implementing protocols for propagation and cultivation, as well as undertaking international breeding programs can make a significant contribution to its protection.

FGBs can be used to conserve most MPs of the different forms of ex situ conservation. FGBs are probably the most costly to establish and maintain, especially in terms of labor and equipment. As a result of economic and practical difficulties in managing FGBs, MPCPs have emerged as the ideal place to establish FGBs and grow live plants of medicinal value, since the urgent need to improve crop yields coincides with economic and practical problems in creating FGBs. Linkages between conservation and use required for the sustainable long-term management of FGBs have been established within MPCPs. Ethnomedical forests, herbarium materials, and raw components of plants are additional strategies. In India, about 3,200 herb gardens have been established in over 300 villages of the southern region; additionally, existing botanic gardens, arboreta, and other sites can be utilized to conserve plants ex situ.

In addition, seeds should be able to withstand desiccation up to 10% so that they can store the genetic diversity of many MPs ex situ as opposed to botanic gardens. Seed banks can offer a better way to store the genetic diversity of many MPs than botanic gardens, and the result is an increase in the biological and genetic diversity of wild plant species. Thus, only species of plants whose seeds could survive the drying process should be retained [24]. In conserving tissues, there are six major steps in the conservation process defined by the conservation use cycle [25] to preserve tissues. In vitro conservation involves collecting, quarantining, propagating, characterization, evaluating, monitoring, storing, and distributing. These techniques should complement other conservation efforts within the greater conservation program of a species or population. Generally, field conservation of MPs involves more space and labor, and it is costly and risky when it comes to damage caused by natural calamities and biotic stresses. Fortunately, recent developments have made it possible to preserve these species in vitro. Tissue culture systems provide some advantages for some species. It must be possible to keep the in vitro collections of species at the same or separate locations but should have clear linkages with FGBs. These factors include (1) very high multiplication rates; (2) aseptic system free of fungus, bacteria, viruses, and pest insects, and the production of pathogen-free stocks; (3) reduced space requirements; (4) reduction in genetic erosion under optimal storage conditions; and (5) reduction of labor costs.

2.9 Cultivation practice

Though wild-harvested resources of MPs are widely attended to be more effective than those that are cultivated, domestic cultivation is a widely applied and commonly accepted practice [16]. Cultivation caters the opportunity to the application of new techniques to solve problems conflicted in the generation of MPs, such as pesticide contamination, toxic components, low amounts of active ingredients, and the misidentification of botanical source. Cultivation under controlled growth situations can ameliorate the yields of active composites, which are almost always SMs, and ensures generation stability. In order to achieve optimal yields of target products, cultivation practices take into account various factors such as water, nutrients, optional additives, temperature, light, and humidity [23]. Moreover, incremented cultivation contributes to declines in the harvest volume of MPs, profits the recovery of their wild resources, and declines their prices to a more sensible range [3]. Various species require different cultivation conditions, and the WHO recommends rotation of crops to prevent insect and disease problems. The cultivation of MAPs can take place through traditional methods or by using conservation agriculture practices such as no-till farming. Since plant characteristics vary greatly depending on the soil type and cropping strategy used, intensive care is required to obtain good yields.

2.9.1 Plant conservation through applied agronomy

In spite of progress made in synthesizing organic chemistry and biotechnology, higher plants still remain a source of drugs and galenic preparations. In developing countries, wild plants are utilized as a main source of higher plants as drugs in traditional and modern medicine. The cultivation of MPs is very rare since so many disadvantages are associated with collecting plants in the wild (Tab. 2.1). Therefore, the need for modern methods of cultivation has grown. Obtaining high-quality MPs requires following cultivation, post-harvest handling, processing, and comprehensive quality control procedures. Determining the genetically enhanced MPs that exhibit high concentrations of active compounds under specific environmental conditions can be achieved. Modern technologies of processing and preserving the raw material of MPs help maintain the quality of the product for a longer period of time.

Cultivating MPs under modern cultural practices represents the primary means of conserving or protecting the genetic material of rare, endangered, or overexploited species, by ensuring that genetic diversity is maintained through selective breeding and selection of the wild flora. The cultivated plants may also provide high-quality raw materials for further processing and/or preparation of galenic products. Agricultural methods are used to domesticate and cultivate MPs in order to preserve and protect them [26].

Tab. 2.1: Collection from the wild versus cultivation of medicinal and aromatic plants.

	Cultivation	Collection
Accessibility	Increasing	Decreasing
Variation of supply	More controlled and quality	Unbalanced
Botanical identification	Not doubtful	Sometimesnot reliable
Quality control	High	Little
Genetic improvement	Appropriate	Not appropriate
Postharvest management	Good	Weak
Agricultural manipulation	Yes	No
Adulteration	Relatively safe	Maybe
Yield	High	Low

2.9.2 Domestication strategy

Domestication of MPs is divided into two phases: at the natural location and at the cultivation site. The natural locations are studied for a range of reasons, including systematic botany, climatic conditions, soil properties, plant development, physiology, propagation, and susceptibility to pests and diseases. The site is also studied for advanced cultural practices, genetic improvement, weed and pest control, the optimal harvesting timing, mechanization, postharvest treatments, raw material inspection and quality control, as well as phytochemical analysis.

2.9.3 Agronomic contribution

Agronomical and chemical traits can be bred into new cultivars using classical and modern breeding methods, thereby preserving highly valuable germplasm in seed banks and botanic gardens. Stable raw materials, with standardized properties, can be produced from MPs and can be used without additional processing. Compared to other categories, MPs have received few studies on agronomic aspects. Various techniques have been investigated for improving the performance of MPs, including genetic improvement, optimal environmental conditions, cultivation under modern cultural practices, and postharvest treatment [26].

2.9.4 Genetic improvement

In future, we will be able to cultivate MPs under conditions outside the current collection sites by breeding improved cultivars that are adapted to different agroecological regions. Recently developed analytical techniques such as the radioimmunoassay and the enzyme-linked immunosorbent assay provide a potential means for breeding cultivars containing more of a specific compound and/or a desired range of SMs. As a result of these analytical methods, both conventional and novel breeding methods no longer take as much time as they once did. Once a suitable cultivar or clone has been identified, it can be propagated using traditional methods like seed division or new methods like micropropagation. There are a number of agronomic and chemical characteristics that make a cultivar suitable for this purpose: uniform seed germination, high biomass yield, resistant lodging, high proportion of desired organs, branching habits, adaptability to mechanical harvest, resistance to pests, diseases, and weeds, as well as an ability to adapt to environmental conditions are agronomic characteristics. Breeding techniques used in genetic manipulation are selection, hybridization, mutation, and polyploidy. The most common phytochemical traits are: a desired spectrum of compounds and a high concentration of the active compound(s) [27].

2.9.5 Selection

There is a wide range of genetic variability in the MPs, so it is relatively easy to select genotypes for desired traits. Selection for suitable cultivars must also take into account both biomass yield and phytochemical content. There is a significant difference between the heritability of these two different traits. The content of the chemicals, in general, has a dominant role; however, the combined yield of biomass and chemical compounds have a much lower heritability. Furthermore, the inherent genetic variability in the wild population allows us to improve thebaine content or produce capsules without shattering.

2.9.6 Crosses and hybridization

Combining desired traits in a hybrid cultivar can result in a stable genotype. Hybrid cultivars offered two benefits: the possibility of increasing yield through heterosis and the protection of seeds under a patent.

2.9.7 Mutation breeding

Chemical or radiation-induced mutation can be an effective breeding method for generating new genotypes; however, limited application of this technique has been achieved for plant breeding largely due to the lack of rapid screen methods for identifying the desired genotype among large populations of plants. As a result, researchers have focused their attention on easily recognizable qualitative traits (e.g., morphological traits such as early flowering and seed germination); only very limited work has been done to identify better genotypes of MPs rich in active compounds; mutagenesis plays an important role in this regard, especially when dealing with vegetative propagated plants with limited natural genetic variability [26].

2.9.8 Polyploidy breeding

There are two basic types of induced polyploidy: autopolyploidy, an augmentation of chromosome number within the species; and alloploidy, an augmentation of chromosome number following interspecific hybridization. Induced artificial polyploids appear to perform differently in different types of crop plants. The results of induced artificial polyploids have consistently been suboptimal both in terms of crop plants in general and MPs in particular.

2.9.9 Optimal environmental conditions

Since most alkaloids are formed in young, growing tissues, factors that affect plant growth can affect the production of the SMs as well. Although alkaloids are largely genes-governed, environmental factors are also important for controlling plant growth and forming SMs. There are many environmental factors that affect the production of MPs, including temperature, light intensity, photoperiodism, mineral and water supply, as well as the altitude above sea level. It is important to realize that not all of these factors act the same way in all MPs. In the highest temperature regime, the levels of gamma-linolenic acid were very low, and on the highest temperature regime, there was a very reduced yield component. Various environmental factors need to be addressed, such as extreme temperatures, water stress, and changes in light interception. Accordingly, in countries with high temperatures, both grain yield and seed quality will be lower than in cooler countries. Although steroid levels increased at a higher altitude for all species, for the *Dioscorea* species, tuber yields were lower at the higher altitude, so diosgenin levels remained the same in both regions [28].

2.9.10 Agrotechnical improvement

The most important forms of cultural practices are propagation, planting dates, irrigation, and fertilization; herbicides and pesticides; and postharvest treatments. The most important forms of cultural practices are plant propagation, sowing dates, irrigation and fertilization; herbicides and pesticides; and postharvest treatments. For manipulating phytochemicals and/or biomass yields, chemical compounds as plant growth regulators can manipulate seed germination, plant growth patterns, flowering, phytochemical quantity and content, leaf defoliation, and postharvest incubation.

2.9.11 Plant establishment

In addition to cultivating plants that thrive in the field, we can experiment with procedures for improving seed germination and emergence, vegetative propagation, and soil moisture conditions that permit better plant growth. Since many MPs were obtained from the wild, their growth rates and germination rates are generally erratic and inadequate. Several seed germination promoters, particularly gibberellins, are known to be very effective upon overcoming dormancy and for speeding the process of germination and emergence, among others. Gibberellic acid (GA3) is by far the most potent of these agents on overcoming dormancy and increasing the speed and uniformity of germination and emergence.

2.9.12 Cultivation procedures

One of the most important agrotechnical and environmental factors is water availability. There are several cases when water stress increased the content of SMs in plants despite a limited water supply. Generally speaking, a limited supply of water adversely impacts plant development.

2.10 Good agricultural practices (GAPs)

Good agricultural practices (GAPs) for MPs have been formulated to adjust generation, make sure quality, and comfort the standardization of herbal drugs. The GAP approach makes sure high-quality, pollution-free, and safe herbal medicines (or crude medicines) by using available science to address different problems [29]. Germplasm, ecological conditions of production sites, plant collection, cultivation, inspection of metal elements, and macroscopic or microscopic production quality are included in GAP. There are a number of countries that actively do their part to implement GAP. For instance, in China, GAP is being promoted for the cultivation of commonly used herbal drugs in areas where these medicines have traditionally been grown [19]. Organic farming has obtained the incrementing regard for its capability to make united, environmentally, humane, and economically sustainable generation systems for MPs [30]. The targets of organic farming of MPs contain generating materials with high productivity and better quality, and ensure the protection and sustainable application of those plants. The defining attributes of organic farming are the nonapplication of synthetic fertilizers, herbicides, and pesticides, which do not let match with many current organic certification standards in North America and Europe [31]. Rigby and Cáceres [30] stated that organic farming is safe to the environment and depends upon farm-taken renewable resources to the ecological balance of habitats and retain biological processes of MPs. The application of organic fertilizers continuously improves the soil stability and supplies soil nutrients, considerably affecting the biosynthesis of essential substances and the growth of MPs. For instance, when organic fertilizers were used, the biomass product of *Chrysanthemum balsamita* was incremented, and its essential oil amount was great relative to those free from organic fertilizers. Organic farming of MPs is becoming incriminatingly main in the sustainability of MPs and the long-term development [32]. Sustainable application of MPs with restricted slow growth and abundance, and destructive harvesting commonly results in resource fatigue and even species extinction [8]. So, good harvesting practices must be formulated, and the sustainable application of MPs should be significant. Root and whole-plant harvesting is more wrecking to MPs (e.g., shrubs, tree, and herbs) than collecting their flowers and leaves or buds. For herbal medicinal made of roots or whole plants, applying their leaves as a treatment can be a benign other. For example, Wang et al. [31] found that ginseng root extract and leaf stem extract have similar pharmacological actions, but leaf or stem ginseng is more sustainable. Conserving (ex situ and in situ conservation) and cultivating MPs should be prioritized in order to prevent further depletion of herbal resources. It is recommended that the government promotes traditional methods for preserving forests, and provides proper assessment of population size, mapping, and biology of threatened species in order to conserve MPs.

The assessment of the biology and population size of threatened plants, the promotion of commercial cultivation of MPs among local farmers, and the dissemination of relevant information (conservation task) through print and electronic media should be undertaken. Factors determining the conservation protocol are range of application, time, capacity for reproduction, storage, infrastructure, security, efficiency, access, sustainability, risks in conservation, regeneration capability, and cost.

2.11 The challenges in conservation and use of MAPs

The number of MPs disappearing or changing rapidly is due to a number of human-induced factors. The majority of MPs are typically obtained from the wild when they are naturally growing; nevertheless, due to overharvesting, deforestation, desertification, and global warming to name a few, MPs are facing the issue of extinction [32]. Amujoyegbe et al. [33] showed that MPs are incriminatingly destruction, not only because they are very favorable for primary healthcare but also because they provide some other purposes such as food, trade, firewood timber, and building poles. Hamilton [3] stated the specific importance of MPs in conservation stems from the main livelihood, and cultural or economic duties that they play in many people's lives. Pimm et al. [4] introduced that basic plant resources of nature are actually threatened by lack of sustenance, overuse, and intensified human development activities. In order to preserve these valuable plant resources, it is paramount to work toward their conservation, not just to preserve nature's bounty but also for the well-being and livelihood of indigenous local communities and society at large.

To follow Dajic-Stevanovic and Pljevljakusic's concluding remarks [34], the most usual challenges with which those who are designing MPs have to deal are the accessibility and abundance of wild populations, market and labor availability, and agrobiotics investment in machine, generation technology, processing capability and costs. Similarly, Saalu [35] noted that the price of modern medicines is steadily increasing with improvements in modern health technology, which often makes them unsuitable to the instant needs of many people living in developing nations. Shingu [36] also noted that the lack of standardization is a challenge to MP conservation, alongside other factors such as loss of MP species and damage to ecosystems. Coherent and direct attempts to conserve plant species have obtained research support with relatively small policy and consideration. MPs have been systematically protected and conserved over the course of history, striving to prevent their extinction. However, this has not been an easy process due to a number of factors that hinder the success [37]. Aside from habitat destruction and overexploitation of MPs, researchers have identified financial difficulties, a lack of proper education of the masses, and a prioritization of species that should be protected as several of the main problems facing MP conservation in Anambra state[38].

Matching to Idu and Onyibe [38], in spite of the incrementing application of MPs, their future, apparently, is being threatened by complacence concerning their conservation. Sinks of MPs and supplies of herbs in developing countries are in hazard of extinction and decreasing. In recent times, lower cost healthcare products, as well as new plant-based therapies, have surpassed more expensive target-specific drugs and biopharmaceuticals as a popular option. These factors have stimulated economic and legal interest in the issue.

Many countries in Africa, as well as the world at large, have a wide array of MAPs. However, the Fourth National Biodiversity Report [39] indicated that many species and habitats are in danger of extinction [40]. Many factors contribute to this, including habitat loss associated with agricultural intensification (including the use of fertilizers and pesticides), increased land drainage, eutrophication caused by watercourse channelization, and reduction of hedgerows. There are also important challenges to overcome so that TMs can be implemented in African countries through regulation, standardization, and integration.

A very serious challenge remains in modern medicine concerning the ethnocentric and medicocentric mentalities of the Western hegemonic system [41]. As noted by Oladele and Alade [42], modern migration in search of social infrastructure poses a threat to TM practice, and younger generations are showing a declining interest in acquiring indigenous knowledge or using plant resources for healthcare. Matching to Dike and Obembe [4], plant conservation has long been overshadowed by conservation attempts straighter toward animals and has also been much divided between attempts focused on distinct generation sections that rely on plant resources – agriculture, forestry, and nonwood forest products and attempts, targeting different kinds of ecosystems.

A second challenge is the lack of standards in standardizing and documenting TM knowledge throughout nature, as summarized by Ikeyi and Omeh [43]. Across the continent, many TM practitioners have become EX before their knowledge has been documented. These natural resources are being rapidly depleted due to overextraction, deforestation, unsustainable land use, urbanization, and industrialization. Egharevba and Ikhatua [25] showed that the loss of conservation measures will increment the number of endangered species resulting in individual extinction of numerous plant taxa that are beneficial as MP remedies. The broad array of herbal medicines need diversity techniques for storage, production, and harvesting, yet seldom are these researched and documented. Shingu [36] stated that safety traditions in sub-Saharan Africa are being missing because they are largely undocumented and oral. Mafimisebi et al. [44] showed that regulations and standards are still not fully developed and usable for the world's MPs, herbal, and aromatic plants market. Earlier, Gideon [45] noted that several efforts have been made in the world with regard to documentation of TM science relating to plant species. In line with these attempts, he saddled the strategic mandate to research, collate, develop, and promote natural medicine, document and nature traditional healthcare system, to complete the same into the national safety care transfer system and to contribute to the socioeconomic development of the countries.

Matching to Abdullahi [41], it is a public belief in medical circle that TM defies scientific procedures in terms of measurement, objectivity, classification, and codification. Even then there are signs that the physical aspects of TM (i.e., the physical ingredients) can be scientifically studied and analyzed. In Yoruba culture, for example, TM comports with the physical and spiritual realms. As the physical aspects can be subjected to scientific analysis applying the contractual scientific methods of research, the mental realm may not. Again, if united, who caters training to medical doctors on the epistemologic ontology and the efficacies of African TM given the racist tendencies in modern medicine? To put it another way: Given the epistemological and ideological differences with TM, who determines its efficacy and effectiveness? The future of MAPs rests on today's ability to resolve the conflicts among conservation and application, and the change to more resource-based agriculture incriminatingly challenged by the globalization of economies [46].

On the other hand, adulteration of herbal medicines is other challenge in conservation of MAPs for generation of herbal drugs. Thillaivanan and Samraj [47] stated medicine adulteration as "mixing or replace the original medicine material with other bogus, faulty, lowly, spoiled, wasteful other parts of like or diversity plant or bad substances or medicine which do not support with the official standards." As MAPs are gathered in the wild, they can contain nontargeted species, intentional adulteration, or even be replaced by accidentally substituting nontargeted species. Common ways of adulterating MAPs include substituting easily accessible or inexpensive species or spiking a generation with synthetic compounds.

Idu and Onyibe [38] observed that genetic diversity of traditional plants and medicinal herbs is constantly threatened by extinction due to the use of environment-hostile harvesting techniques, exploitation of the growth habitat, and unmonitored trade through developed markets. Globally, it is increasingly recognized that biological diversity, including MPs, is of immense value to present and future generations [36].

Thurlaivanan and Samraj [47] outlined the basic challenges in preserving and applying MAPs, namely, managing within risk ranges, communicating uncertainty, toxicological, pharmacological, and clinical documentation, and pharmacovigilance. It is imperative for scientists to understand why excess of harmful additives works, to evaluate medicine interactions, to limit access to people and to clinical examinations, and to standardize assessment, safety, and standardization of clinical exams.

2.12 Paces in assessment of new MAPs

The assessment of new MAPs consists of six paces, namely:
1. Specifications of new substances
2. Pattern and history of application
3. Any determinably reaction
4. Biological action
5. Toxicity and carcinogenicity
6. Clinical exam data

With respect to the paces mentioned, Singh [48] stated that some of the main difficulties in field cultivation of MPs are:
1. Absence/ignorance of cultivation technology
2. Nonavailability of verifiable data on application and access of MPs
3. Land availability due to land roof act and state forest act
4. Nonavailability of planting materials
5. Insufficient irrigation facilities
6. Idiotism of cultivation economics (MPs as a net product may be uneconomical)

2.13 The constraints in herbal medicines

Currently, MPs are applying botanical drugs at the grassroots level in a very different way from what is needed, and there is no apparent trend that this will improve anytime soon [36]. Matching with Erah [49], the main challenges of any pharmaceutical scientist are drastic difficulties with the safety, efficacy of herbal products, and overall quality. Especially in developing countries, dosage measurement and preservation pose a critical challenge. The manufacturer's statements on the label may be far off from what is included in the container of herbal preparations. Many TM practitioners have been reported to have combined orthodox medicines with MP provisions. MPs may be added to orthodox medicine and the user's pain may be treated with one of those increased medicines. Just because a herb is natural does not mean that it is safe, and asserts of considerable healing powers are often not based by reliable evidence. Some limitations include indiscriminate harvesting and poor postharvest remedy practices, needy agriculture and propagation methods, lack of investigation on the development of high-yielding varieties, domestication, and ineffective processing techniques leading to little yields and needy-quality products. Quality control methods are inadequate, good manufacturing practices are lacking, research and development (R&D) on process development is lacking, problems in marketing are prevalent, and manufacturing equipment is not easily accessible locally. With respect to restrictions related to the dealing of herbal drugs, it has been instituted that both the raw herb and the extract including complicated mixtures of organic chemicals which may include alkaloids, saponins, tannins, sterols, fatty acids, lignans, glycosides, terpenes, and flavonoids as well as other small molecules such as oligosaccharides and peptides. Although the component of a herb that is biologically active in humans can often be hard to identify, it is still a problem. According to Dajic-Stevanovic and Pljevljakusic [34], some species have unique biological characteristics or ecological needs (low growth rates, scarce soil nutrients, low germination rates, pest susceptibility, etc.). Additionally, the methods of processing herbs, such as boiling or heating, may change the amount of dissolution of the organic components. Similarly, environmental factors such as altitude, soil moisture, and temperature, and rainfall pattern, length of day, shade, dew, and frost season may affect the constituent levels in any given herb. The use of plants for botanical purposes and plant collection for botanicals are the agents of interest for quality, and other factors include plant density, insect infestations, genetic factors, and competition with other species. On the other hand, the subject of threatened plant species is another challenge. For example, matching to the Fourth National Biodiversity Report [39], a Nigeria report published in 1992 with the Federal Environmental Protection Agency showed that Nigeria contains more than 5,000 species of plants. Further, the manner in which herbs are processed, such as boiling or heating, can affect the process of dissolving the organic constituents. Likewise, environmental factors such as altitude, soil moisture, and temperature, and rainfall pattern, length of day, shade, and frost season may influence their constituent levels. There are 18 species that are threatened and 15 that are CR worldwide. Also, there are about 7,895 plant species specified in 338 and 2,215 families and genera, respectively. Threats to biodiversity in Nature include population pressure, agriculture and habitat destruction, and genetic erosion.

The loss of biodiversity is also another concern: for MAPs, their distribution and sustainable use have revealed interesting insights about an ecosystem's health, and therefore these plants have an important role in the planning of conservation plans for biodiversity. There has been a continual loss of local cultural variety over hundreds of years, and the amount of information about the types, ecology, distribution, management, and techniques for obtaining them in MPs keeps decreasing each year.

Matching to Mamedov [50], as billions of people worldwide count on MPs for sustainability, health, and conservation, it must be our first preference. Each time possible should be done to conserve biodiversity of plant ecosystems, mainly in tropical rainforests. This document confirms that Nigeria is losing biodiversity at a disturbing rate according to the Fourth National Biodiversity Report [39]. The main causes of biodiversity loss are because of the increased quality of life due to industrialization and technological advancement, the interaction between humans and their environment for development, and rapid increase in urbanization. The straight causes of biodiversity loss include rising request for forest products, the following economic policies, weak rules, poor law implement, and cultural practices. Agents such as rapid urbanization have collectively incremented deforestation and biodiversity loss. For instance, incremented export requests for birds and primates for research and trade in nontimber and timber species are direct causes of biodiversity loss in different parts of the world. Exceptionally low budgetary allocations have resulted in a lack of national efforts to reforest large areas that have been decimated. As a result, timber cuts are not replaced in an adequate manner, causing sustained yields of the forests to cease, causing biodiversity loss. Wildlife hunting through forejudgment, wild food harvesting for supplementation of subsistence farming, and bush burning negatively affect biodiversity and the environment, resulting in entanglement in mortality in animal populations, destruction of eggs and plant species, while unlawful grazing of livestock in game reserves poses an imminent threat to wildlife. Cultural practices that cheer the application of specific species for holidays often restrict the population of species mostly occurring under narrow ecological span. The lack of legislation that regulates several species is extremely detrimental to biodiversity loss. In addition, most of the laws that govern the management of several species are outdated and ineffective, leading to overexploitation of resources. Direct causes of biodiversity loss are bush burning, agricultural activities, logging, fuel-wood collection, gathering, and grazing. Since the 1900s, the institution of cash crops like coffee, cocoa, cotton, rubber, oil palm, and groundnut to the farming systems was a big motivation for massive deforestation of ecosystems in Nigeria. The massive rate of deforestation is a direct cause of biodiversity loss.

2.14 Prospects

Gene engineering has enabled the generation of highly efficient and highly favorable bioactive compounds through large-scale biosynthesis of natural products in tissue culture and fermentation of MPs. Tissue culture (including transgenic hairy root culture and plant cell) is a favorable one for the generation of scarce and high-value SMs of medicinal importance [51]. Micropropagation via tissue encapsulation of propagules can not only simplify transportation and storage but also assist higher regeneration rates. Lata et al. [52] stated that artificially encapsulated somatic embryos (or other tissues) can also be used as synthetic seeds when the contents of normal seeds cannot support propagation. Besides, breeding improvements can be carried out using molecular markers, which can greatly reduce the time needed for breeding.

2.14.1 Economic incentives for biodiversity: the need to balance commercial values with conservation values

Research on the economic value of MPs to the pharmaceutical industry can provide valuable insights and data for environmental stewardship. In particular, the studies discussed in this chapter demonstrate that medicinal applications are an important component of biodiversity values, and that ignoring them can result in the undervaluation of biodiversity. Still, there are many omissions in this literature, for example, lists of animals with medicinal value, coverage of coastal and nonforest ecosystems, and an analysis of the value of MPs in pharmaceuticals outside the developed world. Additionally, these studies have yielded contradictory, and sometimes contrasting, conclusions regarding implications for conservation of biodiversity. There is a severe gap in the literature on the matter of MPs, regardless of whether their value is immense or small. The vital concern regarding conservation concerns the extent to which resources and ecosystems can be captured by the people who use and manage them and, ultimately, how far this serves as an economic incentive for conservation and sustainable exploitation. It is unclear whether many of the huge figures cited in valuation studies are actually accurate. If all of these values were derived from the sustainable use of medicinal species as well as those acting on their own actions which may have an impact on biodiversity, then it would make sense (and logic) to postulate a link between economic value and biodiversity conservation. But in reality, this is not the case. Researchers, exporters, and multinational corporations in developed countries tend to receive the vast majority of returns as derived from medicinal species that are employed as chemicals or drugs by the pharmaceutical industry. However, people and countries that live alongside and manage medicinal species receive a much smaller proportion of these returns. In spite of the growing importance of ecological valuation studies, few attempts to examine this distribution of income (or the broader economic, institutional, policy, and legal factors that motivate it) ascertain who captures the rent from exploiting medicinal species, or provide recommendations on how to maximize values in a sustainable way. Ultimately, the total economic value alone would not ensure the preservation of commercially important medicinal species. Instead, how these rents are captured will be the critical factor in helping to ensure sustainability and how far the correlation between values and the economy will extend.

2.15 Recommendations

Our responsibilities are clear:
A number of initiatives and organizations are guiding development of these methods, including CITES, CITES-MAP, the MP Working Group, and the Convention on Biological Diversity. We must commit to the preservation of our remaining wild species and wild places so that the loss of these resources is minimized.

The ISSC-MAP, for example, proposes a series of actions, including:
- wild populations of commercial species are being mapped;
- conservation and restoration of their habitat is being pursued;
- compliance with all applicable laws and regulations is being achieved, such as the endangered species act, public land use policies, harvesting prohibitions, permitting requirements, and quantity limitations;
- learning about the rights, traditions, and practices of local and indigenous groups, and providing them with the resources necessary to sustain their populations;
- a procedure for curbing overexploitation and destructive bioprospecting would be specified in annotated global convention on biological diversity;
- development and implementation of responsible business practices and the development and implementation of sustainable management practices.

It is important to ensure that indigenous and local communities are treated fairly and equally when the use and application of their traditional knowledge takes place.

It is also important to ensure that private and public institutions wishing to employ such knowledge obtain the prior informed approval of indigenous and local communities. In order to ensure that traditional knowledge and its broader applications are respected and preserved, governments must regulate and monitor how impact assessments are conducted regarding any proposed development on sacred sites or on land or waters occupied or used by indigenous populations. Through the adoption of these types of priorities, methodologies, and frameworks, our laws and administration of land and resources must recognize the value of medicinal species and biological diversity in general.

No intelligent species should waste its extinction chance by irresponsibly wasting life-saving treasures. We will lose countless life-saving treasures if we fail to act now.

2.16 Conclusion

As wild lands are destroyed or degraded, we lose unique and precious species of MAPs, which can be used to combat hunger, poverty, natural disasters, and social and economic insecurity. As wild lands are destroyed or degraded, we lose the potential resources needed to combat hunger, poverty, natural disasters, and social and economic insecurity. The loss of diversity may also lead to the disappearance of important cures for diseases both now and in the future. Unchecked commercialization may render these traditionally used products inaccessible and unaffordable to populations, both in the USA and around the world that have depended on them for centuries. This chapter presented some challenges and opportunities in application of medicinal, conservation, and MAPs. Some suggestions and recommendations were made to neutralize and promote conservation and challenges of MAPs. Although there are conservation organizations, more effort is required to take government policies and conservation strategies to rural dwellers and local communities, just to preserve nature's bounty. But the goal of conservation of plant resources should not just be the preservation of nature's bounty. It should also be for the well-being and livelihoods of indigenous communities and society at large. A series of proactive measures and recommendations are urged to be adopted by governments, nongovernmental organizations, business communities, tertiary institutions, research centers, as well as students for the preservation and survival of MAPs. By evaluating the biological activities of MPs, we can preserve our cultural heritage. We can also encourage the populace to cultivate these plants due to the fact that most MAPs are gathered from the wild, and the wild creatures are facing many negative consequences.

References

[1]   S. M. Nalawade, A. P. Sagare, C. Y. Lee, C. L. Kao, H. S. Tsay. Bot Bull Acad Sin. 2003, 44, 79–98.
[2]   M. J. Balunas, A. D. Kinghorn. Life Sci. 2005, 78, 431–441.
[3]   A. C. Hamilton. Biodivers Conserv. 2004, 13, 1477–1517.
[4]   S. Pimm, G. Russell, J. Gittleman, T. Brooks. Science. 1995, 269, 347.
[5]   R. Bentley, (Eds.), Medicinal Plants. Domville-Fife Press, London, 2010, pp. 23–46.
[6]   A. C. Hamilton. Case studies and lessons learned. In: Kala, C. P., (Ed.), Medicinal Plants in Conservation and Development. Plant Life International Publisher, Salisbury, 2008, pp. 1–43.
[7]   S. Zerabruk, G. Yirga. S Afr J Bot. 2012, 78, 165–169.
[8]   H. O. Larsen, C. S. Olsen. Biodivers Conserv. 2007, 16, 1679–1697.
[9]   M. Rafieian-Kopaei. J Herb Med Pharm. 2013, 1, 1–2.
[10]   S. L. Chen, H. Yu, H. M. Luo, Q. Wu, C. Li, F, A. Steinmetz. Chin Med. 2016, 11, 37.
[11]   International Union for Conservation of Nature, Natural Resources. Ecosystems, & Livelihoods Group. Conserving Medicinal Species: Securing a Healthy Future. IUCN. 2006.
[12]   M. E. Soule, J. A. Estes, B. Miller, D. L. Honnold. Bioscience. 2005, 55, 168–176. (Soule et al., 2005).
[13]   K. Sheikh, T. Ahmad, M. A. Khan. Biodivers Conserv. 2002, 11, 715–742.
[14]   H. Huang. Bot J Linn Soc. 2011, 166, 282–300.
[15]   M. S. L. Figueiredo, C. E. V. Grelle. Divers Distrib. 2009, 15, 117–121.
[16]   F. Forest, R. Grenyer, M. Rouget, T. J. Davies, R. M. Cowling, D. P., et al. Nature. 2007, 445, 757–760.
[17]   J. Ma, K. Rong, K. Cheng. Asian J Int Law. 2012, 20, 551–558.
[18]   H. Huang, X. Han, L. Kang, P. Raven, P. W. Jackson, Y. Chen. Science. 2002, 297, 935.
[19]   J. Liu, M. Linderman, Z. Ouyang, L. An, J. Yang, H. Zhang. Science. 2001, 292, 98–101.

[20] C. Liu, H. Yu, S. L. Chen. Framework for sustainable use of medicinal plants in China. Zhi Wu Fen Lei Yu Zi Yuan Xue Bao. 2011, 33, 65–68.

[21] K. Havens, P. Vitt, M. Maunder, E. O. Guerrant, K. Dixon. Bioscience. 2006, 56, 525–531.

[22] Q. J. Yuan, Z. Y. Zhang, J. A. Hu, L. P. Guo, A. J. Shao, L. Q. Huang. BMC Genet. 2010, 11, 52–59.

[23] D. Z. Li, H. W. Pritchard. Trends Plant Sci. 2009, 14, 614–621.

[24] A. Muchugi, G. M. Muluvi, R. Kindt, C. A. C. Kadu, A. J. Simons, R. H. Jamnadass. Tree Genet Genome. 2008, 4, 787–795.

[25] R. K. A. Egharevba, M. I. Ikhatua. J Agric Biol Sci. 2008, 4(1), 58–64.

[26] D. Palevitch. c. 1991, 167.

[27] I. Nwafor, C. Nwafor, I. Manduna. Horticulturae. 2021, 7(12), 531.

[28] B. Ncube, J. F. Finnie, J. Van Staden. S Afr J Bot. 2012, 82, 11–20.

[29] C. Macilwain. Nature. 2004, 428, 792–793.

[30] D. Rigby, D. Cáceres. Agr Syst. 2001, 68, 21–40.

[31] H. W. Wang, D. C. Peng, J. T. Xie. Chin Med. 2009, 4, 20.

[32] S. S. Kankara, M. H. Ibrahim, M. Mustafa, R. Go. S A J Botany. 2015, 97, 165–175.

[33] B. J. Amujoyegbe, J. M. Agbedahunsi, O. O. Amujoyegbe. Int J Med Arom Plants. 2012, 2(2), 345–353.

[34] Z. Dajic-Stevanovic, D. Pljevljakusic. Int J Med Aromat Plants. 2015, 145–164.

[35] L. C. Saalu. Nigeria folklore medicinal plants with potential antifertility activity in males: A scientificappraisal. Res J Med Plant. 2016, 10, 201–227.

[36] G. K. Shingu. Ethnobot Res Appl. 2005, 3, 017–023.

[37] E. C. Orji, G. I. Onwughalu, I. A. Nweke Toxicol Food Technol. 2013, 2(6), 61–63.

[38] M. Idu, H. I. Onyibe. Res J Med Plant. 2007, 1, 32–41.

[39] Federal Republic of Nigeria. Fourth National Biodiversity Report. A publication of Federal Ministry of Environment, Abuja, 2010, pp. 14, 59.

[40] R. N. Okigbo, U. E. Eme, S. Ogbogu. Biotechnol Mol Biol Rev. 2008, 3(6), 127–134.

[41] A. A. Abdullahi. Afr J Tradit Complement Altern Med. 2011, 8(5), 115–123.

[42] A. T. Oladele, G. O. Alade, O. R. Omobuwajo. Medicinal plants conservation and cultivation by traditional medicine practitioners (TMPs) in Aiyedaade local government area of Osun State, Nigeria. Agric Biol J North Am. 2011, 2(3), 476–487.

[43] P. A. Ikeyi, N. Y. Omeh. Int J Curr Microbiol. 2014, 3(1), 675–683.

[44] T. E. Mafimisebi, A. E. Oguntade, I. A. Ajibefun, O. E. Mafimisebi, E. S. Ikuemonisan. Med Aromat Plants. 2013, 2(6), 1–9.

[45] E. C. Gideon. Digitization, Intellectual Property Rights and Access to Traditional Medicine Knowledge in Developing Countries – the Nigerian Experience. A Research paper prepared for International Development Research Centre (IDRC), Ottawa, Canada, 2009.

[46] S. Padulosi, D. Leaman, P. Quek. J Herbs Spices Med Plants. 2002, 9, 243–267.

[47] S. Thillaivanan, K. Samraj. Int J Herb Med. 2014, 2(1), 21–24.

[48] H. Singh. Lead. 2006, 2(2), 198–210.

[49] P. O. Erah. Trop J Pharm Res. 2002, 1(2), 53–54.

[50] N. Mamedov. Med Aromat Plants. 2012, 1(8), 1–2.

[51] S. R. Rao, G. A. Biotechnol Adv. 2002, 20, 101–153.

[52] H. Lata, S. Chandra, I. A. Khan, M. A. Elsohly. Plant Med. 2008, 74, 328.

Sonal Mishra* and Trilok Kumar*

Chapter 3
Culture, tradition, and indigenous practices on medicinal plants

Abstract: Terpenes, alkaloids, flavonoids, and other active phytoconstituents are found in medicinal plants. They can demonstrate potential medical applications as antimicrobial, anti-malaria, antidiabetic, anticancer, antidepressant, and antifungal due to the presence of active ingredients. There is a need for avoidance and conversion of conventional medicinal herbs because of these unique qualities. Knowledge about cultivation, diversity, and local medicinal plants is fundamental in this regard. The current review study focuses mostly on the topics mentioned above from antiquity to modernity.

3.1 Introduction

Plants are the ultimate source of existence of life on Earth. All the living beings depend directly or indirectly on plants for their survival and fulfilling their basic needs. Plants occupy a status of god, they had been worshiped by people since prehistoric times; plants are acting as pillars in the development of human culture and traditions, and play vital role in the evolution of mankind. Living beings depend on plants not only for food and shelters but also has important role in various systems of medicine since ages. Traditional knowledge of medicinal plants is an asset that human race is having from generations. Various countries on map have their own indigenous plants linked with their culture and traditions, and are integral part of their living. China is the largest producer and exporter of medicinal plants in the world followed by India; both countries occupy about 70% of total world trade of medicinal plants and herbs. Plants synthesize large number of chemical compounds called phytochemicals which are used for treating various diseases in living beings and provide defense against insects, microorganisms including fungi, bacteria, and viruses. The Covid pandemic has established the importance of plants and their extracts in each and every single individual and flooded the market with various plant products for boosting the immunity.

*Corresponding author: Sonal Mishra, Department of Botany, Government Digvijay Autonomous Post Graduate College, Rajnandgaon, Chhattisgarh, India, e-mail: sonalmishra2017@gmail.com
*Corresponding author: Trilok Kumar, Department of Botany, Government Digvijay Autonomous Post Graduate College, Rajnandgaon, Chhattisgarh, India, e-mail: trilokdev111@gmail.com

https://doi.org/10.1515/9783110791891-003

Archeological evidences suggest the use of medicinal plants for treating diseases from Paleolithic age about 60,000 years ago. The first documentation of medicinal plants is found in Sumerian civilization where the names and details of various medicinal plants are found written on the clay tablets. Use of plants for medicinal purposes and disease treatments were reported back in prehistoric periods, the use of Sanjeevani buti was mentioned in Hindu epic holy book Ramayana. Ancient Unani writings, Egyptian manuscripts notify the use of plants as medicines. Various cultures like Chinese, Egyptian, and European were using herbs for almost 4,000 years for medicine preparations. Evidences established that Rhishi-Munis, Hakims, Vadis, and Prophets used traditional medicines to cure people.

Countries having Indigenous cultures on earth such as India, Africa, America, Rome, Egypt, Iraq, and Iran documented use of plants in their healing rituals. They had developed various traditional systems of medicine such as Homeopathy, Unani, Ayurveda, Siddha, Folk (tribal), and Chinese Medicine, which traditionally uses plants and their extracts making various decoctions by plants and parts of plants like roots, stems, or leaves which are crushed, meshed, and crude tablets were prepared out of it for varied disease treatments [1]. The World Health Organization (WHO) and Government of India signed agreement for establishing the global center for traditional medicine having aim of exploring the potential of traditional medicine worldwide and connecting it with modern science and technology to improve the health and quality of life of human on earth.

Traditional medicines are difficult to popularize as they have less documentation around the globe; its knowledge is passed orally from one generation to another; there are families treating the same disease from generations learnt the treatments from their forefathers and used indigenous herbs grown in a particular area for treatment [1–3]. The survival of human is not possible without the plants. Before the development of the modern medicine system, the traditional and ancient knowledge of medicines was the only means of cure present for the treatment of diseases. Some ancient literature suggest about performing surgery with the help of articles and instruments made by plants. Plants with medicinal properties are the major source of medicine. Many evidences suggest that herbs are used for treating diseases and restoring normal body functions, in traditional and ancient systems of medicine such as homeopathy, Ayurveda, Unani, folk, and Chinese system of medicine [4]. Additionally, Fig. 3.1 exhibited the traditional system of medicine.

3.2 Involvement of plants in the culture and traditions of mankind

Plants are an integral part of human culture and tradition; since ages plants have helped in human evolution and their cultural, social, and economic development by providing food, shelter, cloths, and medicines even before agricultural practices emerged.

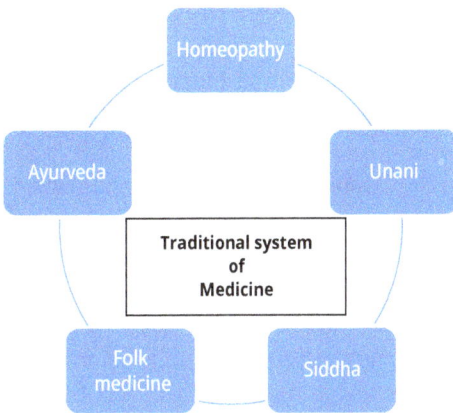

Fig. 3.1: Traditional system of medicine.

People in early ages eat plant products like fruits and roots of many plants and wear cloths made up of leaves and sleep over the trees; their lives revolve around the plants. Human and plants have parallel evolutionary stages. After men learnt the agricultural practices, he began to settle out and live a civilized life; the credit of civilization goes to plants.

Plants play a crucial role in daily life as they provide oxygen for the survival of life on earth, firewood, species, vegetables, fruits, pulses, cereals, beverages, dyes, timber, oils, rubber, cotton, jute and other fibers, coal, petroleum, and medicines, beside these they help to reduce soil erosion and air pollution.

Plants are part of religious rituals from birth to death, philosophically both cradle and grave need wood of plants; they are an important part of the enrichment of living as have share in musical instruments, ornamentation, painting tools, and various organic colors.

We can say that plants and humans have symbiotic association with each other. Worldwide data suggest using about 85% of these ancient and traditional medicines in primary health care. In countries like India and China, they use a lot of spices in food, many of them found to have medicinal properties like ginger, garlic, cinnamon, basil, cardamom, turmeric, coriander, mint, and coconuts.

There is an immense need to document this knowledge of herbs in proper database form in computers; both soft and hard copies should be documented for easy access and understanding. It will open the path for patenting rights. Also the linguistic barriers should be crossed as the traditional and indigenous knowledge is in local languages or difficult one such as Sanskrit and Arabic; this should be translated into easy and understandable language for ordinary laymen to spread the knowledge and practices.

Primitive populations like tribes and folks living in or near forests have abundant knowledge; their knowledge of medicinal herbs and discoveries about wild herbs are must to be documented worldwide. The developing countries had a larger share in traditional medicine and its uses in comparison to developed countries.

3.3 What are medicinal plants

The term "medicinal plant" includes plants with medicinal constituents and bioactive compounds, which can be used for alignments due to their curing capacity; the whole plant or any of its parts is used as herbal medicine or traditional medicine in the treatment of various diseases. The term "herb" defines the nonwoody, small-heighted plant having a herbaceous stem that is generally green in color. The term has been taken from the Latin language, *"herba."*

These plants are precursors for the synthesis of useful compounds. These plants and their parts are time-tested for their healing power and diseases curing abilities in due time period. Generally, they do not have any side effects if consumed in an appropriate quantity; they have their own taste, smell, and aroma, which are specific characteristic features of that particular plant species.

The plants are deeply in coordination with daily life, so they can make easy harmony with the diseased body and create a proper balance between body and mind. So the healing becomes easy and rapid; they work for complete body wellbeing and increase the immunity of the body instead of focusing on one organ, so the healing is slow but the immunity gained lasts longer. Fig. 3.2 shows the cultivation, usage, and applications of medicinal plants.

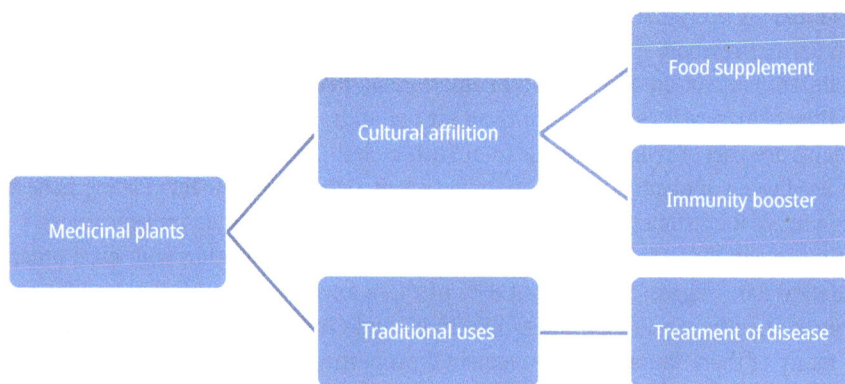

Fig. 3.2: Cultivation, usage, and applications of medicinal plants.

3.4 Characteristic features of medicinal plants

1. Plants are in harmony with the human body so have a holistic approach of treatment.
2. Plants and plant products are easily digestible, and medicinal constituents get easily absorbed by the body.
3. Generally have no side effects in recommended doses.

4. They boost the immune system and beneficial for complete body wellbeing, and act as preventive medicines.
5. Many medicinal plants are known to provide protection against deadly diseases like corona, cancer, and cardiovascular diseases, if taken regularly.
6. It generates the alignment between body, mind, and soul.
7. Traditional herbs have capacity to heal wounds, sores and boils, act as antacids, reduce fever, jaundice, increase appetite, purify blood, and act as an antiseptic.
8. Improve the function of vital organs like brain, heart, kidney, and liver.

3.5 Classification of medicinal plants

Medicinal plants are classified differently depending on the method, criteria, plant part, uses, principal active compounds, and study chosen.

Generally, medicinal plants are classified according to their active compounds or in their particular parts, including roots, stems, barks, leaves, flowers, fruits, and seeds. Medicinal plants are classified to divide them into categories according to their uses. Fig. 3.3 exhibited the classification of medicinal plants according to the usage.

3.5.1 Classification according to the usage

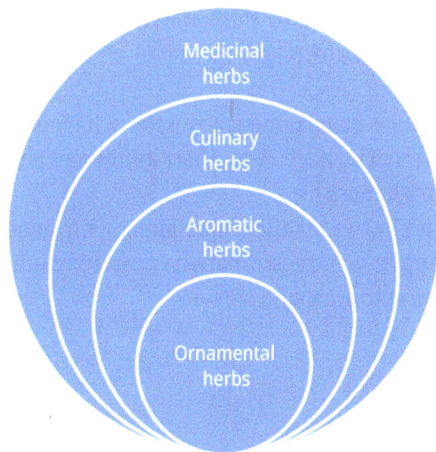

Fig. 3.3: Classification of medicinal plants according to the usage.

Medicinal herbs: The plants which have active compounds and can be used in making medicines because of their healing capacity of disease like serpgandha, ashwagandha, safedmusli, neem, haldi, marigold, lemon, and lavender.

Culinary herbs: They are the herbs which are used in cooking due to their strong flavor and aroma and are good for digestion, mostly used in Thai and Chinese cuisines, e.g., basil, thyme, and oregano.

Aromatic herbs: These are the herbs important for their pleasant smell used in perfumes and air fresheners, and in various therapies, e.g., nilgiri, mint, khas, and rosemary.

Ornamental herbs: They are beautiful in appearance, so used for decorative purposes, e.g., china rose, lavender, lemongrass, and chrysanthemum [5].

3.5.2 Classification based on the active constituents

According to the principle active constituents of herbs, they are classified into different categories. All herbs are divided into five major categories: Bitter, Aromatic, Astringents, Nutritive, and Mucilaginous. Fig. 3.4 exhibit the classification of medicinal plants based on its active constituents.

Bitter herbs: Alkaloids, saponins, and phenols are present in these herbs. They have antibiotic, antiseptic, and blood purifier properties, e.g., parsley, asparagus, chickweed, dog grass, grapevine, and dandelion [6].

Aromatic herbs: Aromatic herbs suggest that the plant is related with aroma, it has a pleasant smell as it contain volatile oils used in perfume-making and in stress-releasing therapies [6].

Astringent herbs: These herbs consist of tannins in them. They have antiseptic, analgesic, antiabortive, and astringent properties used in limited doses as higher doses are harmful and cause vomiting and nausea if taken in higher doses than prescribed.

Mucilaginous herbs: These drugs contain lime mucilaginous exudates which generates sweet taste and slippery texture. These herbs have detoxifier, antibiotic, and antacid properties, e.g., aloe, fenugreek, and Irish moss [6].

Nutritive herbs: They are edible herbs and have nutritive value along with some medicinal effects, e.g., broccoli, cabbage, apple, asparagus, banana, carrot, cauliflower, hibiscus, lemon, onion, orange, papaya, spirulina, and stevia.

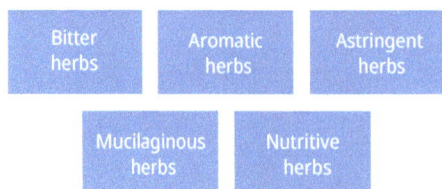

| Bitter herbs | Aromatic herbs | Astringent herbs |
| Mucilaginous herbs | Nutritive herbs | |

Fig. 3.4: Classification of medicinal plants based on active constituents.

3.5.3 Ancient system of medicines

Various systems of medicines are known since ages and all are in practices in different areas around the world. The ancient system of medicines depends on the traditional knowledge, beliefs, customs, and practices which are part of human society for hundreds of years and passed from generation to generation by cultural transmission as it is part of the tradition of each country.

These systems were in practice before before stems were in practiceby hu, and now people are returning to ancient knowledge because of the side effects of synthetic medicines, the high cost of treatments, population rise, inadequate supply of medicines, and development of resistance to the current drugs used. These traditional medicines boost the body's immune system and provide safe and effective treatments for primary health care.

Ancient documentations suggest that in prehistoric times India is the center of a repository of medicinal plants and known as vishwaguru due to its tremendous knowledge in every field. It has always been the capital for the magic herbs, yoga, medicinal plants, culture, and traditional heritage.

Indian forests are always rich in different medicinal herbs to date, but the uses of all the herbs are not identified. Different types of medicinal system have been recognized in India Homeopathy, Ayurvedic, Unani, Yoga, Naturopathy, Siddha, and Folk medicines are the main systems of indigenous medicines in India, which are practiced in different regions of the country; generally, the tribes of the particular region has their knowledge about the use of their indigenous plants which they use for treatments.

World Health Organization confirmed that 75% of people worldwide depend on herbal medicines for their preliminary health care. It has been estimated that the total drug market share of developed countries constitutes about 30% of plant drugs, while developing countries have a larger share of about 75% plant drugs in their total drug production. The medicinal plants are the source of economic development for developing countries like India and China. They have a major share of plant drugs in the world market.

3.5.4 Ayurvedic system of medicines

The details of this system are mentioned in ancient religious writings. This system emphasizes the importance of living in harmony with nature and science. It describes the importance of maintaining a disciplined and proper lifestyle for overall positive health of body and mind. It is assumed that the basic principles of Ayurvedic medicines got organized and arranged in a systemic manner around 1400 BC. The four great Vedas of Indian culture constitute *Atharvaveda*, which has mentioned 114 hymns about the preparation and use of drugs in treating diseases. Ancient writings like Charaka Samhita in the field of Medicine also confirm the use of traditional medicine in which more than

600 plant preparations are mentioned. Sushruta Samhita in the field of Surgery, where basic principles and operative procedure of surgery are mentioned along with the detailed descriptions of instruments and articles required for surgery and postcare in which more than 600 drugs are explained are used in surgery and post care [98].

3.5.5 Siddha system

This system of medicine is practiced in many parts of Southern India, e.g., Tamilnadu. It is closely related to Ayurveda. The term *"Siddha"* has been taken from *"Siddhi"* – describe achievement. *Siddhars* is the title awarded to the men who achieved great knowledge in the medicinal and meditational field [7].

3.5.6 Unani system

This system has its root in Greece. It is believed that this system was established by the great physician Hippocrates (460–377 BC). Another physician Galen contributed for its development (130–201 AD) [98].

3.6 Methods of preparations and uses of indigenous medicine

Traditional herbal medicines were used in different forms, including paste, decoction, juice, powder, and fumes by burning the plants and their parts. Majority of time, plants and plant parts are crushed with the help of pastel and mortar in the form of paste with or without water; depending on the type of plant used, they are swallowed in the form of small balls or tablets and are made and dried to be stored for further uses, successively in the form of juice, decoction, and sometimes the plant materials are dried and grinded to form a powder which is taken with honey generally used in pediatric treatments. The plant products of two or more plants are also used in combination.

3.7 Countries rich in medicinal plants

The growth of any country is measured in terms of their enriched heritage including their traditional knowledge, cultural assets, natural resources, and economic development. Forest is an important natural resource after minerals. Forests are the flag bearer of prosperity as they protect the tribal culture, and provide the basics of living along with the medicinal plants. The countries with large forests are rich in their traditional medical practices and indigenous medicinal plants. The branch of botany known as ethno botany and ethno pharmacology are searching for new methods of preparations of various drugs based on plant origin. Another branch of chemistry and medicine called pharmacochemistry deals with the structure of compounds isolated from plants and their effects as drugs.

Many countries are rich in their medicinal plant's treasures, including China, India, Sri Lanka, Thailand, Australia, Canada, U.S.A, France, Indonesia, and Malaysia and contributing to the overall health and wellbeing of mankind.

3.8 Diversity of medicinal plants in India

India has different climatic conditions in its various parts, so the country is divided into five climatic zones. Further on its basis, the country is broadly classified into 15 agro-climatic zones for resource development and conservation. The agro-climatic zones distribute the agricultural pattern according to the suitable atmospheric conditions of the places and suggest the best plant for cultivation in the particular region. India has about 8% of the total plant biodiversity of the world, around 0.126% million plant species.

India is abundantly rich in its phytodiversity and count of medicinal plant species so-called as botanical garden of the world. India is among the eight major centers of origin of domesticated taxa and one of the 12 mega-diversity countries in the world due to its rich plant diversity, based on its uses of plants as medicine under various indigenous practices and in the traditional medicinal system [8, 9]. India has its strategies for conservation and utilization of its medicinal plant wealth to protect the biodiversity richness. The areas are classified as Botanical gardens, national parks, and biosphere reserves. According to Forest Survey of India, 21.05% of total land area of India is covered with forest cover.

The traditional knowledge of herbal medicines is widely practiced in the country, and use of plants and plant product are an integrated part of people's daily life in India. In India, there are more than 550 tribal communities and 227 ethnic groups living in the forests and nearby villages.

India covers the larger share of the medicinal plant trade in the world. According to The World Health Organization, about 80% of Indian population uses traditional herbs as medicines for preliminary treatments of diseases [10]. According to WHO, more than 40% of the world's population depend on medicinal plants directly for their primary health care [11]. India is the biggest exporter of medicinal herbs, so the medicinal plants contribute to the growth of Indian economy as well as the resident of the forest area are improving their living by earning their livelihood. According to the Botanical Survey of India, in India, more than 7,500 plant species are of medicinal importance.

Proper documentation is a priority for sustaining this knowledge for future generations [12]. To conserve the traditional system of medicines, first step is their proper manuscript formation about their morphological and systematic details and uses in the health care system. In India more than 1,000 herbs are traded, in which about 200 species shows a consumption rate of 95 metric tons yearly, most of the medicinal plants, about 80% are grown in wild in forest cover of India and some are cultivated. In India, 75% of medicinal species are reported from the tropical region in different forest cover, and 25% are from temperate areas of Himalayas. Some important biodiversity centers of India are as follows.

3.8.1 Himalayan region of India

The Himalayas were known as the "Roof of the World" as it encompasses a number of hotspots, center of origin, and repositories of medicinal plants. Different communities live in the Himalayan region where indigenous medicines are the primary means of treatment and health care as the modern medical facility is not so developed in this area. In this area, decoction of leaves and herbaceous stems is most frequently used for treatments. The Himalayan region covers an area of about 591,000 km^2 [13]. It has unique plant diversity, about 18,440 species of plants are found of which 1,748 species are of medicinal importance [14]. It includes the states of Jammu-Kashmir, Himachal Pradesh and Uttarakhand. Himalayas is divided into two zones (a) Trans Himalaya and (b) Himalayas [13]; and these two were further divided in five biotic provinces (i) Tibetan (ii) North-West Himalayas (iii) West Himalayas (iv) Central Himalayas, and (v) East Himalayas.

Some of the important species grown here are *Azadarachta indica, Bacopa monnieri, Boerhaavia diffusa, Pholidotaarticulata, Zanthoxylumarmatum, Aeglemarmelos, Terminaliabellerica, Syzygium cuminii, Rauvolfia serpentina, Withania somnifera, Berberis aristata, Bergenia ligulata, Delphinium denudatum, Potentilla fulgens, Taxus baccata, Taxus wallichiana, Valeriana wallichii, Hyoscymus niger, Polygonatum verticillatum, P. cirrhifolium, Corydalis govaniana, Swertia chirayita, Angelica glauca, Fritillaria roylei,* and *Ephedra gerardiana.*

The importance of medicinal plants increases many folds as the rural people in the Himalayan region mostly depend on these plant-based medicines [15–17]. Traditional knowledge of herbs is gradually decreasing, though some practitioners still use traditional healthcare systems effectively [18].

Primitive people have developed knowledge regarding medicinal plants and their uses through tried and tested methods and develop treatments and therapies using these indigenous herbs of the Himalayan region. Proper documentation of this acquired knowledge in many generations is needed as the culture and traditions are rapidly changing in today's current scenario [15, 19].

3.8.2 North-East India

The northern part of India represents about 8% of the country's total geographical area and consists of eight states, including Assam, Arunachal Pradesh, Mizoram, Manipur, Meghalaya, Nagaland, Tripura, and Sikkim. The tribal communities are dominant in this area. About 225 tribal communities reside in this region [20]. It is one of the hotspots of biodiversity in the world, with the rich phytodiversity in India [21]. According to the Forest Survey of India, the region has approximately 60.02% of forest area, out of which reserves forest has 66.33%, protected forests have 0.03% and unclassified forests represent 33.64% area.

High-valued medicinal plants from the states of North east are *Acorus calamus, Angiopteris evecta, Aconitum heterophyllum, Bacopa monnieri, Berberis aristata, Curcuma caesia, Embelia ribes, Gmelina arborea, Gynocordia odorata, Homalomena aromatica, Mesua ferrea, Paris polyphylla, Picrorhiza kurrooa, Rheum australe, Rubia cordifolia, Smilax china, Solanum anguivi, Taxus wallichiana, Trichosanthis bracteata, Valeriana jatamansi*, and *Zanthoxylum armatum* [22].

3.8.3 Western India

Western Ghats lies parallel to the west coastal area of India. Complete Western Ghats is important in terms of medicinal plant richness and endemism of plants. The Western Ghats has diversified geographical features with hilly and sloppy areas to support the luxuriant growth of varied plant species. It is one of the biodiversity hotspots of the world. It consists of evergreen, moist, dry deciduous and tropical rain forests. The tremendously varied environmental features and the heterogeneity in weather conditions throughout the year make it a heterogeneous biogeography area with rich diversity of flora and fauna [23]. This region is native place to about 700 medicinal plants, many of which are used in folk and traditional medicinal practices.

The important medicinal plants of this region are *Mimosa pudica, Hibiscus angulosus, Leucas aspera, Phyllanthus neruri, Calotropis gigantea, Tridax procumbens, Parthenium hysterophorus, Anona squamosa, Buchanania lanzan, Semecarpus anacardium, Dioscorea bulbifera, Aphanamixis polystachya, Rhincanthus nasuta, Momordica dioica, Cinnamomum zeylanicum, Rauvolfia serpentina, Saraca asoca, Gymnema sylvestre*, and *Gloriosa superba* [24].

3.8.4 Central India

Central India, known as Madhya Bharat, includes Chhattisgarh and Madhya Pradesh. These two states are the pockets of medicinal plants in India. They are the native place of many pharmaceutically important plants. These two are inhabited by various tribal communities like Gond, Santhal, Bharia, Muria, Baiga, and Koru tribes, which carry the glorious history of traditional knowledge of medicinal plants and their importance. Local folklore uses about 400 medicinal plants. The population of age group 50–65 has abundant knowledge about indigenous herbs in comparison to the younger generation [25]. Some important herbs in this area are as follows.

Abrus precatorius, Abutilon indicum, Acanthospermum hispidum, Acacia catechu, Acacia nilotica, Acorus calamus, Achyranthes aspera, Adhatoda vasica, Aegle marmelos, Aloe vera, Alpinia galanga, Andrographis paniculata, Anisomelos indica, Annona squamosa, Anogeissus latifolia, Argemone mexicana, Asparagus racemosus, Azadirachta indica, Bauhinia variegate, Boerhaavia diffusa, Butea monosperma, Catharanthus roseus, Calotropis procera, Cassia tora, Chlorophytum spp., *Citrullus colocynthis, Curculigo orchoides, Curcuma amada, Curcuma aromatic, Diospyros melanoxylon, Eclipta alba, Emblica officinalis, Mucuna pruriens, Nyctanthus arbortistis, Ocimum sanctum, Pongamia pinnata, Rauvolfia serpentine, Sida acuta, Solanum nigrum, Terminalia arjuna, Terminalia bellirica, Terminalia chebula,* and *Withania somnifera* [26].

3.8.5 Southern India

It is also known as Dakshina Bharata, which is rich in its culture heritage and traditional practices. They have their own food, dance, music, and alternative medical practices. Southern India consists of five states: Andhra Pradesh, Karnataka, Telangana, Tamil Nadu, and Kerala; and the union territories of Lakshadweep and Puducherry.

Important medicinal plants of this region are *Areca catechu, Beta vulgaris, Boerhavia diffusa, Bombax ceiba, Camellia sinensis, Capparis decidua, Coccinia indica, Emblica, Ficus bengalensis, Hibiscus rosasinesis, Momordica cymbalaria, Phaseolus vulgaris, Punica granatum, Azadirachta indica, Cardiospermum halicacabum, Erythrina indica, Gloriosa superba, Jatropha curcas, Moringa oleifera, Phyllanthus amarus, Sesbania grandiflora, Tamarindus indica, Tridax procumbens,* and *Vitex negundo* are used in daily life. While *Aloe vera, Azadirachta indica, Curcuma longa, Emblica officinalis, Eucalyptus tereticornis, Gloriosa superba, Moringa oleifera, Ricinus communis, Sesamum indicum, Sesbania grandiflora, Solanum americanum, Tamarindus indica,* and *Zingiber officinale* are commonly cultivated.

3.8.6 Andaman and Nicobar islands

The phytobiodiversity of islands is very rich with endemic flora and fauna. It has Mangroves and a tropical rain forest with rich coral gardens. Traditional medicines are in practice by different ethnic groups. The region has a hot tropical humid climate with uneven rainfall. The tribal population depends on forest cover for food and medicines.

Some important medicinal plants are *Dipterocarpus grandis, Terminalia bialata, Pterocarpus dalbergiodes, Calophyllum inophyllum, Terminalia catappa, Semecarpus kurzii, Strobilanthes andamanensis, Uvaria andamanica, Artabotrys nicobarianus, Calamus andamanicus, Aristolachia tagala, Phyllanthus andamanicus, Glochidion calocarpum* [27]. *Daemonorops manii, Alstonia kurzii, Dichapetalum gelonioides, Dripetes andamanica, Polycarpaea spicata, Suriana maritima, Pleurostylia wightii, Colubrina asiatica, Ochrosia oppositifolia, Cordiasubcordata, Linaria ramosissima, Pisonia alba, Thuarea involuata*, and *Heroandia ovigera* [28].

3.9 Market scenario

India has a rich market for medicinal plants worth Rs. 4.5 billion in 2020 and is predicted to increase by Rs. 14 billion by 2026. The global trade of medicinal herbs, essential oils, and plant extracts is continuously increasing in India as the demand for Indian herbs is increasing in the world market. Traditional herbal remedies have their own importance in the health care system all over the world. The natives of India use varied herbs and extracts to treat diseases.

Recently, the number of medicinal herbs is decreasing in wild due to urbanization, depletion of forest cover, and overexploitation of important herbs. Several plant species are increasing in red data book listing as endangered, vulnerable, and threatened [29, 30]. To preserve the medicinal plant wealth, the forest have to be preserved, its destruction should be prevented while the wild species should be brought under cultivation and their number should be increased; the important medicinal plants can be grown by tissue culture practices. Proper cultivation and conservation of medicinal plants is a need of an hour to preserve, sustain, and flourish the traditional healthcare system. Additionally, Fig. 3.5 shows the market analysis of medicinal plants.

Fig. 3.5: Market analysis of medicinal plants.

3.10 Indigenous medicinal plants of Chhattisgarh state

Chhattisgarh covers a total area of about 1,36,000 km^2 with 58,810 km^2 of cultivable land and 59,772 km^2 of forest cover, which contributes 4.14% of the total forest area of India. The state has 41.33% of its area covered by forests and is one of the richest phytodiverse zones in the country. The state has two major types of forests – Tropical Dry Deciduous Forest (TDDF) and Tropical Moist Deciduous Forest (TMDF). Chhattisgarh occupies the central position of India so-called the heart of India, and is endowed with a rich, geographical, ecological, traditional, cultural heritage, and attractive natural beauty with diverse flora and fauna. Chhattisgarh has been recognized as a herbal state due to its richness in medicinal plants repository. It has a large forest area with abundance of medicinal flora and unique tribal culture that makes it further more prosperous in terms of traditional knowledge of ancient indigenous medicines. The state is rich in endemism with respect to medicinal plants.

The state has favorable weather conditions like temperature and humidity, leading to large production of medicinal plants. The tribes have abundant knowledge about the vast potentiality of indigenous herbs and their use as therapeutics; they uses various plant parts as medicines including root, rhizome, bulb, stem, bark, leaves, flower, fruit, and seed, along with the plant products like sap, gums, tannins, and other exudates in treatments of diseases like malaria, asthma, urinary tract infection, diabetes, and wound healing.

The tribal population of the region includes various communities of tribes like Agariya, Baiga, Gond, Oraon, Mandia, Muriya, Mariya, Bhatra, Bhariya, Bhumiya, Binjhwar, Halwa, Kanwar, Korwa, Khariya, Kamar, Halba, Pardhi, Masturi, and Manglaetc with different cultural, ethnic, religious, and socio-economical and linguistic backgrounds. These communities reside in forests and nearby areas so have full access to forest products and use them for their living. They gathered the medicinal plants and other valuable resources and earn their livelihood through these practices. The tribal populations are carriers of traditional knowledge of medicinal plants found in the nearby vicinity of forests, the use of traditional and local medicines are their customary and this folklore is inherent to their culture from generation to generation, which helps in the conservation of the indigenous herbs. This ethnic pharmacology can be used as a prominent tool for the discovery of new drug [31].

The central part of the state is named as "rice bowl of India," with more than 21,000 rice varieties; the major population depends on agriculture for earning their livelihoods and food security. The large population lacks access to modern physicians due to the lack of infrastructure, knowledge, false beliefs, and travel challenge. So the rural population relies on medicinal herbs and traditional healers for their primary healthcare. Tribals are an integral part of Chhattisgarh population having a rich cultural heritage, they depend on nature for all their primary need. Since ages, tribal communities and forests cover are inter-windily related and dependent on each other. The tribal population is born and brought up in the forests, and their lifelong interactions with nature develop an indigenous knowledge system. This knowledge and belief are passed through generations.

The state has Traditional Healer Association, Chhattisgarh at Bilaspur (THAC), formed to create an organized body of traditional healers to increase and popularize their practice and to make them reachable to remote populations. This association works in accordance with the Chhattisgarh State Medicinal Plant Board, Raipur. These traditional healing practices and drug preparation by herbs are widely practiced in Bastar, Mohala, Manpur, Gariyaband, and Jashpur regions of state. Additionally, Tab. 3.1 provided the list of important medicinal plants of the central India region.

Tab. 3.1: Important medicinal plants of central India region.

Botanical name	Common name	Family	Part used	Medicinal use	References
Abrus precatorius Linn.	Ratti, Ghumchi	Fabaceae	Fruit seeds	Joint pains, paralysis, alopecia, whooping cough	[32, 33]
Abutilon indicum (L.) Sweet.	Kanghi	Malvaceae	Leaves	Facial paralysis, joint disorder, increase	[32]

Tab. 3.1 (continued)

Botanical name	Common name	Family	Part used	Medicinal use	References
Acacia arabica Willd.	Motichoor	Mimosaceae	Leaves	Migraine	[34]
Accacia nilocita (L.) Willd. Ex Del.	Babool	Mimosaceae	Branch twig	Antimutagenic and chemoprotective activity	[35]
Achyranthes aspera Linn.	Chirchita	Amaranthaceae	Whole plant, seeds, roots	Indigestion, cough, asthma, anemia, stomach pain, jaundice, hydrophobia, skin diseases	[32, 33, 36]
Acorus calamus Linn.	Bach	Araceae	Rhizome	Flatulent colic, atonic dyspepsia, ulcers, nervine tonic, antispasmodic	[32, 37, 38]
Adhatoda vasica Ness.	Vasak, Adusa	Acanthaceae	Leaves	Cough, asthma	[33, 34]
Aegle marmalos (L.) Corr.	Bel	Rutaceae	Leaves, fruits	Hypoglycemic, chemopreventive, stomach disorder, diarrhea, dysentery	[36, 39–41]
Albizia lebbeck (Linn.) Benth	Shirish	Mimosaceae	Whole plant parts	Bronchial asthma	[32]
Aloe vera (L.) Burm. F.	Kataban, Ghritkumari	Liliaceae	Leave gel	Skin disorder, blood purifier, constipation, menstrual disorder	[36]
Allium cepa Linn.	Pyaj, onion	Liliaceae	Bulb	Prevents irregular heart beat and abnormal blood pressure. It also prevents cancers, coronary hear tdisease, arteriosclerosis (hardening ofthe arteries), bronchitis, dry orstubborn cough and blood clots	[42]

Tab. 3.1 (continued)

Botanical name	Common name	Family	Part used	Medicinal use	References
Allium sativum Linn.	Lahsun, garlic	Liliaceae	Bulb	Antibacteria and also prevents high blood pressure, anti-inflammatory, antihyperlipidemic, fibrinolytic	[41, 43]
Amorphophallus campanulatus (Roxb.) Blume	Jimikand	Araceae	Leaves, corm	Dysentery, piles, hemorrhoids	[32]
Andrographis paniculata (Burm.f.) Wallich ex Nees	Kalmegh, Bhui neem	Acantaceae	Whole plant	Chronic malaria, jaundice, anemia and loss of appetite	[39, 44]
Argimone mexicana Linn.	Bharbhand	Papavaraceae	Roots	Eczema and skin disease	[33, 39]
Artocarpus heterophyllus Lam.	Kathal	Moraceae	Fruit	Anti-ulcer	[45]
Asparagus racemosus Willd.	Shatawari	Alliaceae	Roots	Infertility, loss of libido, threatened miscarriage	[39]
Azadirachta indica A. Juss.	Neem	Meliaceae	Leaves, root, bark, flower	Arthritis, bronchitis, cough, diabetes, skin diseases	[46]
Bacopa monnieri (Linn.) Pennell.	Brahmi	Scrophulariaceae	Leaves	Nervine tonic	[34]
Barleria prionitis Linn	Vajradanti	Acanthaceae	Leaves	Strengthen teeth, fever, catarrh	[32]
Basella alba Linn.	Poi	Basellaceae	Leaves, fruits, roots	Constipation, gonorrhea, conjunctivitis, intestinal disorder	[36]
Bauhinia purpurea Linn.	Koliyari	Fabaceae	Leaves, flower	Intestinal worm infection, dysentery, diarrhea	[34, 36]
Boerhavia diffusa Linn.	Punarnava	Nyctaginaceae	Roots	Diuretic, anti-inflammatory and anti-arthritic	[39, 47]

Tab. 3.1 (continued)

Botanical name	Common name	Family	Part used	Medicinal use	References
Bombax ceiba Linn.	Silk cotton	Bombacaceae	Roots, fruits, gum	Anti-inflammatory, antidiabetic, anti-obesity, hypotensive, antioxidant, antiangiogenic, antimicrobial, cytotoxicity, aphrodisiac and antipyretic	[48]
Bryophyllumpinnatum (Lam.) Kurtz.	Patharchatta	Crassulaceae	Leaves, roots	Antifungal, antimicrobial, anticancer, antihypertensive activity, prevent and treat Leishmaniasis, protect against chemically induced anaphylact icreactions, wound	[36, 49]
Buchanania lanzan Spr.	Chironji, Chaar	Anacardiaceae	Leaves, roots, fruits	Rich source of Vitamin E, Leprosy, skin diseases, diarrhea, abdominal disorder, constipation, asthma, ulcer, laxative, snake bite, syphilis	[33, 34]
Butea monosperma (Lamk.) Taub	Palash, Tesu	Fabaceae	Bark fruits, gum	Complexion of skin, worm infestations, roundworm, adaptogen, abortifacient, antiestrogenic, anti-gout, anti-ovulatory, dysentery, diarrhea, Scorpion sting, irregular menstruation	[32, 36, 50]
Caesalpinia bonducella Roxb.	Gataran	Fabaceae	Leaves	Cough, asthma	[33]

Tab. 3.1 (continued)

Botanical name	Common name	Family	Part used	Medicinal use	References
Calotropis procera R. Br.	Aak	Asclepiadaceae	Flower, leaves, roots	Whooping cough, asthma, wound, rheumatic pain, wound, bronchitis	[33, 36]
Carissa carandus Linn.	Karonda	Apocynaceae	Fruits, roots	Anemia, purgative, carminative aphrodisiac	[36]
Cassia alata Linn.	Candle bush, ringworm bush	Fabaceae	Leaves, seeds, roots	Antiseptic, laxative and it also inhibits fungal growth	[43, 51]
Cassia fistula Linn	Amaltas	Fabaceae	Leaves, fruits, bark	Ulcer, wound, laxative, antipyretic, worm infestation, rheumatic pain, fever, cough, leprosy	[32, 36, 52]
Cassia tora Linn.	Charota bhaji	Caesalpinaceae	Leaves	Antimicrobial activity, skin disease	[32, 34, 37, 38]
Catharanthusroseus L. G. Don	Sadabahar	Apocynaceae	Leaves	Antileukemic, anticancer, antihypertensive, diarrhea	[53, 54]
Celastrus paniculata Willd.	Malkangni	Celastraceae	Fruit	Sinus problem	[34]
Centella asiatica (L.) Urban	Brahmi	Apiaceae	Whole Plant	Antioxidant, anti-inflammatory, wound healing, memory-enhancing property	[55]
Chlorophytum borivilianum Sant. F	Safed musli	Liliaceae	Roots	Aphrodisiac	[39, 56]
Cissus quadrangularis Linn.	Hadjod	Vitaceae	Whole plant	Bone fracture	[53]
Citrus medica Linn.	Nimboo, Lemon	Rutaceae	Fruits, leaves	Rich in vitamin C, liver disorder, anthelmintic, antiseptic	[36]

Tab. 3.1 (continued)

Botanical name	Common name	Family	Part used	Medicinal use	References
Colocasia esculenta (L.) Schott.	Arbi	Araceae	Leaves, rhizome	Asthma, arthritis, diarrhea, internal hemorrhage, neurological disorders, and skin disorders	[57]
Costus speciosus (Koen. Ex. Rets.) Sm.	Keokand	Zingiberaceae	Rhizome	Snake bite, jaundice	[33, 34]
Curcuma angustifolia Roxb.	Tikhur	Scitanineae	Rhizome	dysentery, dysuria and gonorrhea, antioxidative, anti-inflammatory properties, wound healing, hypoglycemia, anticoagulant, antimicrobial activities	[34, 58]
Cucurma longa Linn.	Haldi, Turmeric	Zingiberaceae	Rhizome	It prevents cancerous growth. It also aids digestion of fats. It is helpful in preventing the blockage of arteries that can gradually cause heart attack or stroke, wound healing, antioxidant	[59–61]
Curculigo orchioides Gaertn.	Kali musli	Amaryllidaceae	Root	Spermatogenesis enhancer	[39, 62]

Tab. 3.1 (continued)

Botanical name	Common name	Family	Part used	Medicinal use	References
Curcuma amada Roxb.	Amahaldi	Zingiberaceae	Rhizome	Appetizer, alexiteric, antipyretic, aphrodisiac, diuretic, emollient, expectorant and laxative and to cure biliousness, itching, skin diseases, bronchitis, asthma, hiccup, and inflammation due to injuries	[39, 63]
Cuscuta reflexa Roxb.	Amarbel	Convolaceae	Twiner	Wound, skin diseases	[33, 39]
Cyperus rotundus Linn.	Nagarmotha	Cyperaceae	Whole plant, roots	Fever, diabetes, solar dermatitis, skin diseases, blood purifier, snake bite	[32, 36]
Dalbergia sissoo Roxb.	Sheesham	Fabaceae	Leaves, flower	Gonorrhea, piles, skin diseases, leprosy, diarrhea	[36]
Datura stramonium Linn.	Dhatura	Solanaceae	Leaves, fruits	Asthma, cardiac pain	[46]
Desmodium gangeticum (Linn.) DC.	Salparni, Balraj	Fabaceae	Whole plant	Contraceptive, headache	[33, 34]
Diospyros melanoxylon Roxb.	Tendu	Ebenaceae	Fruits, bark, leaves flower	Diarrhea, dyspepsia, astringent, dysentery, ulceration of cornea and post-natal pain, urinary and skin troubles, blood diseases, leukorrhea, urinary discharge, anemia, diuretic, scabies, old wounds, laxative and carminative. Fruits are used in stomach disorders	[64]

Tab. 3.1 (continued)

Botanical name	Common name	Family	Part used	Medicinal use	References
Eclipta alba (Linn.)	Bhringraj	Asteraceae	Whole plant	Promotes hair growth, hepatoprotective	[65]
Eucalyptus tereticornis Sm.	Nilgiri	Myrtaceae	Leaves	Antimicrobial activity	[66]
Euphorbia hirta L.	Dudhi plant	Euphobiaceae	Whole plant	Antibacterial, anti-diarrhea, anti-asthmatic	[39, 67]
Ficus bengalensis Linn.	Bargad	Moraceae	Aerial roots, latex of plant	Syphilis, tonic	[33]
Ficus racemosa Linn.	Dumar	Moraceae	Latex, fruits	Piles, diarrhea, dysentery, urinary problem, anti-diabetic	[33]
Ficus religiosa Linn.	Peepal	Moraceae	Bark, leaves, fruits, seeds	Skin disease, constipation, gynecological disease	[46]
Gloriosa superba Linn.	Kalihari	Liliaceae	Tuber	Spasmolytic, oxytocic, source of colchine, leudermic spots	[34, 68]
Glycyrrhiza glabra Linn.	Mulethi	Papillionaceae	Stem	Digestive disorder, ulcer, bronchitis	[32]
Gymnema sylvestre (Retz.) R. Br.	Gurmar	Asclepiadaceae	Leaves, roots	Astringent, stomachic, tonic, anti-diabetic, liver disorder	[44]
Hibiscus rosa-sinensis Linn.	Gurhal	Malvaceae	Flower, roots, stem	Cough, diuretic, kidney problem	[36]

Tab. 3.1 (continued)

Botanical name	Common name	Family	Part used	Medicinal use	References
Ipomea aquatica Forsk.	Swamp morning glory	Convolvulaceae	Leaves	Liver diseases, anthelmintic, central nervous system depression (CNS) depressant, antiepileptic, hypolipidemic effects, antimicrobial, and anti-inflammatory	[69–74]
Madhuca longifolia (L.) Macbride.	Mahua	Sapotaceae	Flower, seed oil	Aching muscle pain relief, heel crack, Scorpion sting, anti-diabetic	[33, 39, 75]
Mangifera indica Linn.	Aam	Anacardiaceae	Leaves, fruits	Anti-emetic, anti-inflammatory, antioxidant, pyorrhea, heart disease	[36, 76]
Momordica charantia Linn.	Karela	Cucurbitaceae	Leaves, root, fruit, seeds	Anti-diabetic	[77]
Momordica dioca Roxb. ex. Willd.	Kheksi, Kakora	Cucurbitaceae	Fruits, leaves, roots	Bleeding piles, urinary infection, antioxidant, hepatoprotective, antibacterial, anti-inflammatory, anti-lipid peroxidative, hypoglycemic, and analgesic properties	[78]
Moringa oleifera Lam.	Munga, Drumstick	Moringaceae	Leaves, Fruits	Antioxidant, antimicrobial, asthma, dysentery, intestinal cancer	[61, 79]
Morus alba Linn.	Shahtut	Moraceae	Bark, roots, fruits	Stomach pain	[53]

Tab. 3.1 (continued)

Botanical name	Common name	Family	Part used	Medicinal use	References
Mucuna puriens L. DC.	Kewanch	Fabaceae	Roots	Gout, joint pain	[33]
Nigella sativa Linn.	Kalonji	Ranunculaceae	Seeds	Diarrhea, dysentery	[46]
Ocimum sanctum Linn.	Tulsi	Lamiaceae	Leaves, fruits	Anti-allergic, anti-diabetic, antioxidant, immunomodulator, hypoglycemic, adaptogen	[46, 80]
Ocimum tenuiflorum Linn.	Krishna tulsi	Lamiaceae	Leaves, flower, stem	Skin diseases, cold, cough, fever, vomiting, anti-cancer, antimicrobial, antiseptic, antispasmodic, antifungal, antiviral, anti-inflammatory, analgesic and immunostimulatory properties	[39, 81]
Oryza sativa Linn.	Rice	Poaceae	Grain	Boils, sores, swellings, skin blemishes, stomach upsets, heart-burn, indigestion, nausea, and diarrhea	[82]
Papaver somniferum Linn.	Aphim	Papavaraceae	Whole plant, roots, stem, bark, leaves, flower, fruit and seeds	Cough, asthma, analgesic, narcotic, sedative, stimulant	[83]

Tab. 3.1 (continued)

Botanical name	Common name	Family	Part used	Medicinal use	References
Passiflora foetida L.	Passion flower	Passifloraceae	Roots, leaves, fruits	Antispasmodic, antibacterial, antihypertension, antiproliferative activity on human breast adenocarcinoma	[84]
Phyllanthus amarus	Bhui amla	Euphobiaceae	Whole plant	Hepatoprotective, jaundice	[39]
Phyllanthus emblica Linn.	Amla	Phyllanthacae	Fruits	Antioxidant, source of vitamin C, anti-inflammatory, analgesic, antipyretic, adaptogenic, hepatoprotective, antitumorand anti-ulcerogenic activities	[39, 85]
Piper longum Linn.	Long pepper	Piperaceae	Fruit, root	Cold, cough, chronic bronchitis, palsy, gout, rheumatism, epilepsy, asthma, piles, leukoderma	[44]
Piper nigrum Linn.	Kali mirch, Black paper	Piperaceae	Fruits seeds	Anti-tumorigenic, immunostimulatory, stomachic, carminative, anticholestrolemic and antioxidant. Antimalarial parasite	[86]
Pongamia pinnata (L.) Pierre K	Karanj	Fabaceae	Seed oil, tender branch	Ezema, toothache, and gum problem	[33]
Rauwolfia serpentina (L.) Benth ex. Kurz	Serpgandha	Apocynaceae	Leaves, roots	Antihypertensive, mental disorders	[32, 87, 88]
Ricinus communis Linn.	Arand	Euphobiaceae	Seeds	Purgative, joint problem	[89]

Tab. 3.1 (continued)

Botanical name	Common name	Family	Part used	Medicinal use	References
Santalum album Linn.	Chandan	Santalaceae	Wood, oil	Disinfectant in genitourinary and bronchial tracts, diuretic, expectorant, and stimulant	[90]
Saraca asoca (Roxb.) de Willd.	Ashoka	Caesalpinaceae	Bark, flower; seed	Astringent used in menorrhagia and uterine, Menstrual irregularities	[44]
Semecarpus anacardium Linn.	Bhelwa	Anacardiaceae	Fruits and nuts	Antiatherogenic, anti-inflammatory, antioxidant, antimicrobial, anti-reproductive, CNS stimulant, hypoglycemic, anticarcinogenic, and hair growth promoter	[39, 91]
Shorea robusta A.W. Roth.	Sal, Sarai	Dipterocarpaceae	Fruits	Dysentery and Scorpion sting	[33]
Sida acuta Burm. F	Horn bean leaf	Malvaceae	Leaves, roots,	Antipyretic, inhibitory activity against bacterial diseases, stomachic, diaphoretic, hepatoprotective	[92, 93]
Sida cordifolia Linn.	Bala	Malvaceae	Whole plant	Bronchial asthma, cold and flu, head ache, nasal congestion, aching joints and bones, edema, nervous and urinary diseases, disorders of the blood and bile, analgesic, anti-inflammatory and hypoglycemic activities, hepatoprotective activity	[94]

Tab. 3.1 (continued)

Botanical name	Common name	Family	Part used	Medicinal use	References
Syzgium cuminii (L.) Skeel.	Jamun	Myrtaceae	Fruits, seeds	Antidiabetic	[33]
Swertia chirata Buch-Ham	Chiraita	Gentianaceae	Whole plant	Tonic, stomachic, laxative, anorexia, biliary disorder, cough, constipation, fever, skin diseases, worms	[44]
Terminalia arjuna (Roxb.) Wight. &Arn.	Arjuna	Combretaceae	Bark	Heart diseases	[95]
Terminella bellirica (Gaertn.) Roxb. B	Bahera	Combretaceae	Fruits	Laxative, antioxidants	[39, 60]
Terminella chebula Retz.	Harra	Combretaceae	Fruits	Laxative, antioxidants	[39, 60]
Thymus vulgaris Linn.	Ajwain	Lamiaceae	Seeds	Antiseptic, antispasmodic	[46]
Tinospora cordifolia (Willd.)Hook. f. and Thoms.	Giloy	Menispermaceae	Stem, leaves	Adaptogen, immunomodulator	[39]
Trigolella foenum-graecum Linn.	Methi	Fabaceae	Seeds	Constipation, diabetes	[46]
Vitex negundo Linn.	Nirgundi	Verbenaceae	Leaves, root, flower, fruit, seeds	Anti-inflammatory, antiarthritic, immunomodulator	[96]
Withania somnifera Linn. Dunal.	Ashwagandha	Solanaceae	Roots, leaves	Tonic, diuretic, narcotic rheumatism	[44]

Tab. 3.1 (continued)

Botanical name	Common name	Family	Part used	Medicinal use	References
Zingiber officinale Rosc. A	Adrak, Ginger	Zingiberaceae	Rhizome	It prevents chilblains and circulatory problems such as Raynaud's disease. It is highly antiseptic, activating immunity, and dispelling a whole variety of bacterial and viral infection. It also inhibits blood clotting. Fever, cough, asthma, anti-emetic	[42, 97]
Ziziphus jujuba Mill.	Ber	Rhamnaceae	Fruit, leaves, bark	Wound, ulcer, fever, abdominal pain, asthma	[36]

3.11 Conclusion

Nowadays, lifestyle is changing and becoming more techno-savvy and materialistic, but recent crises in the world of corona pandemic again explain the importance of nature and natural products in life, and people are again returning to their roots and prioritizing natural products to boost their immunity as well as taking herbal medicines for primary treatments.

The traditional used herbs are of natural origin they are easily locally available and have no side effects so had become the symbol of safety in comparison to synthetic drugs so people are returning back to the natural product which provides safety and security. Its promotion, conservation, and multiplication are need of an hour around the world.

The main reason of depletion of medicinal plants is urbanization, modernization, modification of culture, and industrial expansion which reduces the forest cover. The knowledge of medicinal herbs is not documented properly, and confined to tribal population and local healers through oral tradition; there is a need to document this knowledge for coming generations. Many healers have a belief that training of traditional medicine is effective within a family while providing knowledge to any stranger is culturally illegal.

References

[1]   G. Martin. Ethnobotany: A Method Manual. Chapman and Hall, London, 1995.
[2]   M. Balick, P. Cox. Plants, People, and Culture: Science of Ethnobotany. Scientific American Library, New York, 1996.
[3]   C. Cotton. Ethnobotany: Principles and Applications. Wiley, New York, 1996.
[4]   M. Spinella. The Psychopharmacology of Herbal Medicines. MIT Press, England, 2001, pp. 1–2.
[5]   D. Krishnaiah, S. Rosalam, R. Nithyanandam. A review of the antioxidant potential of medicinal plant species. Food. 2011, 89(3), 217–233.
[6]   N. Farnsworth, D. Soejarto. Global importance of medicinal plants. In: Akerele, O., Heywood, V., Synge, H., (Eds.). Conservation of Medicinal Plants, Cambridge Univerversity Press, Cambridge, 1991, pp. 25–52.
[7]   V. Narayanaswami. Introduction of the Siddha System of Medicine. Anandkumar, A., Pandit, S. S. Anandam Research Institute of Siddha Medicine, T. Nagar, Madras (Chennai), 1975.
[8]   R. L. S. Sikarwar. Ethno-gynaecological uses of plants new to India. Ethno Botany. 2002, 14, 112–115.
[9]   R. Siva. Status of natural dyes and dye-yielding plants in India. Curr Sci. 2007, 32, 916–925.
[10]   M. J. Bhandary, K. R. Chandrashekhar. Glimpses of ethnic herbal medicine of coastal Karnataka. Ethnobotany. 2002, 14, 1–12.
[11]   WHO. WHO Guidelines on Good Agricultural and Collection Practices (GACP) for Medicinal Plants. World Health Organization, Geneva, Switzerland, 2003.
[12]   D. K. Patel. Medicinal plants in G.G.V. Campus, Bilaspur, Chhattisgarh in central India. Int J Med Aromatic Plants. 2012, 2(2), 293–300.
[13]   W. A. Rodgers, H. S. Panwar. Planning a Wildlife Protected Area Network in India, Wildlife Institute of India, Dehradun, 1988.
[14]   Samant, Joshi. Plant diversity and conservation status of Nanda Devi National Park and comparison with highland National Parks of the Indian Himalayan Region. Int J Bio Sci Manage. 2005, 1(1), 65–73. DOI: 10.1080/17451590509618081.
[15]   R. K. Maikhuri, S. Nautiyal, K. S. Rao, K. G. Saxena. Medicinal plant cultivation and biosphere reserve management: A case study from the Nanda Devi Biosphere Reserve, Himalaya. Curr Sci. 1998, 74, 157–163.
[16]   S. Nautiyal, K. S. Rao, R. K. Maikhuri, R. L. Semwal, K. G. Saxena. Traditional knowledge related to medicinal and aromatic plants in tribal societies in a part of Himalaya. J Med Aromat Plant Sci. 2005, 23, 4A and 23/1A, 428–441.
[17]   P. C. Phondani. A study on prioritization and categorization of specific ailments in different high altitude tribal and non-tribal communities and their traditional plant based treatments in Central Himalaya [unpublished Ph.D. thesis], Srinagar, India: HNB Garhwal University, 2010.
[18]   R. K. Maikhuri, S. Nautiyal, K. S. Rao, K. G. Saxena. Role of medicinal plants in the traditional health care system: A case study from Nanda Devi Biosphere Reserve, Himalaya. Curr Sci. 1998, 75(2), 152–157.
[19]   C. P. Kala, N. A. Farooquee, B. S. Majila. Indigenous knowledge and medicinal plants used by vaidyas in Uttarakhand, India. Nat Prod Radiance. 2005, 4, 195–204.
[20]   S. Chatterjee, A. P. Dutta, D. Ghosh, G. Pangging, A. K. Goswami. Biodiversity Significance of North East India. WWF, India, New Delhi, 2006.
[21]   A. A. Mao, T. M. Hyniewta, M. Sanjappa. Plant wealth of North East India with reference to ethnobotany. Indian J Tradit Knowl. 2009, 8, 96–103.
[22]   R. Shankar, M. S. Rawat. Conservation and cultivation of threatened and high valued medicinal plants in North East India. Int J Biodivers Conserv. 2013, 5(9), 584–591.
[23]   M. Gadgil. Western Ghat: A lifescape. Indian Inst Sci. 1996, 76, 495–504.

[24] A. Suja. Environmental Information System, Sahyadri. Medicinal Plants of Western Ghats. Sahyadri E-News. 2005, 2–16.

[25] A. K. Pandey, A. K. Bisaria. Rational utilization of important medicinal plants: A tool for conservation. Indian Forester. 1997, 124(4), 197–206.

[26] A. K. Pandey, A. K. Patra, P. K. Shukla. Medicinal plants in satpura plateau of Madhya Pradesh: Current status and future prospects. Indian Forester. 2005, 131, 857–883.

[27] A. Tewari, A. Srivastava. Economic value of medicinal plants in Indo-Gangetic plain areas of Kanpur: An Urban Environment. Trends Biosci. 2011, 4(2), 201–204.

[28] M. Goyal. Review of medicinal plants used by local community of Jodhpur District of Thar desert. Int J Pharmacol. 2011. DOI: 10.3923/ijp.2011.333.339.

[29] S. K. Uniyal, K. N. Singh, P. Jamwal, B. Lal. Traditional use of medicinal plants among the tribal communities of Chhota Bhangal, Western Himalaya. J Ethnobiol Ethnomed. 2006, 2, 14. DOI: 10.1186/1746-4269-2-14

[30] R. C. Uniyal, M. R. Uniyal, P. Jain. Cultivation of Medicinal Plants in India.A Reference book.New. TRAFFIC India & WWF India, Delhi, India, 2000.

[31] P. A. Cox, M. J. Balick. The ethnobotanical approach to drug discovery. Sci Am. 1994, 270(6), 82–7.

[32] A. Sofowora, E. Ogunbodede, A. Onayade. The role and place of medicinal plants in the strategies for disease prevention. Afr J Tradit Complementary Altern Med. 2013, 10(5), 210–229.

[33] M. Kujur, R. K. Ahirwar. Folklore claims on some ethno medicinal plants used by various tribes of district Jashpur, Chhattisgarh, India. Int J Curr Microbiol Appl Sci. 2015, 4(9), 860–867.

[34] S. Kumar. Study of forest based indigenous technology at Bastar region of Chhattisgarh. Int J Res Anal Rev. 2016, 3(3), 304–311.

[35] K. Kaur, S. Arora, M. E. Hawthorne, S. Kaur, S. Kumar, R. G. Mehta. A correlative study on antimutagenic and chemopreventive activity of *Acacia auriculiformis* A. Cunn. and *Acacia nilotica* (L.) Willd. Ex Del. Drug Chem Toxicol. 2002, 25(1), 39–64. DOI: 10.1081/dct-100108471.

[36] P. K. Sahu, A. Tiwari, S. Banerjee, P. Pandey. Ethnomedicinal plants used in the healthcare sysytems of tribal people in Chhattisgarh India. Indian J Sci Res. 2017, 13(2), 119–124.

[37] G. V. Satyavati, M. K. Raina, M. Sharma, (Eds.),. *Acorus*.Medicinal Plants of India-Vol. 1. New Delhi Indian Council of Medical Research 1976, pp. 18–22.

[38] B. C. Bose, R. Vijayavargiya, A. Q. Safi, S. K. Sharma. Some aspects of chemical and pharmacological studies of *Acorus calamus* Linn. J Am Pharm Assoc. 1960, 49, 32.

[39] P. Shah, A. Maitry, R. Naidu, P. Netam, A. Mandavia, S. Singh. Common Medicinal Plants of Chhattisgarh, India, 2020. 10.13140/RG.2.2.29954.30400.

[40] D. S. Vyas, V. N. Sharma, S. A. H. Naqvi, S. Ahmad, Khanna. Preliminary study on anti-diabetic properties of *Aegle marmelos* and *Enicostemma littorale*. J Res Indian Med Yoga Homoeopathy. 1979, 14(3), 63–66.

[41] P. P. Dixit, J. S. Londhe, S. K. Ghaskadbi, T. P. A. Devasagayam. Anti-diabetic and related beneficial properties of Indian medicinal plants. In: Sharma, R. K., Arora, R. (Eds.), Herbal Drugs: A Twenty First Century Perspective. JAYPEE Brothers, New Delhi, 2006, 377–395.

[42] Ozougwu, Eyo. Evaluation of the activity of *Zingiber Officinale* (Ginger) aqueous extracts on alloxan-induced diabetic rats. Pharmacol Online. 2011, 1, 258–269.

[43] L. S. Gill. Ethnomedical Uses of Plants in Nigeria, University of Benin press, Nigeria, 1992.

[44] S. Tike. Review of region wise distribution of medicinal plants of India. Int J Res Indian Med. 2016, 1(1), 67–72.

[45] I. O. Lawal, N. E. Uzokwe, A. B. I. Igboanugo, A. F. Adio, E. A. Awosan, J. O. Nwogwugwu, B. Faloye, B. P. Olatunji, A. A. Adesoga. Ethnomedicinal information on collation and identification of some medicinal plants in Research Institutes of South-west Nigeria. Afr J Pharm Pharmacol. 2010, 4(1), 001–007.

[46] S. Gangola, P. Khati, P. Bhatt, Parul, A. Sharma. India as the heritage of medicinal plants and their uses. Curr Trends Biomed Eng Biosci. 2017, 4(4), 0050–0051.

[47] P. C. Sharma, M. B. Yelne, T. J. Dennis, (Eds.), Data base on medicinal plants used in ayurveda volume 1. Central Council for Research in Ayurveda and Siddha. In: Raktapunarnava (*Boerhavia diffusa*). New Delhi, 2000a, pp. 360–377.

[48] R. Verma, K. Devre, T. Gangrade, S. Gore, S. Gour. A Pharmacognostic and pharmacological overview on *Bombax ceiba*. Scholars Acad J Pharmacy. 2014, 3(2), 100–107.

[49] S. Ghasi, C. Egwuibe, P. U. Achukwu, J. C. Onyeanusi. Assessment of the medical benefit in the folkloric use of *Bryophyllum pinnatum* leaf among the Igbos of Nigeria for the treatment of hypertension. Afr J Pharm Pharmacol. 2011, 5(1), 83–92.

[50] P. C. Sharma, M. B. Yelne, T. J. Dennis, (Eds.), Data Base on Medicinal Plants Used in Ayurveda Volume- 1. New Delhi, Central Council for Research in Ayurveda and Siddha, Palasha (Butea monosperma). 2000b, pp. 336–347.

[51] W. F. Sule, I. O. Okonko, T. A. Joseph, M. O. Ojezele, J. C. Nwanze, J. A. Alli, O. G. Adewale. In vitro antifungal activity of *Senna alata* Linn.crude leaf extract. Res J Biol Sci. 2010, 5(3), 275–284.

[52] P. K. Joshi. Pharmaco-clinical study of Argavadha with special reference to vicharchika. M.D. (Ayu) Dissertation submitted to Gujarat Ayurved University, Jamnagar. India, 1998.

[53] V. K. Painkara, M. K. Jhariya, A. Raj. Assessment of knowledge of medicinal plants and their use in tribal region of Jashpur district of Chhattisgarh, India. J Appl Nat Sci. 2015, 7(1), 434–442.

[54] A. A. Elujoba, O. M. Odeleye, C. M. Ogunyemi. Traditional medical development for medical and dental primary healthcare delivery system in Africa. Afr J Tradit Complementary Altern Med. 2005, 2(1), 46–61.

[55] V. Seevaratnam, P. Banumathi, M. R. Premlatha, S. P. Sundaram, T. Arumugam. Functional properties of *Centella asiatica* (L.): A Review. Int J Pharmacy Pharm Sci. 2012, 4(5), 8–14.

[56] A. A. Farooqi, M. M. Khan, M. Vasundhara. Natural Remedies. Production Technology of Medicinal and Aromatic Crops, Bangalore, India, 2001, pp. 90–91.

[57] P. Sudhakar, V. Thenmozhi, S. Srivignesh, M. Dhanalakshmi. *Colocasia esculenta* (L.) Schott: Pharmacognostic and pharmacological review. J Pharmacogn Phytochem. 2020, 9(4), 1382–1386.

[58] S. Sharma, S. K. Ghataury, A. Sarathe, G. Dubey, G. Parkhe. *Curcuma angustifolia* Roxb, (Zingiberaceae): Ethnobotany, phytochemistry and pharmacology: A review. J Pharmacogn Phytochem. 2019, 8(2), 1535–1540.

[59] R. M. Tripathi, S. S. Gupta, D. Chandra. Antitrypsin and Anti-hyaluronidase activity of *Curcuma longa* (Haldi). Indian J Pharmacol. 1973, 5, 260–261.

[60] S. Narasimhan, G. Anitha, K. Illango, K. R. Mohan. Bio-assay guided isolation of active principles from medicinally important plants. In: Sharma, R. K., Arora, R., (Eds.), Herbal Drugs: A Twenty First Century Perspective. JAYPEE Brothers, New Delhi, 2006, 70–76.

[61] T. Odugbemi. Outlines and Pictures of Medicinal Plants from Nigeria. University of Lagos Press, Lagos, 2006, pp. 283.

[62] P. K. Joshi. A comparative pharmacognostical, phytochemical and pharmacotherapeutic study of Sweta Musali (*Curculigo orchioides*) with special reference to Vrsya karma. Ph.D. (Ayu) Thesis, submitted to Gujarat Ayurved University, Jamnagar. India, 2005.

[63] R. S. Policegoudra, S. M. Aradhya, L. Singh. Mango ginger (*Curcuma amada* Roxb.) – A promising spice for phytochemicals and biological activities. J Biosci. 2011, 36(4), 739–748.

[64] V. Gupta, V. Maitili, P. K. Vishwakarma. Comparative study of analgesic activity of *Diospyros Melanoxylon* (Roxb.) Bark and Root Bark. J Nat Remedies. 2013, 13(1), 15–18.

[65] T. Chandra, J. Sadique, S. Somasundaram. Effect of *Eclipta alba* on inflammation and liver injury. Fitoterapia. 1987, 58(1), 23–32.

[66] P. Jain, S. Nimbrana, G. Kalia. Antimicrobila activity and phytochemical analysis of Eucalyptus tereticornis bark and leaf methanolic extracts. Int J Pharm Sci Rev Res. 2010, 4(2), 126–128.

[67] N. D. Onwukaeme. Medicinal plant of Nigeria Natural medicine development agency federal ministry of science and technology. Phytother Res. 1995, 9, 306–308.

[68] P. C. Sharma, M. B. Yelne, T. J. Dennis, (Eds.), Data Base on Medicinal Plants Used in Ayurveda Volume 4. In: Langli (*Gloriosa superba*). Central Council for Research in Ayurveda and Siddha, New Delhi, 2002, pp. 341–357.

[69] C. Malakar, P. P. N. Choudhury. Pharmacological potentiality and medicinal uses of *Ipomoea aquatic* Forsk: A review. Asian J Pharm Clin Res. 2015, 8(2), 60–63.

[70] S. M. Badruzzaman, W. Husain. Some aquatic and marshy land medicinal plants from Hardoi district of Uttar Pradesh. Fitoterapia. 1992, 63, 245–7.

[71] S. C. Datta, A. K. Banerjee. Useful weeds of West Bengal rice fields. Econ Bot. 1978, 32, 297–310.

[72] A. K. Nadkarni. Indian Materia Medica. 3rd ed, Popular Books, Bombay.

[73] S. Dhanasekaran, P. Muralidaran. CNS depressant and antiepileptic activities of the methanol extract of the leaves of *Ipomoea aquatic*a Forsk. Eur J Chem. 2010, 7, 1555–61.

[74] S. Dhanasekaran, M. Palaya, K. S. Shantha. Evaluation of antimicrobial and anti-inflammatory activity of methanol leaf extract of *Ipomoea aquatica* Forsk. Res J Pharm Biol Chem Sci. 2010, 1, 258–64.

[75] J. B. Jain, S. C. Kumane, S. Bhattacharya. Medicinal flora of Madhya Prasesh and Chattisgarh- A review. Indian J Tradit Knowl. 2006, 5(2), 237–242.

[76] T. F. Okujagu, S. O. Etatuvie, I. Eze, B. Jimoh, C. Nwokereke, C. Mbaoji, Z. Mohammed. Medicinal plant of Nigeria, north-west Nigeria. Vol 1. Nigerian J Nat Prod Med. 2008, 17/(1(1)), 32–34.

[77] I. Ahmed, M. S. Lakhani, M. Gillet, A. John, H. Raza. Hypotriglyceridemic and hypocholesterolemic effects of antidiabetic *Momordica charantia* (Karela) fruit extract in streptozocin induced diabetic rats. Diabetes Res Clin Pract. 2001, 51(3), 155–161.

[78] B. Bawara, M. Dixit, N. S. Chauhan, V. K. Dixit, D. K. Saraf. Phyto-pharmacology of *Momordica dioica* Roxb. ex. Willd. Int J Phytomed. 2010, 2, 01–09. 10.5138/ijpm.2010.0975.0185.02001.

[79] S. Gupta, R. Jain, S. Kachhwaha, S. L. Kothari. Nutritional and medicinal applications of *Moringa oleifera* Lam. – Review of current status and future possibilities. J Herb Med. 2018, 11, 1–11.

[80] P. Uma Devi. Radioprotective, anticinogenic and antioxidant properties of the Indian holy basil, *Ocimum sanctum* (Tulsi). Indian J Exp Biol. 2001, 39, 185–190.

[81] R. Palla, A. Elumalai, M. C. Eswaraiah, K. Raju. A review on Krishna tulsi, *Ocimum tenuiflorum* Linn. Int J Res Ayurveda Pharmacy. 2012, 3(2), 291–293.

[82] M. Umadevi, R. Pushpa, K. P. Sampathkumar, D. Bhowmik. Rice-traditional medicinal plant in India. J Pharmacogn Phytochem. 2012, 1(1), 6–12.

[83] Masihuddin, M. A. Jafri, A. Siddiqui, S. Shahid Chaudhary. Traditional uses, phytochemistry and pharmacological activities of *Papaver somniferum* with special reference of unani medicine: An updated review. J Durg Delivery Ther. 2018, 8(5), 110–114.

[84] P. Moongkarndi, N. Kosem, O. Luanratana, S. Jongsomboonkusol, N. Pongpan. Antiproliferative activity of Thai medicinal plant extracts on human breast adenocarcinoma. Fitoterapia. 2004, 75, 375–377.

[85] M. D. Khurshid, V. Shukla, B. Bhupendra Kumar, Amandeep. A review paper on medicinal properties of *Phyllanthus emblica*. Int J Pharm Biol Sci. 2020, 10(3), 102–109.

[86] P. Shaba, N. N. Pandey, O. P. Sharma, J. R. Rao, R. K. Singh. Anti-trypanosomal activity of *Piper Nigrum* L (Black pepper) Against *Trypanosoma Evansi*. J Vet Adv. 2012, 2(4), 161–167.

[87] J. O. Kokwaro. Current status of utilization and conservation of medicinal plants in Africa, South of the Sahara. Acta Horticulturae (ISHS). 1993, 332, 121–130.

[88] S. M. S. Chauhan, B. Ambika, T. Singh, P. P. author. Bio-active compounds from Himalyan medicinal plan. In: Sharma, R. K., Arora, R. (Eds.), Herbal Drugs: A Twenty First Century Perspective. JAYPEE Brothers, New Delhi, 2006, 190–199.

[89] I. Hingora, A. Sharma. Traditional uses of medicinal plants of Gariaband District Chhattisgarh. Int J Recent Res Life Sci. 2017, 4(1), 9–11.
[90] R. K. Sindhu, Upma, A. Kumar, Arora, S. *Santalum album*Linn: A review on morphology, phytochemistry and pharmacological aspects. Int J Pharm Tech Res. 2010, 2(1): 914–919.
[91] M. Semalty, A. Semalty, A. Badola, G. P. Joshi, M. S. Rawat. *Semecarpus anacardium*Linn.: A review. Pharmacogn Rev. 2010, 4(7), 88–94.
[92] I. R. Iroha, E. S. Amadi, A. C. Nwuzo, F. N. Afiukwa. Evaluation of the antibacterial activity of extracts of *Sida acuta* against clinical isolates of *Staphylococcus aureus* isolated from Human Immunodeficiency Virus/Acquired immunodeficiency syndrome patients. Res J Pharmacol. 2009, 3 (2), 22–25.
[93] C. D. Sreedevi, P. G. Latha, P. Ancy, S. R. Suja, S. Shyamal, V. J. Shine, S. Sini, G. I. Anuja, S. Rajasekharan. Hepatoprotective studies on *Sida acuta*. Burm F J Ethnopharmacol. 2009, 124(2), 171–175.
[94] A. Jain, S. Choubey, P. K. Singour, H. Rajak, R. S. Pawar. *Sida cordifolia* (Linn) – An overview. J Appl Pharm Sci. 2011, 01(02), 23–31.
[95] G. Karunakaran, L. C. Mishra. Ischemic heart diseases. In: Mishra, L. C., (Eds.), Scientific Basis for Ayurvedic Therapies. CRC Press, New York, 2004, 512–531.
[96] A. M. Nair, M. N. Saraf. Inhibition of antigen and compound 48/80 induced contractions of guinea pig trachea by the ethanolic extract of the leaves of *Vitex negundo* Linn. Indian J Pharmacol. 1995, 27, 230–233.
[97] P. C. Sharma, M. B. Yelne, T. J. Dennis, (Eds.), Data Base on Medicinal Plants Used in Ayurveda Volume 5. Central Council for Research in Ayurveda and Siddha, New Delhi, Shunthi (*Zingiber officinale* Rosc), pp. 315–390.
[98] *http://www.indianmedicine.nac.in.*

Tarun Kumar Patle, Pramod Kumar Mahish
and Ravishankar Chauhan*

Chapter 4
Plants alkaloids and flavonoids: biosynthesis, classification, and medicinal uses

Abstract: Alkaloids and flavonoids are vital natural pharmacological active secondary metabolites that have long been concern because of their significant health benefits for the human being and treating many ailments. This chapter summarizes the types, biosynthesis, sources, and health benefits of alkaloids and flavonoids as fascinating substitute sources for medicinal and pharmaceutical applications. Biosynthesis pathways and classification of secondary metabolites, particularly alkaloids and flavonoids have been demonstrated briefly here with their molecular structures. The presence of these phytoconstituents in different medicinally important plants and their applications in medical and pharmaceutical aspects, particularly for health-promoting, e.g., free radical inhibitors, antiviral, antitumor, antibacterial, anti-inflammatory, antidiabetic, and so forth are highlighted. Conclusively, an effort was made to précis the plant-derived alkaloids and flavonoids with useful biological activities to increase an understanding of their effects on the health of the human being.

4.1 Introduction

Plants synthesize a vast diversity of naturally occurring chemical compounds known as phytochemicals/plant metabolites. These organic chemical compounds have low molecular weight with various therapeutic benefits as well as attributed nutritional benefits [1]. Phytochemicals are categorized into two major parts, primary metabolites and secondary metabolites. Primary metabolites are responsible for the growth and development of plants, whereas secondary metabolites are specialized metabolites or natural products having several health benefits such as antioxidant, antiviral, antimicrobial, anticancer, enzyme detoxification regulation, immune system modulation, anti-arthritis, reduced platelet aggregation, antidiabetic, and hormone metabolism property [2–5]. Primary metabolites include carbohydrates, proteins, vitamins, and

---

*Corresponding author: Ravishankar Chauhan, Department of Botany, Pandit Ravishankar Tripathi Government College, Bhaiyathan, 497231, Surajpur, Chhattisgarh, India, e-mail: ravi18bt@gmail.com
Tarun Kumar Patle, Department of Chemistry, Pt. Sundarlal Sharma (Open) University, Bilaspur, 495009, Chhattisgarh, India
Pramod Kumar Mahish, Department of Biotechnology, Government Digvijay (Autonomous) PG College, Rajnandgaon, 491441, Chhattisgarh, India

https://doi.org/10.1515/9783110791891-004

amino acids, whereas secondary metabolites such as phenolic acids, flavonoids, alkaloids, and terpenoids protect the plant from different threats [6].

Phenolics are considered to be a very important class of bioactive compounds due to their diverse medicinal properties [7]. The antioxidant property of phenolics is primarily associated with the hydroxyl group within the compounds [8]. Figure 4.1 shows a flow diagram of different class of phytochemicals. Flavonoids are an enormous class of bioactive constituents containing 10,000 well-known chemical structures [9]. The major significance of the flavonoids is their low toxicity in humans; many are used as a medicine [10]. Flavonoids show many pharmacological activities such as antioxidant, antiulcer, anti-inflammatory, anticancer, anti-hepatotoxic, and antidiabetic [11, 12]. Alkaloids are often noxious to humans but have numerous biological activities such as antimicrobial, anti-inflammatory, and anticancer; hence, they are broadly used for pharmaceutical purposes [13, 14].

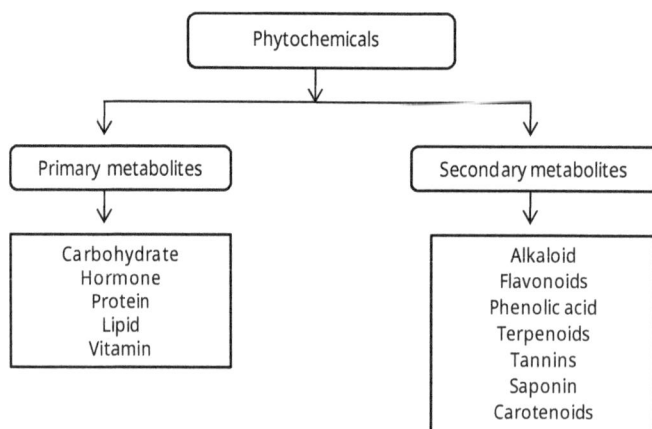

Fig. 4.1: A flow diagram of classification of plants primary and secondary metabolites.

Plants synthesized a huge variety of organic compounds. A bioactive compound can be defined as compounds that have biological activity and benefits to human health. Plants synthesize several bioactive compounds during their metabolic pathways to effectively treat stress conditions associated with different factors. These bioactive compounds are protective against herbivores, viruses, and microbes, and release chemical substance to attract the attention of insects and animals for seed-dispersing or pollinating, as well as defending the plant from oxidants or harmful ultraviolet rays [15]. Ethnobotanical data show that tribal people use plant material as ethnomedicine to treat several diseases. They are taking either whole, raw, or cooked material to make a tonic/paste to treat several diseases [16]. This chapter provides a summarized form of two major secondary metabolites, i.e., alkaloids and flavonoids. The chapter included the synthesis pathway, basic structures, and some examples of health benefits of these two phytochemicals.

4.2 Alkaloids of plants

Alkaloids are one of the most important classes of phytochemicals. They are nitrogen-containing molecules in their structure [17]. Amino acids are a key precursor for the biosynthesis of alkaloids in plants and its derivative, which contains the heterocyclic ring or another ring system. Due to the several biological actions of alkaloids like anti-cancer, anti-inflammatory, and antimicrobial, they are broadly used in the pharmaceutical industries [14].

4.2.1 Biosynthesis of alkaloids in plants

The primary steps in an alkaloid synthesis are mostly critical; thus, the alkaloids can be synthesized in plants through various pathways, including the shikimate pathway, amino acid pathway, terpenoid, and polyketide pathway [18]. The precursors change from primary to specific compounds by several chemical and enzymatic processes. There are four important steps involved in the biosynthesis of an alkaloid: (i) accumulation of an amine precursor, (ii) accumulation of an aldehyde precursor, (iii) formation of an iminium cation, and (iv) a Mannich-like reaction [19].

4.2.2 Classification of alkaloids and its importance

To date, more than 40,000 alkaloids have been identified from natural sources and classified according to their plant origin, natural origin, synthesis pathway, biochemical origin, and chemical structure. The naming and classification of alkaloids based on structure is frequently used, and it is based on CN skeleton [20]. Alkaloids are divided into three major types on the basis of their structure and synthesis: (a) true alkaloids (heterocyclics), (b) protoalkaloids (non-heterocyclics), and (c) pseudo alkaloids [20].

4.2.2.1 True alkaloid (heterocyclic)

True alkaloids are physiologically active and chemically complex derivatives of cyclic amino acids. The basic skeleton of true alkaloid has intracyclic nitrogen with some aliphatic acid [21, 22]. The major amino acid involves in the basic structure of true alkaloids are L-arginine, L-ornithine, L-phenylalanine, L-tyrosine, L-lysine, L-tryptophan, L-histidine,

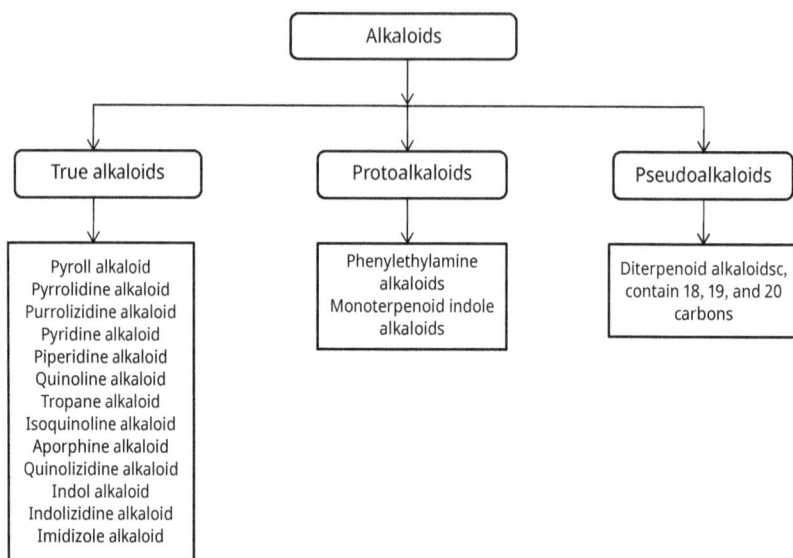

Fig. 4.2: A flow diagram of classification of plant alkaloids.

and anthranilic acid/glycine [23]. Indole, quinoline, and pyrroloindole alkaloids have basic skeleton of L-tryptophan; isoquinoline alkaloids have basic skeleton of the amino acid L-tyrosine, similarly, tropane, pyrrolizidine, and pyrrolidine alkaloids have base of L-ornithine; quinolizidine and piperidine alkaloids have base of L-lysine; pyridine has base of aspartate, quinoline has base of anthranilic acid, and imidazole alkaloids are derivatives of L-histidine [24]. The heterocyclic alkaloids are divided into several types shown in Fig. 4.2. Within these groups, we can find alkaloids, such as berberine, salsoline, geissospermine, piperine, nicotine, lobeline, nantenine, cocaine, quinine, dopamine, and morphine [25] (Fig. 4.3).

4.2.2.2 Protoalkaloid (non-heterocyclic)

The protoalkaloids are also synthesized from amines or amino acids. These non-heterocyclic alkaloids contain the nitrogen atom in the side chain of structure [26]. Some of them are colchicine, mescaline, ephedrine, and cathinon [27]. L-tryptophan and L-tyrosine are the main amino acids involved in the biosynthesis of protoalkaloids, further converted into phenylethylamine and monoterpenoid alkaloids [28]. A phenylethylamine alkaloid called mescaline was obtained from *Lophophora williamsii*, and indole alkaloids have been found in plants belonging to Loganiaceae, Apocynaceae, and Rubiaceae family [29, 30].

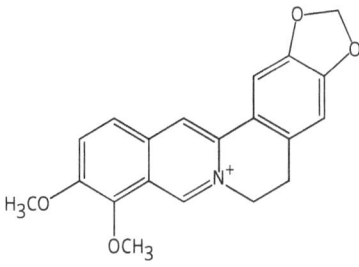

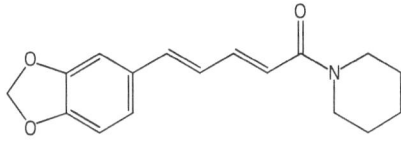

Piperine

Berberine

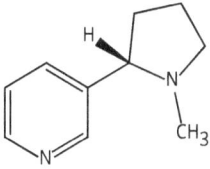

Nicotine

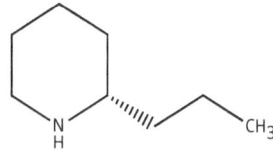

Coniine

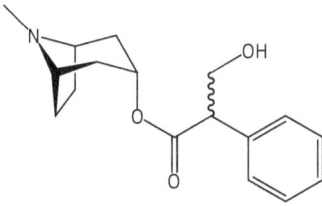

Atropine

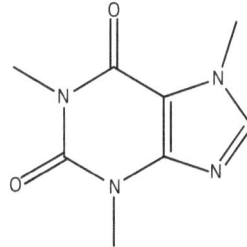

Caffein

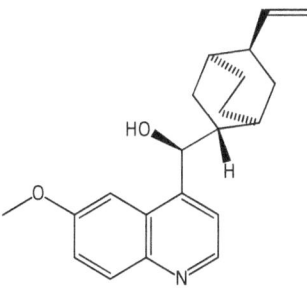

Quinine

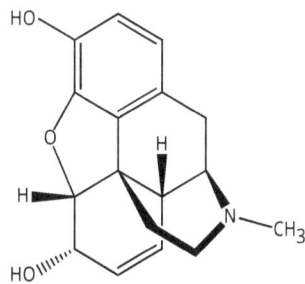

Morphine

Fig. 4.3: Several examples of alkaloids and its structure.

4.2.2.3 Psedualkaloids

Pseudoalkaloids are derived from pyruvic acid, acetate, adenine/guanine, or geraniol [31]. These are heterocyclic in nature, but due to biosynthesis by non-amino acid compounds it is called pseudoalkaloids [31]. The main example of pseudoalkaloid is diterpenoid alkaloidsc, contains 18, 19, and 20 carbons, obtained from the genus *Consolida, Aconitum*, and *Delphinium* [32].

4.2.3 Pharmacological importance of alkaloids

Plant alkaloids have several pharmacological functions such as antimicrobial, antimalarial, anticancer, anti-Alzheimer, anti-inflammatory, anesthetic, cardiotonic, and migraine preventive [17]. Isoquinoline alkaloids like morphine are a painkiller drug, anticancer, and antidiarrheal drug. Morphin and papaverine obtained from *Papaver somniferum* are used for chronic pain in terminal cancer treatment [33]. Atropin is a tropane alkaloid present in *Atropa belladonna* that can lower blood pressure and heart rate, is used to relief in gastrointestinal disorders, and is also useful in Parkinson's disease [34]. Caffeine is purine alkaloids present in CaffeaArabica and Caffeacanephora, has antidiarrheal, strong painkiller property, antiasthmatic, and metabolism enhancer [35]. Some indol alkaloids such as ajmaline, vincamine, vinblastine, and reseprine used as antidiabetic, blood flow enhancer in ischemia, anticancer, and central sedative activity, respectively [36]. Some alkaloids such as nicotine and anabasine can be used as biopesticides to protect crops from pest. Alkaloids are a key class of phytochemicals that have tremendous medicinal uses. Table 4.1 shows the functions of alkaloids obtained from plant sources.

4.3 Flavonoids of plants

A group of polyphenolic compounds is known as flavonoids are broadly distributed in plants with thousands of well-known varieties [9]. These are nontoxic for humans and are usually used as medicine due to its several pharmacological activities such as antioxidant, anti-inflammatory, antidiabetic, anti-hepatotoxic, anticancer, and anti-ulcer [49]. Several have antiviral actions, and many of them are very useful in cardiovascular diseases. Many in vitro studies revealed that flavonoids have the potential to inhibit the growth and development of tumors of various cancer cell lines [50–52].

Tab. 4.1: Plant-derived alkaloids and its pharmacological activities.

Alkaloids	Examples	Source plant	Pharmacological functions	References
Isoquinoline	Berberine, Jatrorrhizine, Palmatine, Coptisine	*Berberis vulgaris*	Cholagogue, antileishmanial, and antibacterial	Hostalkova et al. [37]
Tropane	Atropine	*Atropa belladonna, Datura stramonium, Lyciumbarbarum*	Parasympatholytic, mydriatic, and spasmolytic	Jakabová et al. [38], Adams et al. [39], Dimitrov et al. [40]
	Cocaine, Hygrine, Truxilline Cinnamylcocaine,	*Erythroxylon coca*	Anesthetic, CNS stimulant, and parasympathetic	Dos Santos et al. [41]
Pyrrolizidine	Gamma-linolenicacid, Lycopsamine, Amabiline, Thesinine	*Borago officinalis*	Cardiotonic, diuretic, gastrointestinal regulator, asthmatic	Avila et al. [42]
	Intermedine, Simphytine, Lycopsamine	*Symphytum officinale*	Wound healing and anti-allergic	Trifan et al. [43]
Purin	Caffeine, Theobromine, Theophylline	*Cola acuminate, Cola nitida*	Antioxidant, relaxing in tiredness	Ekalu et al. [44]
Indol	Quinine, Quinidine, Cinchonine, Cinchonidine	*Cinchona*	Antimalarial, antipyretic, and antiarrhythmic	Nair [45]
	Ajmaline, Reserpine, Aricine, Rescinnamine, Serpentinine,Yohimbine	*Rauwolfia serpentina*	Antiarrhythmic, antihypertensive, and temperature-decreasing properties	Itoh et al. [46]
Quinolizidine	Piperine, Piperettine	*Piper nigrum*	Biopestecide, antibacterial, and antirheumatic, and antiphlogistics	Li et al. [47]
	Nicotine, Anatabine, Anabasine, Tabacine, Choline	*Nicotiana tabacum*	Act as nicotinic receptors	Wei et al. [48]

4.3.1 Biosynthesis of flavonoids in plants

The biosynthesis of flavonoids is associated with the phenolic compounds that are synthesized in plants through the shikimic acid pathway [53]. Briefly, the primary step is the conversion of glucose into the glucose-6-phosphate in the pentose phosphate pathway in the presence of glucose-6-phosphate dehydrogenase. In this process, nicotinamide adenine dinucleotide phosphate (NADP) is produced as reducing equivalents. The glucose-6-phosphate is further converted into ribulose-5-phosphate in the presence of 6-phosphogluconolactone dehydrogenase. This process also produces nicotinamide adenine dinucleotide phosphate as reducing equivalents. Further, in the pentose phosphate pathway, the ribulose-5-phosphate is converted into erythrose-4-phosphate and phosphoenol pyruvate from glycolysis. These two chemical compounds then participate in the phenylpropanoid pathway to synthesize phenolic compounds. Further, the phenylpropanoid pathway leads to the synthesis flavonoids in plants. A study on a berry suggested that the cytosolic multienzyme complex present in the endoplasmic reticulum also known as flavonoid metabolon started the biosynthesis of flavonoids in plants. In particular, some of these enzymes belong to the cytochrome-p450 family in vacuoles, plastids, and nuclei [53]. Similarly, a study of the synthesis pathway of flavonoids in tobacco revealed that several enzymes are associated with the pathways [54]. These researchers might put forward a result that a multi-branching distribution of the enzymes involved in flavonoid biosynthesis [55].

4.3.2 Classification of flavonoids and its importance

4.3.2.1 Flavonols

Flavonols are the most plentiful flavonoids, include quercetin, rutin, kaempferol, isorhamnetin, and myricetin [56, 57]. Among them, quercetin is one of the most abundant flavonoids which form skeletons of others. The major natural sources of quercetin are tea (*Camellia sinensis*), lavages (*Levisticum officinale*), berries, cherries, apples, grapes, red onions, coriander, pepper, and citrus fruits [58]. Quercetin has several health-beneficial properties such as free radical scavenger, controlling hypertension, and acting as an anticancer agent [59, 60]. Additionally, some glycoside derivatives of quercetin can act as antidiabetic agents [61]. Another important flavonol is rutin which is also known as quercetin-3-O-rutinoside, naturally present in oranges, grapes, peaches, lemons, and berries [58]. Rutin was also reported to have several biological activities such as antioxidants, antidiabetic, anti-obesity, neuroprotective, and antidiabetic effects [62]. Antidiabetic studies show that 25 and 50 mg/kg rutin can reduce the cognitive impairment and hyperalgesia induced by hyperglycemia in diabetic rates [62]. Isorhamnetin diglucoside is usually found in herbal plants such as mustard leaf (*Brassica juncea*) reducing the oxidative stress in streptozotocin-induced diabetes rates [63]. Additionally, isorhamnetin has numerous medicinal properties and shows several pharmacological activities, such as antidiabetic and anti-obesity. Kaempferol is found in tomato, apple, potato, grape, tea, spinach, broccoli, and berries. Kaempferol can also show several biological functions like other flavonols [64]. Myricetin found in fruits, vegetables, wines, teas, and berries also exhibit antioxidant, anti-obesity, anti-inflammatory, and antidiabetic properties [65]. Some example of structure of flavonols is given below (Fig. 4.4).

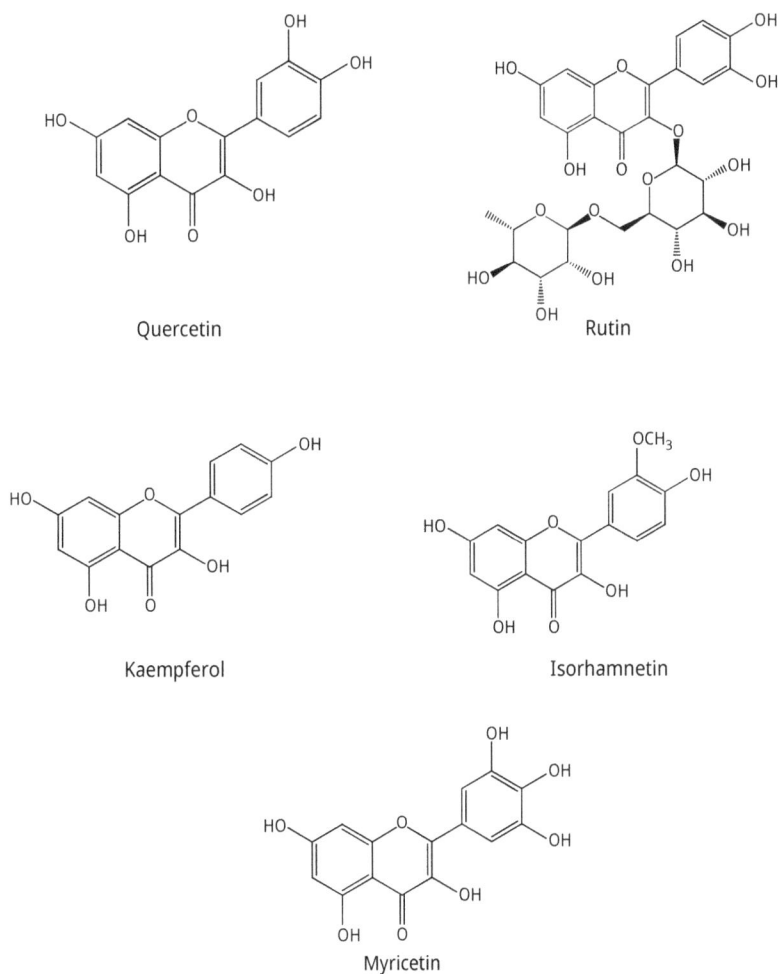

Quercetin

Rutin

Kaempferol

Isorhamnetin

Myricetin

Fig. 4.4: Several examples of flavonols and its structure.

4.3.2.2 Flavanones

Naringenin, a flavanone plentiful in many citrus fruits, has tremendous biological functions including anticancer, antioxidant, antibacterial, antiviral, anti-inflammatory, and cardioprotective effects [58, 66]. Naringin, a flavanone is also naturally found in citrus fruits with several health benefits such as anticancer, anti-inflammatory, antioxidant, bone regeneration, and antidiabetic [67]. A bioflavonoid called hesperidin is also found in citrus fruits such as lemons and limes with many pharmacological activities such as antimicrobial, antioxidant, anticancer, and anti-inflammatory [68]. According to Pyrzynska [68], the by-products, such as peels, seeds, and cell and membrane residues of citrus fruits are a good source of hesperidin. So, the appropriate extraction procedure is required to extract hesperidin from these by-products. Pyrzynska [68] reported a review article on the methodology for extracting hesperidin from plants. The structures of some flavones are given below (Fig. 4.5).

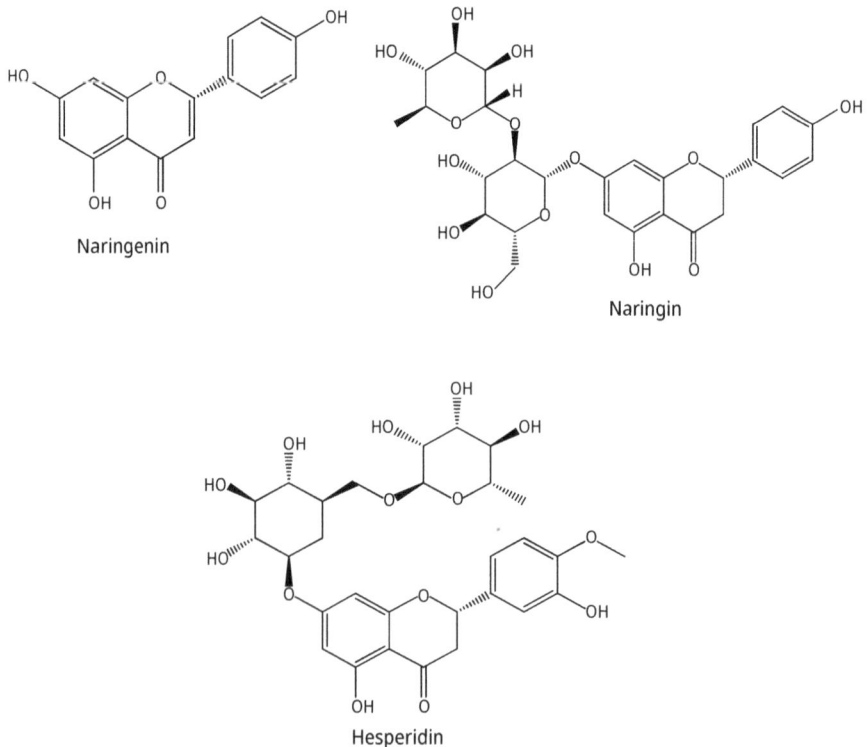

Naringenin

Naringin

Hesperidin

Fig. 4.5: Several examples of flavanones and its structure.

4.3.2.3 Isoflavones

Isoflavones are an important class of bioflavonoids usually present in leguminous plants. The major dietary sources of isoflavones are soybean and its products, which have mostly two types of isoflavones called daidzein and genistein [58, 69]. Křížová et al. [69] reported that the isoflavones were measured as chemoprotective for breast and prostate cancer, also beneficial in some hormonal disorders and have valuable effects on key risk factors of heart disease. Examples of some isoflavones are given below (Fig. 4.6).

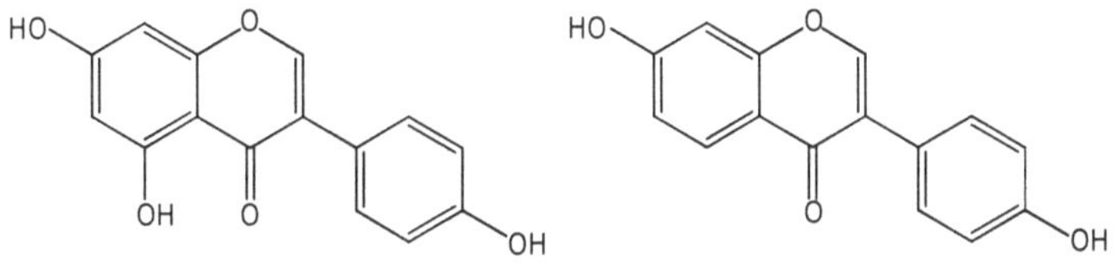

Genistein   Daidzein

Fig. 4.6: Several examples of isoflavones and its structure.

4.3.2.4 Flavones

Flavones are another class of bioflavonoids found in parsley, celery, and several other herbal plants [58]. The most important nutritional flavones are apigenin and luteolin, and they have tremendous medicinal uses including anti-inflammatory and immuno-modulatory effects [70]. Some herbal plants such as chamomile and passion flower contain natural apigenin, and these plants were sued as conventional medicines to cure several ailments. Luteolin naturally occurs in edible herbal plants, and fruits like onion, cabbage, parsley, celery, peppers, carrots, and apple have potent antidiabetic, anti-inflammatory, and antidiabetic activities [71]. Examples of some flavones are given below (Fig. 4.7).

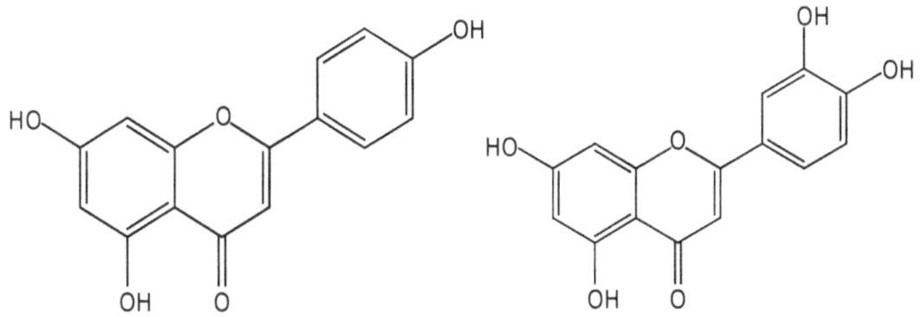

Apigenin   Luteolin

Fig. 4.7: Several examples of flavones and its structure.

4.3.2.5 Flavan-3-ol

Flavan-3-ols such as catechin, epicatechin, epicatechingallate, gallocatechin, epigallo-catechin, and epigallocatechin gallate are also referred to as flavanols and are found in various fruits, vegetables, and tea [58]. Regular consumption of green tea has been revealed to be efficient for curing metabolic and cardiovascular diseases. Aron and Kennedy [72] reported that flavan-3-ols exhibit could act as antioxidant, anticancer, cardiopreventive, antimicrobial, antiviral, and neuro-protective agents. Examples of some flavan-3-ol are given below (Fig. 4.8).

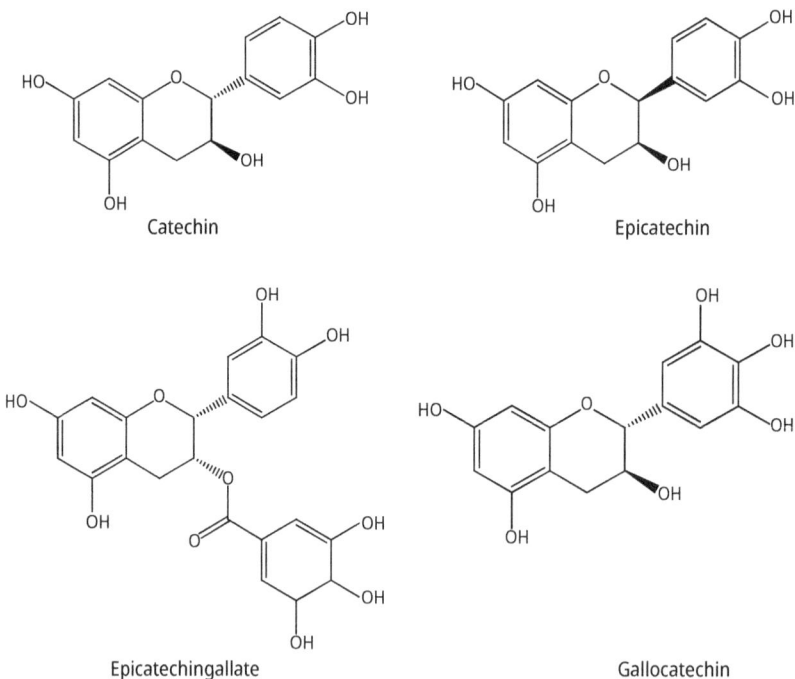

Catechin

Epicatechin

Epicatechingallate

Gallocatechin

Fig. 4.8: Several examples of flavan-3-ol and its structure.

4.3.2.6 Anthocyanidins

Anthocyanidins are another class of flavonoids broadly present in vegetables, fruits, and berries. Significant consideration has been specified for anthocyanins because of their latent wellbeing benefits, such as reduce the risk of chronic disorders, anticancer, cardiovascular disorders, and antidiabetic [73]. Several anthocyanin compounds are well-known, and the most widespread of these compounds are delphinidin, cyanidin, pelargonidin, malvidin, peonidin, and petunidin [58, 74]. Structures of some anthocyanidins present in medicinal plants are given below (Fig. 4.9).

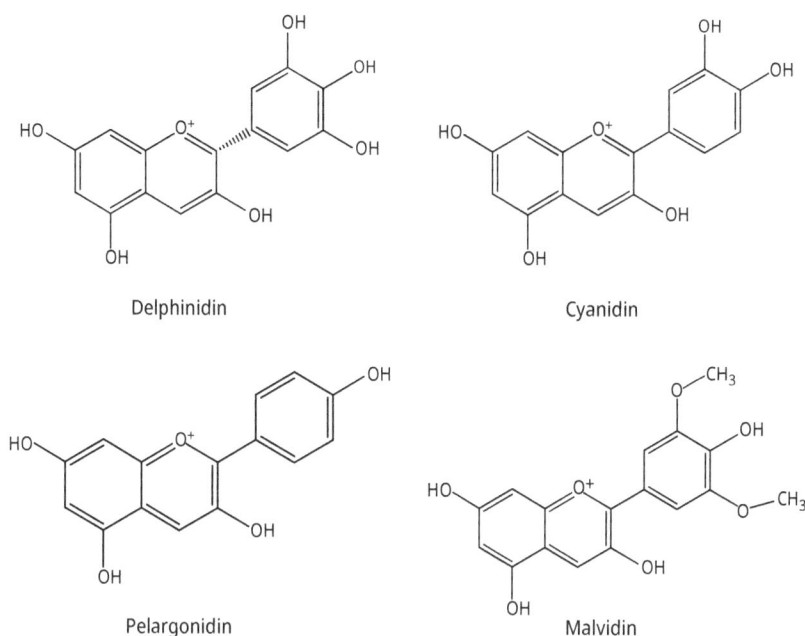

Fig. 4.9: Several examples of anthocyanidins and its structure.

4.3.3 Pharmacological activity of flavonoids

Flavonoids show tremendous antioxidant activity and prevent oxidative damage caused by free radicals [9]. Besides the radical inhibiting property, flavonoids also have various biological actions that are beneficial for human health. These activities are antiviral, anti-inflammatory, anticancer, antiulcer, cytotoxic, antidiabetic, and antimicrobial [75]. Table 4.2 represents the biological function of plant-derived flavonoids.

4.4 Conclusions

The curative effects of alkaloids and flavonoids have been proved in most pre-clinical trials in model organisms. Thus, their production should be enhanced in plants by expressing their biosynthetic pathway enzymes.. Further, the conjugates of these phytochemicals with other important drugs may improve the potency of those active compounds. Conclusively, more study is required to resolve more structures of alkaloids and flavonoids; and to investigate their therapeutic applications. Moreover, the structures of these phytoconstituents will always inspire researchers to design and synthesize new effective drugs for various diseases.

Tab. 4.2: Plant-derived flavonoids and its pharmacological activities.

Alkaloids	Examples	Source plant	Pharmacological functions	References
Flavan-3-ol	Catechin	*Cola acuminata*	Tonic, physical and mental tiredness, stimulation of metabolism, and antitrypanosomal activity	Atawodi et al. [76]
Flavan-3-ol Flavonols Flavones	Epicatechin, Epicatechin-3-O-gallate, Epigallocatechin, Epigallocatechin-3-O-gallate Quercetin, Myricetin Kaempferol Apigenin, Luteolin	*Camellia sinensis*	Tonic, physical and mentaltiredness, stimulation of metabolism, antimicrobial, and antiviral	Anesini et al. [77]
Falvon	Vitexin	*Passifloraincarnata*	Irritation, mental stress, headaches, heartbeat disorders, anxiety, sleeplessness, mild depression, and effective in Parkinson's disease	Pagassini et al. [78]
Flavonols Flavon-3-ol	Quercetin, Myricetin Routine, Kaempferol Galangin, Catechin, Epicatechin, Quercetagetin	*Capsicum frutensens Amburanacearensis Caesalpiniapulcherrima*	Antioxidants, chemoprotectants, cardioprotectors, antiviral, and antiaging	Obohand Rocha [79], Leal et al. [80], Srinivas et al. [81]
Flavones	Luteolin, Apigenin, Routine, Crisine, Baicalein	*Citrus* spp.	Antiviral and antifungal	Wollenweber [71]
Flavanones	Naringin, Naringenin Eriodictyol	*Citrus* spp.	Chemoprotective effects	Musumeci et al. [82]
Isoflavonoids	Daidzein, Genistein, Coumestrol, Glycytheine	*Glycine max*	Antiestrogenic effects	McCue and Shetty [83]
Anthocyanins	Cyanidin, Delfinidin, Petunidin,	*Ipomoea batatas*	Atherosclerosis, antioxidant, antibacterial, and cytoprotectors	Rupasinghe and Arumuggam [73]

References

[1]   T. K. Patle, K. Shrivas, R. Kurrey, S. Upadhyay, R. Jangde, R. Chauhan. Spectrochim Acta Part A: Mol Biomol Spectrosc. 2020, 242, 118717.

[2]   R. Chauhan, S. Keshavkant, A. Quraishi. Ind Crops Prod. 2018, 113, 234–239.

[3]   R. Chauhan, A. Quraishi, S. K. Jadhav, S. Keshavkant. Acta Physiologiae Plantarum. 2016, 38, 116.

[4]   M. C. Dias, D. C. Pinto, A. M. Silva. Molecules. 2021, 26, 5377.

[5]   P. K. Mahish, S. Singh, R. Chauhan, Bioactive secondary metabolites from endophytic phomaspp. In: M. Rai, B. Zimowska, G. J. Kövics (Eds.), Phoma: Diversity, Taxonomy, Bioactivities, and Nanotechnology. Springer, Cham, 2022, pp. 205–219.

[6]   G. Velu, V. Palanichamy, A. P. Rajan, Phytochemical and pharmacological importance of plant secondary metabolites in modern medicine. In: Bioorganic Phase in Natural Food: An Overview. Springer, Cham, 2018, pp. 135–156.

[7]   D. Lin, M. Xiao, J. Zhao, Z. Li, B. Xing, X. Li, M. Kong, L. Li, Q. Zhang, Y. Liu, H. Chen. Molecules. 2016, 21, 1374.

[8]   A. Christ-Ribeiro, C. S. Graça, L. Kupski, E. Badiale-Furlong, L. A. de Souza-soares. Process Biochem. 2019, 80, 190–196.

[9]   S. Mitra, M. S. Lami, T. M. Uddin, R. Das, F. Islam, J. Anjum, M. J. Hossain, T. B. Emran. Biomed Pharmacother. 2022, 150, 112932.

[10] S. Kunjam, S. S. Chauhan, R. Chauhan, P. K. Mahish, S. K. Jadhav. Int J Bot Stud. 2021, 6, 394–397.

[11] S. S. Chauhan, P. K. Mahish. Res J Pharmacutical Technol. 2020, 13, 5647–5653.

[12] R. L. Nagula, S. Wairkar. J Control Release. 2019, 296, 190–201.

[13] T. Fukuda, F. Ishibashi, M. Iwao. LamellarinAlkaloids: Isolation, Synthesis, and Biological Activity. Alkaloids: Chem Biol. 2020, 83, 1–12.

[14] Z. H. Pu, M. Dai, L. Xiong, C. Peng. Nat Product Res. 2022, 36, 3489–3506.

[15] M. A. Ashraf, M. Iqbal, R. Rasheed, I. Hussain, M. Riaz, M. S. Arif. Environmental Stress and Secondary Metabolites in Plants: An Overview. In: Plant Metabolites Regul under Environ Stress. 2018, 153–167.

[16] M. P. Goh, A. M. Basri, H. Yasin, H. Taha, N. Ahmad. Asian Pac J Trop Biomed. 2017, 7, 173–180.

[17] N. Bribi. Asian J Bot. 2018, 1, 1–6.

[18] I. Desgagné-Penix. Phytochem Rev. 2021, 20, 409–431.

[19] B. R. Lichman. Nat Prod Rep. 2021, 38, 103–129.

[20] E. P. Gutiérrez-Grijalva, L. X. López-Martínez, L. A. Contreras-Angulo, C. A. Elizalde-Romero, J. B. Heredia, Plant alkaloids: Structures and bioactive properties. In: M. Swamy (Ed.), Plant-Derived Bioactives. Springer, Singapore, 2020, pp. 85–117.

[21] S. K. Talapatra, B. Talapatra, Alkaloids – General introduction. In: S. K. Talapatra, B. Talapatra (Eds.), Chemistry of Plant Natural Products: Stereochemistry, Conformation, Synthesis, Biology, and Medicine. Springer, Berlin Heidelberg, 2015, pp. 717–724.

[22] T. Xu, S. Liu, L. Meng, Z. Pi, F. Song, Z. Liu. J Chromatogr B. 2016, 1026, 56–66.

[23] Y. X. Xiong, Z. S. Huang, J. H. Tan. Eur J Med Chem. 2015, 97, 538–551.

[24] R. Kaur, T. Matta, H. Kaur. Saudi J Life Sci. 2019, 2, 158–189.

[25] G. Hussain, A. Rasul, H. Anwar, N. Aziz, A. Razzaq, W. Wei, M. Ali, J. Li, X. Li. Int J Biol Sci. 2018, 14, 341.

[26] A. C. Lindsay, S. H. Kim, J. Sperry. Nat Prod Rep. 2018, 35, 1347–1382.

[27] T. T. Cushnie, B. Cushnie, A. J. Lamb. Int J Antimicrob Agents. 2014, 44, 377–386.

[28] A. C. Alves de Almeida, F. M. De-faria, R. J. Dunder, L. P. B. M, A. R. M. S-b, A. Luiz-Ferreira. Evidence Based Complementary Altern Med. 2017, 2017, 8528210. doi: 10.1155/2017/8528210. Epub 2017 Jan 9. PMID: 28191024; PMCID: PMC5278565.

[29] J. Beyer, O. H. Drummer, H. H. Maurer. Forensic Sci Int. 2009, 185, 1–9.

[30] Q. Pan, N. R. Mustafa, K. Tang, Y. H. Choi, R. Verpoorte. Phytochem Rev. 2016, 15, 221–250.
[31] S. Nakamura, M. Hongo, S. Sugimoto, H. Matsuda, M. Yoshikawa. Phytochemistry. 2008, 69, 1565–1572.
[32] F. Gao, -Y.-Y. Li, D. Wang, X. Huang, Q. Liu. Molecules. 2012, 17, 5187–5194.
[33] U. Rinner, T. Hudlicky. Topics Curr Chem. 2012, 309, 33–66.
[34] T. Zhao, S. Li, J. Wang, Q. Zhou, C. Yang, F. Bai, X. Lan, M. Chen, Z. Liao. ACS Synthetic Biology. 2020, 9, 437–448.
[35] H. Ashihara, A. Crozier. Trends Plant Sci. 2001, 6, 407–413.
[36] A. Ndagijimana, X. Wang, G. Pan, F. Zhang, H. Feng, O. Olaleye. Fitoterapia. 2013, 86, 35–47.
[37] A. Hostalkova, J. Marikova, L. Opletal, J. Korabecny, D. Hulcova, J. Kunes, L. Novakova, D. I. Perez, D. Jun, T. Kucera, V. Andrisano. J Nat Prod. 2019, 82, 239–248.
[38] S. Jakabová, L. Vincze, Á. Farkas, F. Kilár, B. Boros, A. Felinger. J Chromatogr A. 2012, 1232, 295–301.
[39] M. Adams, M. Wiedenmann, G. Tittel, R. Bauer. Phytochem Anal. 2006, 17, 279–283.
[40] K. Dimitrov, D. Metcheva, L. Boyadzhiev. Sep Purif Technol. 2005, 46, 41–45.
[41] N. A. Dos Santos, C. M. de Almeida, F. F. Gonçalves, R. S. Ortiz, R. M. Kuster, D. Saquetto, W. Romão. J Am Soc Mass Spectrom. 2021, 32, 946–955.
[42] C. Avila, I. Breakspear, J. Hawrelak, S. Salmond, S. Evans. Fitoterapia. 2020, 142, 104519.
[43] A. Trifan, S. E. Opitz, R. Josuran, A. Grubelnik, N. Esslinger, S. Peter, S. Bräm, N. Meier, E. Wolfram. Food Chem Toxicol. 2018, 112, 178–187.
[44] A. Ekalu, J. D. Habila. Med Chem Res. 2020, 29, 2089–2105.
[45] K. P. Nair, Cinchona (Cinchona sp.). In: K. P. Nair (Ed.), Tree Crops. Springer, Cham, 2021, pp. 129–151.
[46] A. Itoh, T. Kumashiro, M. Yamaguchi, N. Nagakura, Y. Mizushina, T. Nishi, T. Tanahashi. J Nat Prod. 2005, 68, 848–852.
[47] S. Li, Y. Lei, Y. Jia, N. Li, M. Wink, Y. Ma. Phytomedicine. 2011, 19, 83–87.
[48] X. Wei, S. P. Sumithran, A. G. Deaciuc, H. R. Burton, L. P. Bush, L. P. Dwoskin, P. A. Crooks. Life Sci. 2005, 78, 495–505.
[49] A. Ullah, S. Munir, S. L. Badshah, N. Khan, L. Ghani, B. G. Poulson, A. H. Emwas, M. Jaremko. Molecules. 2020, 25, 5243.
[50] P. O. de Souza, S. E. Bianchi, F. Figueiró, L. Heimfarth, K. S. Moresco, R. M. Gonçalves, J. B. Hoppe, C. P. Klein, C. G. Salbego, D. P. Gelain, V. L. Bassani. Toxicology In Vitro. 2018, 51, 23–33.
[51] H. Kikuchi, B. Yuan, X. Hu, M. Okazaki. Am J Cancer Res. 2019, 9, 1517.
[52] K. Muniyandi, E. George, S. Sathyanarayanan, B. P. George, H. Abrahamse, S. Thamburaj, P. Thangaraj. Food Sci Hum Wellness. 2019, 8, 73–81.
[53] E. Petrussa, E. Braidot, M. Zancani, C. Peresson, A. Bertolini, S. Patui, A. Vianello. Int J Mol Sci. 2013, 14, 14950–14973.
[54] Q. Zhu, S. Sui, X. Lei, Z. Yang, K. Lu, G. Liu, Y. G. Liu, M. Li. PLoS One. 2015, 10, e0139392.
[55] F. R. Quattrocchio, A. N. Baudry, L. O. Lepiniec, E. R. Grotewold, The regulation of flavonoid biosynthesis. In: The Science of Flavonoids. Springer, New York, 2006, pp. 97–122.
[56] A. Crozier, I. B. Jaganath, M. N. Clifford. Nat Prod Rep. 2009, 26, 1001–1043.
[57] A. I. Oraibi, M. N. Hamad. JPharmSci Res. 2018, 10, 2407–2411.
[58] M. Kawser Hossain, A. Abdal Dayem, J. Han, Y. Yin, K. Kim, S. Kumar Saha, G. M. Yang, H. Y. Choi, S. G. Cho. Int J Mol Sci. 2016, 17, 569.
[59] W. M. Dabeek, M. V. Marra. Nutrients. 2019, 11, 2288.
[60] X. K. Fang, J. Gao, D. N. Zhu. Life Sci. 2008, 82, 615–622.
[61] M. Bule, A. Abdurahman, S. Nikfar, M. Abdollahi, M. Amini. Food Chem Toxicol. 2019, 125, 494–502.
[62] P. Hasanein, A. Emamjomeh, N. Chenarani, M. Bohlooli. Nutr Neurosci. 2020, 23, 563–574.
[63] T. Yokozawa, H. Y. Kim, E. J. Cho, J. S. Choi, H. Y. Chung. J Agri Food Chem. 2002, 50, 5490–5495.
[64] S. Cid-Ortega, J. A. Monroy-Rivera. Food Technol Biotechnol. 2018, 56, 480–493.

[65] Y. Taheri, H. A. Suleria, N. Martins, O. Sytar, A. Beyatli, B. Yeskaliyeva, G. Seitimova, B. Salehi, P. Semwal, S. Painuli, A. Kumar. BMC Complementary Med Ther. 2020, 20, 1–4.

[66] B. Salehi, P. V. Fokou, M. Sharifi-Rad, P. Zucca, R. Pezzani, N. Martins, J. Sharifi-Rad. Pharmaceuticals. 2019, 12, 11.

[67] R. Chen, Q. L. Qi, M. T. Wang, Q. Y. Li. Pharm Biol. 2016, 54, 3203–3210.

[68] K. Pyrzynska. Nutrients. 2022, 14, 2387.

[69] L. Křížová, K. Dadáková, J. Kašparovská, T. Kašparovský. Molecules. 2019, 24, 1076.

[70] C. Zaragozá, L. Villaescusa, J. Monserrat, F. Zaragozá, M. Álvarez-Mon. Molecules. 2020, 25, 1017.

[71] E. Wollenweber, Flavones and flavonols. In: The Flavonoids. Routledge, 2017, pp. 259–335.

[72] P. M. Aron, J. A. Kennedy. Mol Nutr Food Res. 2008, 52, 79–104.

[73] H. P. V. Rupasinghe, N. Arumuggam. Health Benefits of Anthocyanins. In: M. S. Ling (Ed.), Anthocyanins from Natural Sources. RSC 2019, pp. 121–158.

[74] P. Mena, R. Domínguez Perles, A. Gironésvilaplana, N. Baenas, C. García Viguera, D. Villaño. IUBMB Life. 2014, 66, 745–758.

[75] M. M. Jucá, F. M. Cysne Filho, J. C. de Almeida, D. D. Mesquita, J. R. Barriga, K. C. Dias, T. M. Barbosa, L. C. Vasconcelos, L. K. Leal, J. E. Ribeiro, S. M. Vasconcelos. Nat Product Res. 2020, 34, 692–705.

[76] S. E. Atawodi, B. Pfundstein, R. Haubner, B. Spiegelhalder, H. Bartsch, R. W. Owen. J Agri Food Chem. 2007, 55, 9824–9828.

[77] C. Anesini, G. E. Ferraro, R. Filip. J Agri Food Chem. 2008, 56, 9225–9229.

[78] J. A. Pagassini, L. J. de Godoy, F. G. Campos, G. R. Barzotto, M. A. Vieira, C. S. Boaro. Sci Rep. 2021, 11, 22064.

[79] G. Oboh, J. B. Rocha. Eur Food Res Tech. 2007, 225, 239–247.

[80] L. K. Leal, K. M. Canuto, K. C. da Silva Costa, H. V. NobreJúnior, S. M. Vasconcelos, E. R. Silveira, M. V. Ferreira, J. B. Fontenele, G. M. Andrade, G. S. de Barros Viana. Basic Clin Pharmacol Toxicol. 2009, 104, 198–205.

[81] K. V. Srinivas, Y. K. Rao, I. Mahender, B. Das, K. R. Krishna, K. H. Kishore, U. S. Murty. Phytochemistry. 2003, 63, 789–793.

[82] L. Musumeci, A. Maugeri, S. Cirmi, G. E. Lombardo, C. Russo, S. Gangemi, G. Calapai, M. Navarra. Nat Product Res. 2020, 34, 122–136.

[83] P. McCue, K. Shetty. Crit Rev Food Sci Nutr. 2004, 44, 361–367.

Maryam Mohammadi-Cheraghabadi* and Saeid Hazrati

Chapter 5
Terpenoids, steroids, and phenolic compounds of medicinal plants

Abstract: Phytochemical compounds are largely responsible for the multidimensional and wide medicinal effects of pharmaceutical plants. Phytochemical compounds are generally divided into two classes based on their roles in basic metabolic processes, that is, primitive and secondary metabolites (SMs). In this chapter, secondary chemical compounds from plants are discussed as important sources of medical benefits from plants. Terpenoids (or terpenes) are one of the most diverse natural product families, with over 40,000 individual compounds that occur both primary and secondary in metabolism. Terpenoids are anti-inflammatory, antitumor, antibacterial, antimalarial, and antiviral, inhibit and remedy cardiovascular diseases, promote transdermal absorption, and have hypoglycemic effects. It is well-known that flavonoids and phenolic compounds are the most important bioactive factors and antioxidants that have enjoyed verbose importance due to their advantages for curing and preventing plenty of illnesses for humans. Several bioactivities of phenolic compounds are involved in their chemopreventive properties, including antimutagenic or anticarcinogenic, anti-inflammatory and antioxidant properties, as well as their ability to boost apoptosis by inhibiting cell proliferation, inhibition of DNA binding, differentiation, migration, and blocking signaling. Conversely, plant steroids are mainly composed of sugars to form glycosides. Steroidal factors used to counter inflammatory disorders are glucocorticoids; long-term treatment causes intense side effects. As a result, it is imperative that research be undertaken to identify new phytochemicals that have a remedial potential without or with remarkably decreased side effects. In general, we argue about terpenoids, phenolic compounds, and steroids. This chapter aims to provide an overview of terpenoids and phenolic compounds as the interesting alternative originals for pharmaceutical and medicinal plants with anti-inflammatory effects and containing chemical constituents.

*Corresponding author: Maryam Mohammadi-Cheraghabadi, Department of Agronomy,
Faculty of Agriculture, TarbiatModares University, PO Box 14115-336, Tehran, Iran,
e-mail: mohammadi.maryam@modares.ac.ir
Saeid Hazrati, Department of Agronomy, Faculty of Agriculture, Azarbaijan Shahid Madani University,
Tabriz, Iran, e-mail: saeid.hazrati@azaruniv.ac.ir

https://doi.org/10.1515/9783110791891-005

5.1 Introduction

Traditional medical plants have been applied for medicinal and dietary therapy for multiple millennia in East Asia, for instance, in Thailand, Japan, China, and India, and are currently broadly used in cancer therapy. Natural composites separated from medical plants, as affluent origins of new anticancer medicines have been of incrementing concern since then. During long-term people exercise, a large number of anticancer medical plants and many related provisions have been screened and applied for preventing and curing different cancers. In fact, natural products play a chief role in cancer cure and prevention.

Efferth et al. [1] reported that a remarkable number of antitumor factors commonly applied in the clinic are of natural source. For instance, more than 50% of anticancer prescription drugs approved internationally were natural products or their derivatives. Moreover, the National Cancer Institute discovered during the 1960s (in the United States) that plant extracts can inhibit cancer cell growth.

Phytochemicals are specified as bioactive; and do not require nutrients from plants. They have a diversity of man-healthy impacts, such as having supposed chemoretentive attributes (antimutagenic and anticarcinogenic) and interjecting with tumor promotion and progression. The National Cancer Institute, based on plenty of reports describing an anticancer activity, specified about 40 edible plants possessing cancer-preventive attributes. Plus, there are more than 400 species of traditional Chinese medical plants related to anticancer.

In particular, terpenoids are included in multiple medicinal plants, and many terpenoids have been indicated to be extant for medicinal applications, for instance, taxol and artemisinin as cancer and malaria drugs, respectively. Different terpenoids are included in several plants for not just plant medicinal application but also dietary application. Terpenoids are mainly existed in the shape of fugacious oils in higher medical plants and mostly present in the following medicinal plant groups: Labiatae, Celastraceae, Ranunculaceae, Taxaceae, Magnoliaceae, Compositae, Lauraceae, Oleaceae, Pinaceae, Rutaceae, Aristolochiaceae, Umbelliferae, Acanthaceae, and Araliaceaeetc.

Plus, universal metabolic, physiological, and structural functions, and several special terpenoids function in different conditions, including defense and communication. Organs of the isoprenoid group also contain industrially helpful agrochemicals (e.g., azadirachtin and pyrethrins) and polymers (e.g., chicle and rubber). Also many terpenoids also have immunomodulatory, anti-oxidant, anti-aging, insecticidal, and neuroprotective effects; the terpenoids paclitaxel and artemisinin have been broadly applied in clinical practice. So, the probe on the biological activity of the terpenoids will help to the choice of the recovery of medicines and cure methods supply a scientific basis for the expansion of novel medicines, which researchers pay many regard [2].

Secondary metabolites (SMs) are generally understood as compounds containing at least one hydroxyl group per aromatic ring present in plant tissues. Also, based on reports of Kumar and Pandey [3], it is many absorbing to note that half of these phenolic composites are flavonoids rendering as aglycone, glycosides, and methylated derivatives. These phytochemical materials exist in plant drugs and nutrients. Both flavonoids and phenolic compounds have been reported for their efficient antioxidants, antibacteria, anticancer, cardio-protective agents, immune system promoting, anti-inflammation, skin conservation from UV radiation, and concerning candidacy for drug and medicinal usage.

Phenolic compounds supply required functions in the propagation and growth of plants; act as advocacy mechanisms versus parasites, predators, and pathogens; and chip into the color of plants. Plus, phenolics plenty in fruits and vegetables are reported to play a main role as chemopreventive factors; for instance, the phenolic components of apples have been connected by the prohibition of colon cancer in vitro.

Several phenolic compounds have been reported to possess the strong antioxidant activity and to have anticancer or antimutagenic/anticarcinogenic, anti-atherosclerotic, antibacterial, antiviral, and anti-inflammatory activities to a higher or lesser limit. Hundreds of natural phenolic composites have been specified from Tungmunnithum et al. [4] tested medical plants and dietary plants, mainly including flavonoids, tannins, coumarins, phenolic acids, stilbenes, quinones, lignans, curcuminoids, and phenolic mixtures and other phenylpropanoids and phenylethanoids. These attributes arise from their role in signaling detoxifying enzymes as well as in scavenging free radicals and providing antioxidant protection. Inflammation, a process unfavorable and familiar to everyone, happens in reaction to wounds, allergens, autoimmune, and infection situations. Inflammation is specified with edema, redness, heat, pain, and alteration of function of affected tissue. Creating an inflammatory reaction is necessary for length. Excessive and uncontrolled inflammation results in a vast rank of illnesses that contains the highly current situation of asthma, allergic, inflammatory bowel diseases, Crohn's disease, rheumatoid arthritis, and upper airway diseases such as allergic rhinitis, chronic sinusitis, and allergic conjunctivitis. Two main groups of drugs are applied in controlling inflammation: steroidal and nonsteroidal anti-inflammatory factors. Glucocorticoids (steroidal anti-inflammatoryfactors) are broadly applied for alleviating inflammation in chronic inflammatory diseases, which are related to incremented expression of inflammatory genes by binding to glucocorticoid receptors on many signaling routes. Medicinal herbs include chemical components which chemically look like in organize with steroids, and modern clinical studies have confirmed their role as anti-inflammatory factors.

Generally, this study helps to survey flavonoids, phenolic phytochems, terpenoids, and steroids as potential sources of medical applications and pharmaceuticals. And the recent documented studies help several interesting orientations for future investigation.

The boundary among primary and SMs is blurred.

Based on their biosynthetic sources, natural plant products can be divided into three main groups: the alkaloids, the terpenoids, and allied phenolic and phenylpropanoids composites. The chemical precursor of terpenoids is isopentenyl diphosphate (IPP), which comprises five carbons with more than 25,000 secondary compounds attached. About 8,000 or so phenolic composites are shaped by way of either the malonate/acetate or the shikimic acid route. Primary and SMs cannot readily be distinct based on chemical structures, precursor molecules, or biosynthetic sources. For instance, primary and SMs are detected between the diterpenes (C_{20}) and triterpenes (C_{30}). There are many similar enzyme reactions that are required to synthesize kaurenoic acid and abietic acid, which are the intermediates for the synthesis of gibberellins growth hormones detected in all plants, while the late is a resin composite largely limited to members of the Pinaceae and Fabaceae. In the lack of a reliable difference based on biochemistry or structure, we return to a functional description, with primary generates participating in essential metabolic processes and nutrition inside the plant, and SMs generate influencing ecological interplay among the plant and its environment.

5.2 Terpenoids

There are many structurally diverse classes of natural plant products, but terpenoids may be the most structurally diverse. The roots of the term terpenoid go back to the origins of turpentine, a mineral known as terpenin in German; terpenoids are most commonly made up of duplicate combinations of branched five-carbon units (C5) in accordance with isopentane skeletons. These monomers are commonly introduced to as isoprene units because thermal analysis of multiple terpenoid substances yields the alkene gas isoprene as a product, and because proper chemical situations can contain isoprene to polymerize many of five carbons, producing many terpenoid skeletons. For these causes, the terpenoids are often named isoprenoids, though scholars have known for well over 100 years that isoprene itself is not the biological pioneer of this family of metabolites [5].

5.2.1 The number of five-carbon units in terpenoids determines their classification

The five-carbon (isoprene) units that form the terpenoids are often connected in a "head-to-tail" procedure, but head-to-head fusions are also usual, and some generates are made up by head-to-middle fusions. Therefore, because wide structural modifications with carbon–carbon bond rearrangements can happen, detection of the source template of isoprene units is sometimes hard. The smallest terpenes include a single isoprene unit; as a group, they are called hemiterpenes (half-terpenes). The best-known hemiterpene is isoprene itself, a transient generate released from photosynthetically active tissues. The enzyme isoprene synthase exists in the leaf plastids of several C3 plant species, but the metabolic principle for the light-dependent generation of isoprene is unknown. Estimated annual foliar emissions of isoprene are fully substantial (5×10^8 metric tons of carbon), and the gas is an original reactant in the NO_x radical-induced foundation of tropospheric ozone. C10 terpenoids, though they include two isoprene units, are named monoterpenes; as the first terpenoids separated from turpentine in the 1850s, they were attended to be the base unit from which the further nominations are taken. In addition to essential oils and volatile essences, monoterpenes account for 5% of the botanical dry weight. Monoterpenes are separated by extraction and/or distillation and find significant industrial applications in perfumes and flavors. The terpenoids deduced from three isoprene units possess 15 carbon atoms known as sesquiterpenes (i.e., one and one-half terpenes). Like monoterpenes, multiple sesquiterpenes are detected in essential oils. Also, many sesquiterpenoids act as antibiotic composites generated by plants in reaction to microbial compete and as material antifeedants that dispirit opportunistic herbivory. Though the plant hormone abscisic acid is structurally a sesquiterpene, its C_{15} pioneer, xanthoxin, is not synthesized straight from three isoprene units but rather is generated by asymmetric cleavage of a C_{40} carotenoid. The triterpenes, which include 30 carbon atoms, are produced by the head-to-head connecting of two C_{15} chains, each forming three isoprene units connected head-to-tail. This large class of molecules contains brassinosteroids, certain phytoalexins, phytosterol membrane composites, composites of level waxes, several toxins, and feeding deterrents, such as oleanolic acid of grapes. Tetraterpenes are a class of secondary pigments that have 40 carbon atoms and 8 isoprene units, which are necessary for photosynthesis. The polyterpenes (those having more than eight isoprene units) include long-chain polyprenols involved in sugar transfer reactions, the prenylated quinone electron carriers, and enormously long polymers such as rubber (mean molecular mass greater than 106 Da), often detected in latex. Natural generations of blended biosynthetic sources partly isolated from terpenoids are often named meroterpenes. For instance, cytokinins and multiple phenylpropanoid composites include C5 isoprenoid side chains. Certain alkaloids, containing the anticancer medicines vinblastine and vincristine, include terpenoid fragments in their structures. Also, some modified proteins contain a 15- or 20-carbon terpenoid side chain that anchors the protein in a membrane [6].

5.2.2 Numerous conserved reaction mechanisms are used to synthesize terpenoid compounds

As a consequence of structural considerations of multiple terpenoids, Otto Wallach formed the idea that most terpenoids could be theoretically constructed by repetitively connecting isoprene units at the turn of the twentieth century. This principle provided that first imaginary framework for a common building relationship between terpenoid natural products. In the 1930s, Ruzicka excavated Wallach's idea by formulating the "biogenesis of isoprene rule," emphasizing mechanistic investigations of terpenoids in terms of cyclizations, rearrangements, and electrophilic elongation. This hypothesis rejects the precise nature of the biological forerunner and assumes just that they are "isoprenoid" in structure. As a working model for terpenoid biosynthesis, the biogenetic isoprene law has proved necessary right. Despite the great variety in function and form, the terpenoids are unified in their common biosynthetic source. The biosynthesis of all terpenoids based upon simple components can be divided into six different phases: (1) synthesis of the primary precursor IPP; (2) repetitive steps of IPP to produce homologs of the prenyl diphosphate; (3) transesterification of these homologs to produce deuterated terpenoids; (4) transesterification of these homologs to form the instant precursors of the various classes of terpenoids; (5) complexity of these allylic prenyl diphosphates with special terpenoid synthases to product terpenoid skeletons; and (6) secondary enzymatic modifications to the skeletons (largely redox reactions) to give rise to the functional attributes and large chemical diversity of this family of secondary metabolites.

5.2.3 Biosynthesis of terpenoids is compartmentalized, as is production of the terpenoid precursor IPP

Specifically, plants generate a much wider diversity of terpenoids than microbes or animals, diversity reflected in the complex formation of plant terpenoid biosynthesis at the cellular, subcellular, tissue, and genetic surfaces. The generation of large quantities of terpenoid natural products and their subsequent release, secretion, or accumulation is approximately always related to the attendance of anatomically highly determined structures. A terpenoid essential oil as part of the terpenoid essential oils emitted by the glandular epidermis, glandular trichomes, and secretory cavities of foliage greatly encourages insects to pollinate. Terpene biosynthesis occurs in resin blisters of coniferous trees and ducts in the coniferous trees. A more basic, and maybe universal, feature of the formation of terpenoid metabolism exists at the subcellular level. Triterpenes (C30), sesquiterpenes (C15), and polyterpenes are generated primarily in plastids, whereas monoterpenes (C10), diterpenes (C20), and certain prenylated quinones appear to originate primarily in the cytosol and endoplasmic reticulum (ER) of cells. The evidence now indicates that the biosynthetic routes for the foundation of the basic forerunner IPP differ markedly in these compartments, with the glyceraldehyde phosphate/pyruvate routes in the plastids and the classical acetate/mevalonate pathway being active in the cytosol and ER. Regulation of these dual routes may be difficult to determine, given that plastids may supply IPP to the cytosol for use in biosynthesis, and vice versa. Mitochondria, a third compartment, may produce the ubiquinone prenyl group by the acetate/mevalonate route, though small is known about the ability of these organelles for terpenoid biosynthesis.

5.2.3.1 Acetate/mevalonate is made of two enzymes: hydroxymethylglutaryl-CoA reductase and hydroxymethylglutaryl-CoA synthase

Acetate/mevalonate biosynthesis is widely acknowledged to have fundamental enzymology. This cytosolic IPP route includes the two-phase condensation of three molecules of acetyl-CoA catalyzed by hydroxymethylglutaryl-CoA synthase and thiolase.

This secreted substance is later converted into 3-hydroxy-3-methylglutaryl-CoA (HMG CoA). These gene families exhibit complex patterns of expression, with individual genes expressing fundamental, tissue-specific, or hormone-induced activity. Specific HMG-CoA reductase genes are induced by wounding or pathogen infection. HMG-CoA reductase activity may be altered post-translationally, such as by phosphorylating and, as a result, inactivating the enzyme by a protein kinase cascade. Allosteric modulation maybe plays a regulative role. Proteolytic decay of HMG-CoA reductase protein and the rate of turnover of the corresponding mRNA transcripts may also affect enzyme activity. Scholars have not achieved a common plan that illustrates how the

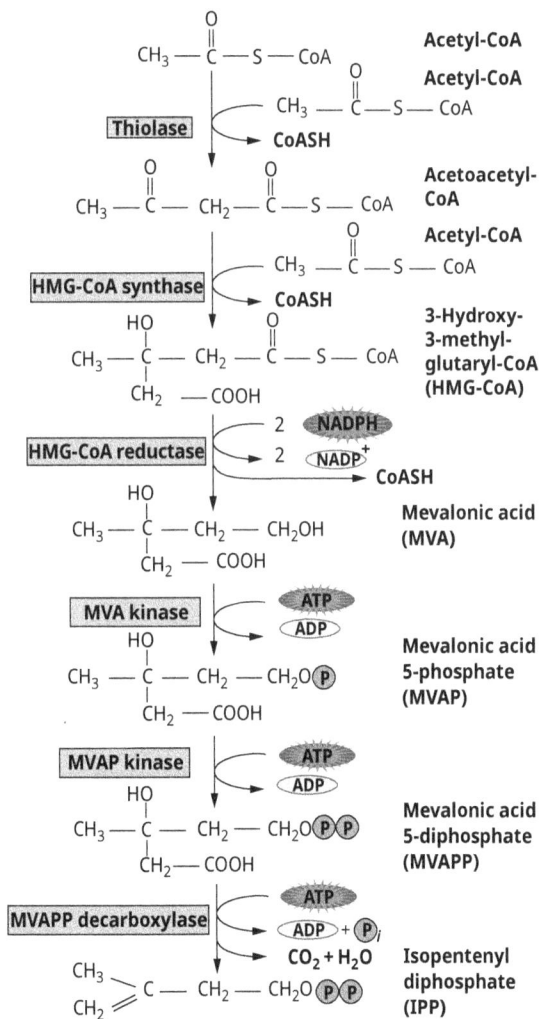

Fig. 5.1: It requires three molecules of acetyl-CoA to synthesize each IPP unit in the acetate/mevalonate pathway.

different mechanisms that regulate HMG-CoA reductase facilitate the generation of various terpenoid families. The accurate biochemical controls that affect activity have been hard to recognize in vitro because the enzyme is related to the ER membrane.

5.2.3.2 In plastids, IPP is synthesized from glyceraldehyde 3-phosphate and pyruvate

The plastid-localized pathway to IPP involves various routes, indicated in green algae and multiple eubacteria plants. As thiamine pyrophosphate reacts with hydroxyethyl-TPP, pyruvate forms a two-carbon fragment, which constricts with glyceraldehyde-3-phosphate. 1-Deoxy-D-xylulose 5-phosphate is formed from TPP. This medium can be degraded by rearranging and decreasing to form 2-C-methyl-D-erythritol 4-phosphate and finally transformed into IPP (Fig. 5.2, upper pathway). The detection of this new route for IPP foundation in plastids proposes that these organelles, assumed to have emanated as prokaryotic endosymbionts, have retained the bacterial machinery for the generation of this key middle of terpenoid biosynthesis. The details of the glyceraldehyde 3-phosphate/pyruvate route and the enzymes accountable have not yet been totally determined. However, generations of the two IPP biosynthesis routes can be easily distinguished in experiments that employ [1–13C] glucose as a forerunner for terpenoid biosynthesis. Nuclear magnetic resonance (NMR) spectroscopy can be applied to specify the 13C-labeling template of each isoprene unit in a terpenoid composite, allowing researchers to derive the labeling pattern of the corresponding IPP units [7] (Fig. 5.2).

5.2.4 Pharmacological activities of terpenoids

5.2.4.1 Antibacterial activity

Terpenoids (patchouli alcohol, andrograpolide, oleanolic acid) have strong antibacterial effects. The monoterpenoids in terpenoids are primarily detected in the genus *Mentha*, and other checks have detected that most of the composites extracted from plants of the genus *Mentha* have potent antimicrobial activity. Menthol is a cyclic monoterpene that has antibacterial activity. Also, scholars detected that menthol indicated considerable suppressor activity of biofilm when checking the effects of plant-derived terpenoids on *Candida albicans*.

Patchouli alcohol (PA) is a tricyclic sesquiterpenoid composite detected in *Pogostemoncablin* (Blanco) Benth. Scholars detected that it had an anti-*Helicobacter pylori* activity in vivo and in vitro. Andrographolide is a diterpene lactone composite in the Chinese medicine *Andrographis paniculata* Nees. Scholars detected that andrographolide had a considerable inhibitory efficacy on synergistic antibacterial efficacy with

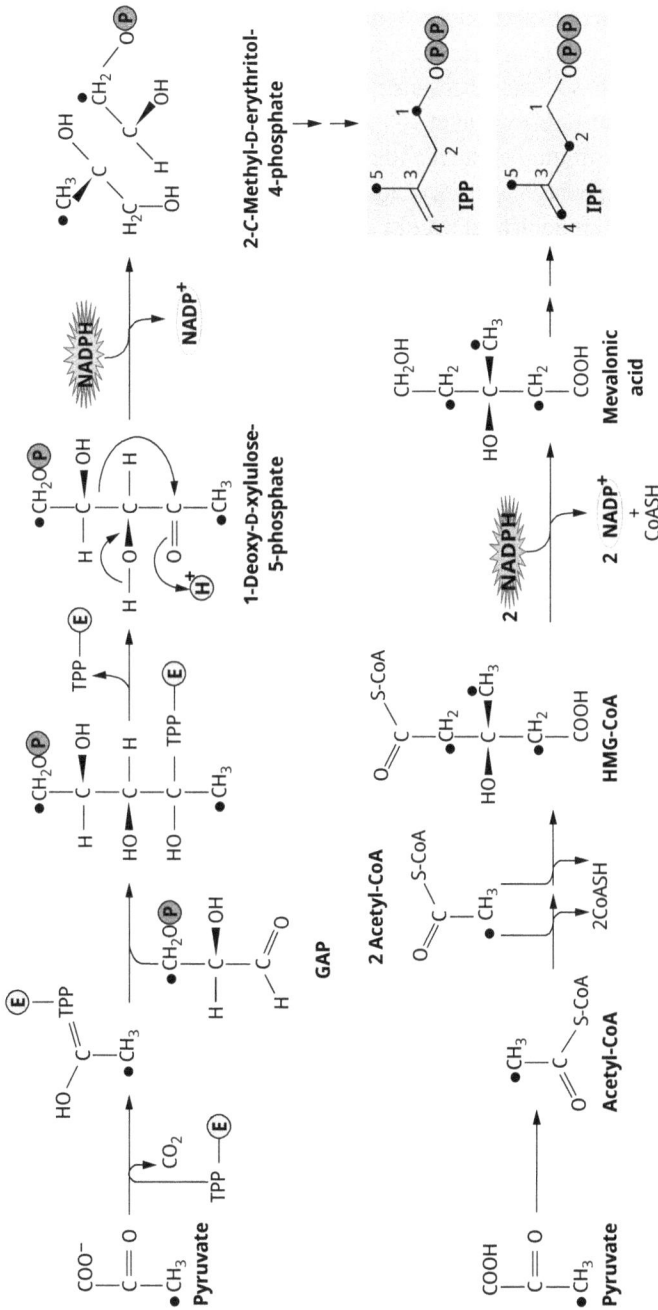

Fig. 5.2: Feeding studies distinguish two pathways of isoprenoid biosynthesis. Glyceraldehyde-3-phosphate (GAP) is isotopically labeled at C-3, which appears in acetyl-CoA and pyruvate after it is degraded by glycolytic enzymes and pyruvate dehydrogenase. IPP synthesized from labeled pyruvate and GAP by the plastid-localized pathway will be labeled at C-1 and C-5 (upper panel), whereas IPP formed from labeled acetyl-CoA by way of the cytosolic acetate/mevalonate pathway will be labeled at C-2, C-4, and C-5 (lower panel).

azithromycin and the biofilm of *P. aeruginosa*. Banerjee et al. [8] detected that androg-rapholide indicated potential antibacterial activity versus most of the tested Gram-positive bacteria, between that it was most sensitive to *Staphylococcus aureus* with a minimum inhibitory content (MIC) value of 100 µg/mL and was detected to have a suppressor effect on the constitution of the biofilm of *S. aureus*.

Oleanolic acid (OA) is a pentacyclic triterpenoid composite separated from plants. The results of one investigation indicated OA has a determined inhibitory efficacy on *Streptococcus mutans*, methicillin-resistant *S. aureus* and *S. aureus*. Kim et al. [9] also detected that OA can kill *Enterococcus faecium*, *Enterococcus faecalis*, and *Listeria monocytogenes*, by the MICs were 15–30 µg/mL for *L. monocytogenes*, 34–68 µg/mL for *E. faecium* and *E. faecalis*, destroying cell membranes of the bacteria and bacterial cell viability reduced after exposure to 2 × MIC of OA.

5.2.4.2 Anti-inflammatory activity

Terpenoids with anti-inflammatory activities are Paeoniflorin, 4-O-methyl paeoni-florin, paeoniflorin (MPΓ), 4-O-methylbenzoyl paeoniflorin (MBPF), IVSE, JEUD-38, Triptolidenol, Triptolide, Tripterine, Triptonide, and Ginsenoside.

Inula japonica Thunb (inula flower) is a conventional Chinese medicinal plant of the genus *Asteraceae*. Researchers detected that the sesquiterpene lactone composite IVSE can stop the generation of NO impelled with LPS, therewith displaying the anti-inflammatory activity in the inula flower. Scholars separated a new sesquiterpene lactone composite JEUD-38 in the inula flower; this composite remarkably attenuated LPS-induced NO generation and had the efficacy of treating and preventing inflammatory illness. In this study, the results indicated that under the excitation of LPS, the content of NO incremented with about 11 times compared with the blank group. The suppression efficacy of suppressor cell lines associated with dose of JEUD-38 (the cells were exposed to varying JEUD-38 contents) showed a remarkable suppression of NO generation after the supplementation with JEUD-38. In addition, being a key bioactive compound in *Tripterygium wilfordii*, it is the most effective natural product for inflammation and immune regulation, which has been shown to cure a wide range of inflammatory autoimmune disorders. Its chief mechanism is to prevent the generation of inflammatory cytokines. Scholars also have indicated that tripterine, tripto-nide, and triptolide all have clear anti-inflammatory effects. In one study, it was detected that both triptonide and triptolide remarkably mitigated the signs of torrid lung injury (ALI) in mice. In ALI mice, triptonide can alleviate pathological aspects of the lung tissues, while triptonide can also significantly suppresses the negative effects of LPS on chemokines, such as MIP-1 β, MIP-1 α, MCP-1, IP-10, and RANTES.

Ginsenoside-Rb1 (G-Rb1) is a possible anti-inflammatory factor. G-Rb1 can remark-ably prevent the activation of NF-κB (NF-κB is a key agent in inflammation, and is also a regulative agent of the generation of TNF-α). Plus, it was detected that ginseno-side-Rd (G-Rd) and ginsenoside-Rb2 (G-Rb2) indicated neuroprotective effects.

5.2.4.3 Antimalarial affect

Artemisinin is a sesquiterpene lactone composite separated from *Artemisia annua* L. in the 1970s. It is the most efficient antimalarial medicine after chloroquine, prima-quine, and pyrimethamine, and has the specifications of high performance and low toxicity. Later, antimalarial medicines such as arteether, artemether, and artesunate have been synthesized with changing the chemical structure of artemisinin. These medicines have low detrimental responses and can effectively mortify the plasmo-dium in the red blood cell stage. Usual research indicates that when red blood cells are phagocytosed by plasmodium, heme molecules in high contents are liberated.

Artemisinin is activated by heme in the place where the plasmodium metabolized potently and binds to the parasite proteins in the plasmodium body to make the pro-teins inactivated, performing the target of mortifying plasmodium. Plus, translational controlled tumor protein and glutathione S transferase, sarcoplasmic endoplasmic re-ticulum calcium ATPase (PfATP6) have also been recognized as non-heme protein sub-stances that interact with artemisinin in the plasmodium. Artemisinin is generally a successful antimalarial drug, provided it is regularly reviewed.

5.2.4.4 Antitumor activity

Terpenoids are fascinated by concern of multiple medicinal chemists with their good antitumor activity and specific structural attributes, and have the potential to be Ser-bian composites to expand secure and effective antitumor medicines. In this section, the antitumor effects of terpenoids, it was detected that geraniol, paclitaxel, and per-illyl alcohol are the terpenoids-containing antitumor activities.

Its strong antitumor activity has given it considerable interest due to its monocyclic monoterpene and its presence in medical plants' essential oils. It is effective at eliminat-ing tumor cells at a high output, and it is non-toxic. The growth of tumor cells was re-markably forbidden after perillyl alcohol was appended in the culture of tumor cells in plenty of animals, and perillyl alcohol plays a repellent and therapeutic role in cancer. Also, results indicated that execution of perillyl alcohol at an amount of 1–2 g/kg in rats can remarkably decrease the plurality and occurrence of colonic invasive adenocarci-noma effected with the infusion of carcinogen azomethane. Intraperitoneal infusion of perillyl alcohol at a dose of 75 mg/kg per week remarkably interdicted lung tumor foun-dation affected with the concurrent infusion of the carcinogen 4-(methyl-nitrosamino)-1-(3-pyridyl)-1-butanone. In addition to developing high cytotoxicity against HCT-116, SF-295, and OVCAR-8 man tumor cell lines, perillyl alcohol also showed high growth inhibition properties through the MTT test. The cell growth blockage percentage contents ranged from 90.94% to 95.92%. Last investigations have disclosed that the tumor growth inhibition rate of perillyl alcohol measured in mice was 45.4% and 35.3% at doses of 200 and 100 mg/kg/day, respectively. No toxicologically considerable effect was detected in kidney and liver parameters.

Geraniol is broadly detected in essential oils of perfumed medical plants. Current experiential document displays that geraniol has prophylactic or therapeutic effect on various kinds of cancers, for example, lung cancer. It has been showed that geraniol can adjust a diversity of signaling molecules and takes part in a diversity of life processes, such as cell proliferation, cell cycle, metabolism, autophagy, and apoptosis. It can be applied as a multi-target medicine for the cure of cancer; the effect is remarkable and is not affected by the adaptive persistence. For instance, geraniol inhibits tumor cell growth by blocking the G1 phase of the Michigan Cancer Foundation (MCF)-7 breast cancer cell cycle.

Hassan et al. [11] investigated the health-promoting attributes of *Thymusbovei* Benth. The anti-herpes simplex virus kind 2 (HSV-2), antihypertensive, and cytotoxic composites of TB-EO may be of profit in clinical use. TB-EO indicated very weak cytotoxicity on the healthful man embryonic lung fibroblast cells with an IC50 amount of 118.34 μg/mL compared with that of cisplatin (IC50 = 10.08 μg/mL). TB-EO and its important part, geraniol has indicated anti-HSV-2 with the half-maximum efficient content (EC50) amounts of 2.13 and 1.92 μg/mL, respectively. Geraniol and TB-EO at a content of 15 μg/mL indicated eminent inhibitory activities versus angiotensin-converting enzyme with the percent of inhibition 92.2% and 95.4%, respectively, compared with that of standard suppressor captopril (99.8% and 15 μg/mL). The results may be a new finding during an investigation for active and new anticancer, antihypertensive and anti-HSV-2 factors.

Costunolide is a sesquiterpene lactone composite and is one of the main chemical compounds of the medical plant *Aucklandia lappa* Decne. Studies have indicated that Costunolide has effects on anti-bladder cancer, prostate cancer, leukemia, and ovarian cancer mostly by forbidding cancer cell reproduction, separation of cancer cells, and inducing apoptosis, incursion of cancer cells, and forbidding metastasis, inhibiting angiogenesis, and reversing multidrug resistance.

Artemisinin and its derivatives, well-known as antimalarial medicines have particular inhibitory effects on tumors in vivo and in vitro, by distinct effects, small side effects, and low costs. For instance, artesunate is a semisynthetic derived from artemisinin that displays anticancer activity versus different kinds of tumors, such as colon cancer, melanoma, leukemia, lung cancer, no small cell lung cancer, breast cancer, ovarian cancer, and prostate cancer. Das investigates that semisynthetic artemisinin derivatives can apply anticancer effects by causing cell cycle G0/G1 arrest, inducing participatiion in oxidative stress and cell apoptosis, which result in more effective antitumor activity than monomeric composites, though this mechanism of action has not been explained.

Paclitaxel is a type of tetracyclic diterpenoid separated from taxus plants and has a good remedial effect on cancers such as breast and ovarian cancer.

The UA could stimulate tumor cells to apoptosis and reduce HBx-mediated autophagy via synthesis of Ras homolog gene family member A (RhoA). The widespread research indicated that UA could function by targeting glycolysis even in phenotypically different cancer cells. UA is known to have anticancer properties. UA can also prevent reproduction and induce apoptosis of man osteosarcoma 143B cells with inactivating Wnt/β-catenin signaling.

Wang et al. [12] have indicated that UA activates cell apoptosis in prostate cancer via rho-related protein kinase/phosphatase and tensin homolog-mediated mitochondrial translocation of cofilin-1.29. And UA is made of nanoparticles which can ameliorate anticancer bioavailability and trace.

5.2.4.5 Antiviral effect

Terpenes are between the very promising original of novel antimicrobials agents that have activity versus bacteria, viruses, protozoa, and fungi. Terpenoids mentioned here include Isoborneol, Encalyptole, Borneol, Artesunate, Glycyrrhizin, Betulinic acid, Dammarenolic acid, Tereticomate A, Notoginsensenoside ST-4i, Putranjivain A, Moronic acid, and Betulonic acid.

Several monoterpenes present in various plant essential oils have anti-herpes simplex virus-1 (HSV-1) properties, including isoborneol, borneol, and monoterpenoids. This composite indicated total prohibition of HSV-1 repetition at a content of 0.06%. Other monoterpene composites such as α-terpinene, thymol, γ-terpinene, α-terpineol, 1,8-cineole, and citral separated from thyme, eucalyptus, and tea tree indicates in vitro antiviral activity versus HSV-1 with inhibition of virus greater than 80%.

Hassan et al [13] separated 12 pure composites from the twigs and leaves of *Eucalyptus globulus* and all of them were tested for anti-herpetic activity versus the repetition of antigen types HSV-2 and HSV-1. Based on results in the anti-HSV-1 evaluation, we found that the composite tereticornate A (IC50: 0.98 g/mL; selectivity index CC50/IC50: 216.6) had the greatest potency, even exceeding acyclovir (IC50: 1.94 g/mL; selectivity index CC50/IC50: 111.5), a standard antiviral medicine. And the composite also demonstrated modest anti-inflammatory activity and antibacterial effects.

Putranjivain A, a diterpene separated from *Euphorbia jolkini*, indicated an antiviral effect versus HSV-2 in Vero cells with an IC_{50} value of 6.3 μM. *Rhus javanica* extracts, a beta-ceramide and a monoterpene-containing triterpenes exhibit potent suppressor properties against HSV-1, with EC50 contents of 2.8 and 3.8 g/mL, respectively. Notoginsenoside ST-4i, separated from the Chinese herb *Panax notoginseng*, indicated in vitro significant suppressor activities versus HSV-2 and HSV-1 with EC50 values of 19.44 and 16.4 μM, respectively. The results of suppressor activity versus HSV-1 of glycyrrhizin (GR) showed that GR removed the enhanced sensitivity of thermally harmed mice to HSV infectivity on the excitation of CD4 + contrasuppressor T cells [14].

Artemisinin and its derivatives demonstrated remarkable suppressor activity on the hepatitis B and C viruses (HBV and HCV, respectively). Artesunate is the semisynthetic derivation of artemisinin, which has a more vocalized suppressor efficacy on hepatitis B level antigen (HBsAg). A synergistic efficacy was noticed by merging with lamivudine (a medicine for treating HBV) and artesunate.

Past studies have showen that andrographolide has a remarkable suppressor effect on the repetition process of the chikungunya virus (CHIKV) (a mosquito-borne alphavirus). Andrographolide inhibited CHIKV putridity and decreased virus generation with almost 3log10 with a 50% EC 50 of 77 μM without cytotoxicity. RNA transfection and time-of-addition studies indicate that andrographolide can directly affect CHIKV replication and that the drug is independent of cell type in its activity. Dengue fever is the most common viral illness transmitted by arthropods in men, but there are commonly no medicines for the dengue virus (DENV). Panraksa exhibited impressive anti-DENV activity in both HeLa and HepG2 cell lines, tremendously reducing both viral yield as well as the level of infection in both, with 50% EC 50 values for DENV of 22.739 M and 21.304 M, respectively. Harada [15] detected that GR (a triterpenoid saponin) can prevent the repetition of intense keen respiratory syndrome-related virus and adjust the movability of the cover of the man immunodeficiency virus (HIV). A pentacyclic triterpene compound derived from Syzigiumclaviforum and betulinic acid (BA) contains pentacyclic triterpenoids. BA and its derivatives have anti-HIV activity, and it is attended to be the earliest detected pentacyclic triterpenoid composite by anti-HIV activity, which can affect the merger among cells and virus, and also prevent the activity of the assembly of the virus and reverse transcriptase. Ursolic acid, Dammarenolic acid, and Oleanolic acid also have anti-HIV activity.

5.3 Flavonoids comprise a diverse set of compounds and perform a wide range of functions

It is well known that flavonoids play a critical role in many plant–animal interactions. The colors of fruits and flowers that often function to attract pollinators and seed dispersers, result primarily from vacuolar anthocyanins such as the cyanidins (magenta and crimson), the pelargonidins (salmon, orange, red, and pink), and the delphinidins (blue mauve, and purple). Related flavonoids, such as flavones, flavonols, aurones, and chalcones also contribute to the color definition. Manipulating flower color by targeting different enzymatic phases and genes in flavonoid biosynthesis has been fully successful, solely in petunia. Specific flavonoids can also support plants versus UV-B irradiation, a role sometimes imputed to kaempferol. Others can act as insect-feeding attractants, such as isoquercetin in mulberry, anagent involved in silkworm assessment of its host species. In contrast, condensed tannins such as the proanthocyanidins increase a distinct bitterness or astringency to the taste of specified plant tissues and function as antifeedants. Apigenin and luteolin are communication molecules between rhizobium bacteria and legumes that facilitate nitrogen fixation (Fig. 5.3). In an associated function, isoflavonoids are involved in inducible defense versus fungal assault in alfalfa (e.g., medicarpin; Fig. 5.3) and other plant species. Maybe the most poorly checked and minimum understood classes of flavonoids are the oligomeric and polymeric substances related with formation of specified heartwood and bark tissues. These composites contain proanthocyanidins and their congeners in woody gymnosperms and isoflavonoids in woody legumes from the tropics. In both cases, their great deposition during the heartwood foundation contributes remarkably and characteristically to the overall color, quality, and rot resistor of wood. These metabolites can be misidentified as lignins because some constituents are not readily solubilized and are mostly dissolved only under the same situations that affect lignin dissolution. Different flavonoids have also been studied extensively from the perspectives of healthy conservation and pharmacological usefulness, for which mammalian enzyme systems have been applied to recognize flavonoid activity. Flavonoids have been analyzed as modulators of immune and inflammatory responses for their effect on smooth muscle function and as antiviral, anticancer, hepatoprotective, and antitoxic factors. There is a considerable current concern in the consumption of isoflavonoids in cancer prohibition. Dietary use of the isoflavonoids daidzein and genistein, which are present in soybeans, is ideal to decrease substantially the occurrence of prostate and breast cancers in men [16].

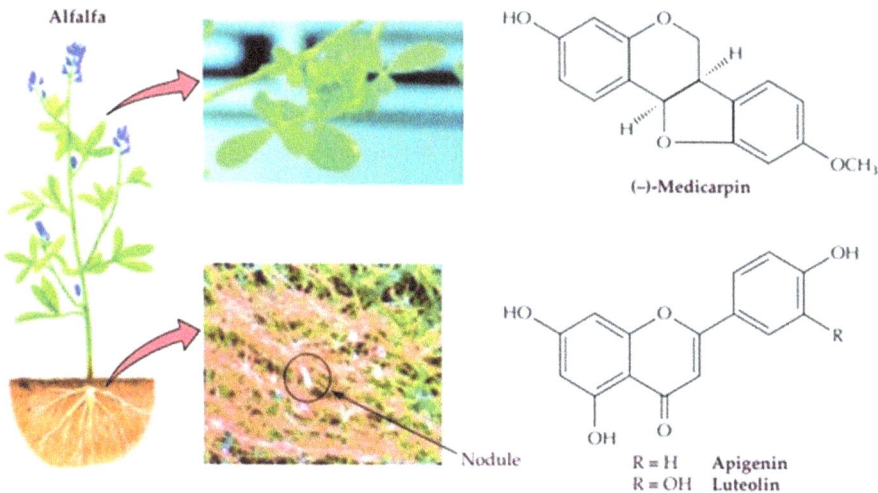

Fig. 5.3: Flavonoids perform diverse functions in alfalfa. The flavonoids apigenin and luteolin function as signaling molecules that induce Nod gene expression in compatible Rhizobium bacteria, facilitating the development of nitrogen-fixing root nodules. The phytoalexin isoflavonoid medicarpin participates in inducible plant defense.

5.3.1 The biosynthesis pathway of flavonoid has numerous important branch points

Anthocyanins, a flavonoid which is generated by the oxidation of malonyl-CoA by acetate, are the most common anthocyanins. The flavonoids contain various plant metabolites such as aurones, chalcones, leucoanthocyanidins (flavan-3,4-diols), flavonones, flavones, isoflavonoids, and flavonols. CHS, a dimeric polyketide synthase with each subunit at about 42 kDa, has no cofactor needs. In certain species, the consonant action of CHS and an NADPH-dependent reductase produces a 6-deoxychalcone (isoliquiritigenin). Chalcones can then be transformed to aurones, a subclass of flavonoids detected in the determined plant species. On the way to CHS, most of the flavonoid biosynthesis routes undergo a chalcone isomerization step by a chalcone isomerase (CHI). This isomerization step is responsible for producing 2S flavanones, naringenin, and (less commonly) liquiritigenin. The flavanones may demonstrate the main ramified point in flavonoid metabolism, because isomerization of these composites yields the phytoalexin isoflavonoids, presentation of a C-2–C-3 double bond affords flavonols and flavones, and hydroxylation of the 3-position produces dihydroflavonols. Entrance to the isoflavonoid branch point happens by way of two enzymes. The first, isoflavone synthase (IFS), catalyzes an unusual C-2 to C-3 aryl immigration and hydroxylation to give the 2-hydroxyisoflavanones and has lately been demonstrated to be an NADPH-dependent cytochrome P450 enzyme. Dehydration of the 2-hydroxyisoflavanones, catalyzed by 2-hydroxyisoflavanone dehydratase (IFD), shapes the isoflavonoids genistein and daidzein. The isoflavonoids can be further metabolized, primarily in the Fabaceae, to yield phytoalexins or to produce isoflavonoid-derived substances known as rotenoids in tropical legumes. The rotenoids, which are separated mostly from *Derris elliptica* and relevant species, are applied extensively as insecticidal factors but have other use as well. For instance, rotenone is applied as a rat poison and a suppressor of NADH dehydrogenase. Interestingly, the NADPH-dependent isoflavone reductase (IFR) step involved in isoflavonoid formation demonstrates remarkable homology to pinoresinol/lariciresinol reductase, suggesting a phylogenetic link among both lignan and isoflavonoidroutes for plant defense. The second branching point in common flavonoid metabolism involves that of dehydration of naringenin at the C-2/C-3 positions to give such plenty flavones as apigenin. This variation is catalyzed by flavone synthase (FNS), which changes in enzymatic kind depending on the plant species. For instance, in parsley cell cultures, flavone formation is catalyzed by an α-ketoglutarate-dependent dioxygenase, whereas an NADPH-dependent microsomal preparation generates this reaction in Antirrhinum flowers. The third main branch point in flavonoid metabolism is stereospecific 3-hydroxylation of naringenin (or its 3'-hydroxylated analog) to give dihydroflavonols such as dihydrokaempferol (or dihydroquercetin). The enzyme involved, flavanone 3-hydroxylase, is an Fe^{2+}-requiring, α-ketoglutarate-dependent dioxygenase. Specific hydroxylation involving an NADPH-dependent cytochrome P450 monooxygenase of naringenin can also directly give dihydroquercetin, which can be converted to quercetin (a flavanol) by flavonol synthase (FLS)-catalyzed C-2–C-3 double bond foundation; FLS is an α-ketoglutarate-dependent dioxygenase. Alternatively, dihydroquercetin can be decreased by an NADPH-dependent dihydroflavonol reductase (DFR) to give the corresponding flavan-3,4-diols. Subsequent species- and tissue-specific enzymatic protects can construct vast arrays of structurally diverse groups of flavonoids. For instance, in flower petals and the leucoanthocyanidins (e.g., leucopelargonidin) can be transformed into the colored anthocyanins (e.g., pelargonidin) via the action of a dehydratase, anthocyanidin synthase (ANS), which is an α-ketoglutarate-dependent dioxygenase. Leucoanthocyanidins can also render as precursors of the (epi-)catechins and condensed tannins. The enzymology related with those coupling processes, chain extension mechanisms, and oxidative modifications, however, is not yet appointed [17, 18].

5.3.2 Effects of plant phenolics on human health promoting, diseases curing, and preventing

5.3.2.1 Antibacterial effect

The mechanisms of antibacterial action of phenolic composites are not yet quite detected, but these composites are well known to involve several sites of action at the cellular surface. Many scholars illustrated this activity by the correction in permeance of cell membranes, the shifts in different intracellular functions compelled by hydrogen binding of the phenolic composites to enzymes, or by modifying the cell wall stiffness with totality losses due to various interactions with the cell membrane.

Thus, promoting the lipophilic character of phenolic composites increases their antimicrobial activity by favoring their interplay with the cell membrane. This may compel immutable harms of the coagulation and cytoplasmic membrane of the cell amount that can even lead to the prohibition of intracellular enzymes. For instance, condensed phenylpropanoids – tannins may compel harm at the cell membrane and even inactivate the metabolism by binding to enzymes, while phenolic acids have been indicated to interrupt membrane totality, as they cause resultful leakage of critical intracellular constituents. Flavonoids may connect to the soluble proteins determined outside the cells and with bacterial cell walls; thus, promoting the constitution of intricates. Flavonoids also may act via prohibition of both DNA synthesis energy and metabolism, thus affecting protein and RNA syntheses. In the case of Gram-positive bacteria, intracellular pH modification, as well as interposition with the energy (ATP) generating system, was reported.

Most of these investigations are about essential oils (i.e., volatile phenolic compounds due to their low molecular weight). Bouarab-Chibane et al. 2019 indicated that the significance of the share of the octanol-water partition coefficient (Log P) in relation to the hydrophobic and amphiphilic character of the molecule.

Interestingly, a major number of phenolics display antibacterial effect; such composites can be broadly detected in flowering medical plants to the non-flowering ones. Against the three pathogens, *Proteus vulgaris* Hauser, *Pseudomonas aeruginosa,* and *Proteus mirabilis* Hauser, the *Aspleniumnidus nidus* L. contained quercetin-7-O-rutinoside and gliricidin 7-O-hexoside. Plus, phenolic composites are synthesized by different plant groups containing many medical plant species that are applied in dietary consumption or traditional drug. A distinct instance is a nutmeg or *Myristica fragrans*; this plant is mainly applied traditionally as a flavoring factor in countries in South East Asia as Indonesia. However, ethanolic extract of the nutmeg seed, which included 30,40,7-trihydroxyflavone indicated efficient potential versus MDR gram-negative bacteria, for example, *Escherichia coli* Castellani, *Providencia stuartii* Ewing, and Chalmers. Additionally, *Pseudarthria hookeri* Wight and Arn, a traditional African plant medicine used to treat abdominal pain, pneumonia, diarrhea, and cough, has been reported as being effective in treating these conditions. Matching to the antibacterial study of this medical species, Dzoyem and his team detected that flavonoids from this plant indicated the highest antibacterial effect versus both gram-negative and gram-positive bacteria, for example, *E. coli, Pseudomonas aeruginosa* Migula, *Kilpper-Blazand, Klebsiella pneumonia, Staphylococcus aureus* Rosenbach, and *Enterococcus faecalis* (Andrew and Horder) Schleifer; the maximum antibacterial activities detected in 6-prenylpinocembrin and pseudarflavone. Some of antibacterial activities of terpenoids were introduced by Wenqiang et al. [14] that are including compounds of monoterpenes (1,8-eucalyptus, limonene, geranialdehyde, sabinene, menthol, sabinol, carvone), sesquiterpene (patchouli alcohol), diterpene (artemisinin and andrographolide), and triterpene (oleanolic acid).

5.3.2.2 Anticancer effect

Purification of phenolic composites from natural sources to homogeneity is challenging work. So, many studies have experimented with the capability of either raw extracts affluent in phenolic composites or fractions, including a mixture of phenolic composites for forbidding cancers in vivo and in vitro.

Currently, Flavopiridol is the scientific name for a flavonoid-based medicine from *Dysoxylum binectariferum*, where it derives is an instance of anticancer medicines emanating from a phytochemical composite for leukemia treatments and lymphomas. Based on reports, phenolic composites, particularly flavonoids, have long been announced as chemopreventive versus cancer cure.

Danciu and colleagues [21] have also analyzed the biological activities and compounds of ethanolic extracts collected from the roots of *Curcuma longa* and *Zingiber officinale* Roscoe, the species core representative of the Zingiberaceae family. This study team proposed the extract of *C. longa* rhizome as the undertaking source of natural active composites to fight virulent melanoma due to its potential anticancer exclusivity on the B164A5 murine melanoma cell line. The scholars also proposed that the increment in anticancer activity was associated with the increment in the content of polyphenol composites. Furthermore, the results from multiple biomedical research groups showed that different species of flavonoids could promulgate apoptosis in different cancer cells. Quercetin, a flavonol part, is examined as an interesting anticancer substance versus breast and prostate cancers. On the other hand, Quercetin 7-O-rutinoside and Glirici-din7-O-hexoside, which were the flavonoids separated from the medicinal fern (*Asplenium nidus*) were also contemplated as the potential chemopreventive man carcinoma HeLa cells and man hepatoma HepG2, matching to the severe studies of Hashemzaei and his research group on quercetin and apoptosis-inducing capability both in vivo and in vitro surfaces. Anticancer activity of quercetin was examined by Hashemzaei et al. [22] in several cancer cell categories. Generally, results indicated that quercetin could remarkably compel apoptosis of every experimented cell lines compared with control group. The in vivo experiments were carried out in mouse models (i.e., mice bearing CT-26 tumors and mice bearing MCF-7 tumors). The results of the experiment indicated a significant decline in tumor volume and size in the quercetin-cure groups compared with controls. Clifford's team also found that quercetin is able to infect miR-200b-3p, so it is able to adjust the self-renewing partitions of the pancreatic cancer. The severe examined on genistein and its molecular effects on prostate cancer by scholars indicated that a soy isoflavone genistein prohibited the activity of nuclear factor kappa B (NF-κB) signaling route that is engaged to the balance cell survival and apoptosis, this soy isoflavone could also take its action to warfare versus apoptotic, cell growth, and metastasis processes containing epigenetic modifications in prostate cancer. One of phenolic composites indicating anticancer effects to peel cancers is named curcumin, this phenolic can affect the cell cycle by acting as a pro-apoptotic factor. According to Dzialo et al. [23], curcumin inhibits the proliferation of melanoma cancer cells on the surface of B16F10 murine melanoma cells in an in vitro study. Also, they indicated that curcumin reacted as a non-selective cyclic nucleotide phosphodiesterases (PDE) stopper to prevent melanoma cell multiplication, which is relevant to epigenetic integrator UHRF1; these scholars also proposed that curcumin occurring in diets might aim to forbid this cancer and chip in the gene expression through epigenetic control. Research carried out by Hisamitsu's team on human prostate cancer cells, such as LNCaP and 22Rv1, showed that curcumin inhibited intracrine androgen synthesis in both vivo and in vitro. In vitro experiments performed on human prostate cancer cells, LNCaP and 22Rv1, indicated curcumin reduced the expression of genes in steroidogenic acute regulatory proteins, supporting the hypothesis that testosterone synthesis was suppressed. Curcumin prohibited multiplication of the elected cell lines in this experiment and compelled apoptosis of the cancer cells with dose-dependent reply. There was also an in vivo study on transgenic adenocarcinoma of the mouse prostate (TRAMP) mice with 1 month oral curcumin administration that managed to reduce the level of testosterone in prostate tissue by lowering the expression of steroidogenic enzymes, including the enzyme AKR1C2.

5.3.2.3 Antioxidant effects

During the generation of adenosine triphosphate (ATP) to produce energy for cells by applying oxygen, reactive oxygen species (ROS) and reactive nitrogen species (RNS) are generated as the by-crops from these cellular redox reactions. At the parity level, ROS and RNS are helpful composites for cellular functions and immune reactions, but the unbalance content of ROS and RNS will lead to oxidative stress, which can motive chronic and degenerative disturbances. The number of naturally occurring antioxidant molecules has increased considerably both in terms of applications and research studies; many natural antioxidant composites have been used in pharmaceuticals and medical crops as an alternative to artificial antioxidant compounds, which have been found to be one of the main causes of cancer. Medicinal plants have long been announced as a futuristic pole of natural antioxidant composites, especially plant secondary metabolites, that is, flavonoids and phenolic compounds which are produced by plant to vindicate itself or to promulgate the growth under unfavorable situations. In addition, Heim et al. [24] reported that functional group succession, adjustment of the number of hydroxyl groups, and configuration were also influenced by the antioxidant activity of flavonoids (example of metal ion chelation capability and/or ROS). Flavonoids and phenolics are generally well known as the highest phytochemical molecules with antioxidant attributes from plants.

Oki et al. [25] reported antioxidant activity of anthocyanins and other phenolic composites from different cultivars of *Ipomoea batatas*, an eatable and economic medicinal species in Japan by diphenyl-2-picrylhydrazyl (DPPH) radical-scavenging activity; the acquired results indicated the positive solidarity among the activity of free-radical scavenging and phenolic amount, also results of similar were obtained by Mohammadi et al. [26, 27]. Plus, chlorogenic acid (as phenolic composites) that operated as predominant DPPH radical-scavenger in "Bise" and "Miyanou-36" (cultivars of I. batatas), while anthocyanins were the predominant DPPH radical-scavengers of "Kyushu-132" and "Ayamurasaki" cultivars. A medicinal plant, *Bauhinia variegate*, that was used in traditional medicine in Asian countries as India and Pakistan, was examined by researchers, and they detected that leaf extracts of *Bauhinia variegate* included flavonoid compounds and presented antioxidant attributes versus oxidative harm by iron binding, radical neutralization, and reducing power abilities. The antioxidant activity of skin of the pulp and fruits, six Spanish Mediterranean cultivars of *Opuntia ficusindica* (L.) Mill. and leaf of *Salvia officinalis*, were examined by Andreu et al. [28] and Mohammadi et al. [26, 27], respectively. This investigation group detected that the considerable surfaces of total phenolic compounds in the best antioxidant cultivar played a considerable role in oxidative stress. The antioxidant attribute and bioactive composites from the fruits of *Aesculus indica* (Wall. ex Cambess.) Hook, a medicinal plant from moderate regions of Asia, that is, India, Pakistan, Nepal, and Afghanistan, were examined by the research group of Zahoor; their results showed that 2-(3,4-dihydroxy phenyl)-3,5,7-trihydroxy-4 H-Chromen-4-one (quercetin) and 2-hydroxy-2-phenyle acetic acid (mandelic acid) were the chief bioactive molecules with considerable antioxidant attribute to reduce oxidative stress reasoned by ROS. Furthermore, the rhizomes extract of *Polygonatum verticillatum* (L.) All., a medicine plant from India, has also been shown to possess antioxidant activity that is related to its surface phenolic compound. The investigation group of Meng [29] estimated the phytochemical profiling and biological activity from the leaf extract of *Camellia fangchengensis* Y. C. Zhong and S. Ye Liang, a wild tea species, which local public have been applied for green tea or black tea generation, that is a domestic tea species in Guangxi province, Republic of China. The obtained results demonstrated that monomers and flavan-3-ol oligomers were strong antioxidant compounds and were plentifully detected in this species.

Ustun et al. [30] showed that needle and spring extracts and essential oils of *P. pinea* L., J. F. Arnold, Tenore, *P. sylvestris* L., *P. brutia*, *P. nigra*, pycnogenol, and *P. halepensis* which is the bark extract from P. *pinaster*, until study their phytochemical composites and antioxidant activities by applying DPPH, ferric-reducing antioxidant power (FRAP), *N,N*-dimethyl *p*-phenylendiamine (DMPD) radical scavenging, and metal-chelating assays. Their results showed that pycnogenol had the richest total phenol amount and disclosed effective antioxidant effects. Like, Apetrei and his collaborators [31] managed their study on biological activity and phytochemical compounds of *Pinus cembra* L., a local species of Central European, the Carpathian, and Alps Mountains; they found that the hydromethanolic extract from bark supplied a higher amount of flavonoids and total phenolics than that of needle extract. Also, the bark extract indicated better ability as free ROS.

5.4 Steroids

Plants and plant parts are major sources of food and constituents of traditional medicines. The active ingredients of these plant-derived traditional medicines, practiced in many countries, are the secondary metabolites present in them. As steroids are ubiquitous in plants, they play a major role in pharmacological activities of these medicines. With the advancement of high-throughput screening methods and introduction of new bioassays, these plants have been subjected to various chemical and biological investigations. These studies have led to the recognition of steroids with a diversity of biological activities.

5.4.1 Phytosterols

All vegetable foods are known to contain appreciable quantities of phytosterols. The most abundant sources of phytosterols are vegetable oils. Except for highly refined carbohydrates and animal products, nearly all foods contribute considerably to phytosterol intake by humans. The major sterols present in edible fats of plant origin are β-sitosterol (~65% of sterol content) and campesterol (~30% of sterol content). Phytosterols are reported to interfere with the intestinal absorption of cholesterol and lower blood cholesterol levels. The mechanism of action has been attributed to the bulkiness of phytosterols compared to cholesterol. Recent work comparing phytosterol-free corn oil and commercial corn oil in a single test meal on human subjects reported that cholesterol absorption increased when corn oil was free of phytosterols. A reduction in cholesterol absorption was observed when phytosterols were added back to the phytosterol-free corn oil, confirming that this effect was due to the presence of phytosterols in natural corn oil. Many studies have been carried out on this subject, and it was found that, on average, a 13% reduction of blood low-density lipoprotein-cholesterol (LDL cholesterol) and a 10% reduction of total blood cholesterol occurred without affecting the high-density lipoprotein-cholesterol (HDL-cholesterol) and total lipid content in humans. Sitostanol, the saturated analog of β-sitosterol, which is present in minute amounts in plants, is found to be more potent in the inhibition of intestinal cholesterol absorption. Stanols (saturated sterols) can be acquired by catalytic hydrogenation of sterols. A subsequent investigation by Jones et al. [32] has revealed that a mixture of β-sitosterol and sitostanol was as effective as sitostanol alone in inhibiting cholesterol absorption.

Apart from the hypocholesterolemic activity, phytosterols show in vitro anticancer properties and are claimed to inhibit the development of various cancers in humans. It has been reported that β-sitosterol is effective in the symptomatic treatment of benign prostatic hyperplasia (BPH) (enlargement of the prostate), although the prostatic size remained unchanged during the treatment period. Further, it has been shown that β-sitosterol exhibits antihelminthic and antimutagenic activities, whereas both β-sitosterol and β-sitosteryl-β-d-glucoside encountered in the leaves of *Mentha cordifolia* Opiz (Lamiaceae) have exhibited analgesic activities.

5.4.2 Withanolides

Biological activities of withanolides have been extensively reviewed. Plants containing withanolides as major constituents such as *W. somnifera* (L.) Dunal, *Physalis angulata* L., and other *Physalis* species are widely used in traditional medicines in various countries. Roots of *W. somnifera* (L.) Dunal have been used in Ayurvedic medicine for over 3,000 years as Ashwagandha or Indian ginseng. In the course of extensive research on these plants, it has been revealed that their constituent withanolides exhibit antitumor, immunomodulatory, anti-inflammatory, antileishmanial, and antimicrobial properties. Significant biological activities of a few selected withanolides are summarized by Kamal et al. [33] are as follows: *withaferin A* (anticancer, cytotoxic, Immunosuppresive cholinesterase inhibitory, anti-inflammatory) *withanolide A* (cholinesterase inhibitory), *withanolide D* (antitumor, anti-metastatic), *withangulatin A* (immunosuppresive), physagulin H (trypanocidal), 4-Deoxywithaperuvin (antimicrobial), bracteosin A (cholinesterase inhibitory), viscosalactone B (anti-inflammatory), and witharistatin (diuretic).

5.4.3 Brassinosteroids

Brassinosteroids are plant growth regulators occurring in very low concentrations in plants and are required for the normal plant growth and development. Brassinosteroids may also protect plants against the adverse effects of harsh environments, including drought, high temperatures, herbicidal injury, heavy metals, and salinity. The presence of 24-epibrassinolide increases the thermotolerance of tomato plants by causing mitochondrial small heat shock proteins to increase. It is noteworthy that 24-epibrassinolide has shown anti-genotoxicity effect in a chromosomal aberration assay using *Allium cepa* L., and antioxidant and neuroprotective properties in a mammalian neuronal cell culture model.

5.4.4 Phytoecdysteroids

Plants containing phytoecdysteroids are used in folk medicines in many countries, while some herbal preparations containing these plants have been approved by certain countries as tonics in phytomedicine. Some of the preparations containing phytoecdysteroids are also used for the purpose of body building by athletes. Among the natural phytoecdysteroids, 20-hydroxyecdysone has shown anabolic, hypoglycemic, anti-arrhythmic, immunostimulant, and hepatoprotective activities on animal test models. It is noteworthy that the anabolic effects exhibited by phytoecdysteroids are not associated with androgenic, antigonadotropic, and thymolytic (causing destruction of thymic tissue) side effects. Pharmacological effects of some selected phytoecdysteroids are summarized by Kamal et al. [33] are 20-hydroxyecdysone (anabolic, anti-arrhythmic, antioxidant, hypoglycemic hepatoprotective, hypoazatemic, and immunostimulant), polypodine B (hypoazatemic), cyasterone (anabolic, hepatoprotective, and hypoazatemic), 2-deoxyecdysone (immunostimulant), 2-deoxy-20-hydroxyecdysone (immunostimulant), ecdysone (hypocholesterolemic), pterosterone (anabolic), and turkesterone (anabolic, hypoazatemic).

5.4.5 Steroidal alkaloids

Steroidal glycoalkaloids of solanidine and spirosolane types are known to possess antitumor, antifungal, antiviral, and embryotoxic activities. Jervine and 11-deoxyjervine (cyclopamine) are known to be natural teratogens that induce holoprosencephaly (a cephalic disorder in which the forebrain of the embryo fails to develop into two hemispheres) in animals. Structure–activity relationship (SAR) studies of these C27 steroidal alkaloids have revealed that the presence of Δ5-double bond is a critical structural feature for their teratogenicity. Some selected C27 steroidal alkaloids with significant biological activities are summarized by Kamal et al. [33] are lycioside A (α-Glucosidase inhibitor), 22α,23α-epoxysolanida-1,4,9 trien-3-one (anti-ophidic), ebeinone (cholinesterase inhibitor), ebeiensine (anti-asthmatic), Jervinone (anti-hypertensive), and stenophylline B (antifungal).

Steroidal alkaloids bearing pregnane skeleton show antibacterial, antifungal, anti-viral, antimalarial, glutathione S-transferase inhibitory, and cholinesterase inhibitory activities. Glutathione S-transferases are detoxification isozymes that can form adducts with electrophilic substances. Anticancer medicines with electrophilic centers readily form adducts with glutathione S-transferase, which are easily excreted from the body, thus lowering the performance of the drug. Hence, glutathione S-transferase inhibitors are important in cancer chemotherapy. Cholinesterases including acetyl- and butyl-cholinesterases are found to be responsible for the hydrolysis of acetylcholine. Deficiency of acetylcholine in the brain has been identified as a cause for Alzheimer's disease. Hence, one of the treatment strategies for Alzheimer's disease is to use potent cholinesterase inhibitors.

Plants belonging to families Asclepiadaceae, Apocynaceae, and Buxaceae that contain pregnane alkaloids are used in traditional medicines in many parts of the world. For example, Holarrhena anti-dysenterica (Apocynaceae) is a plant used in traditional medicine in the eastern Asian countries as a remedy for diseases of liver, amoebic dysentery, diarrhea, and other intestinal ailments as well as asthma, bronchopneumonia, and malaria.

A study of a methanol extract of *H. anti-dysenterica* Wall. ex A.DC. as well as its alkaloid constituents has shown strong acetylcholinesterase inhibitory activity [34]. Significant biological activities of some selected steroidal alkaloids having pregnane skeleton introduced by Kamal et al. [33] are conessine (cholinesteraseinhibitor), ire-hine (glutathione, s-transferase inhibitor), paravallarine (cytotoxic), terminamine A (anti-metastasis), salignenamide E (cholinesteraseinhibitor), sarcovagine D (cytotoxic), and hookerianamide J (cholinesteraseinhibitor).

5.5 Conclusions

The primary metabolic processes of plant growth and development are not directly involved in the synthesis and composition of organic compounds. Until recently, it was not appreciated in an analytical context just what these natural products or secondary metabolites did in plants. A natural product's main role appears to be in providing defense against predators and pathogens and in aiding reproduction by attracting pollinators and spreading seed. They may also act to create competitive advantage as poisons of rival species. Medical plants indicated major ability to synthesize various kinds of terpenoids applying secondary metabolic routes. Terpenoids are reported in most of the herbal plants and multitude terpenoid derivatives have been indicated to be available for medicinal applications, for example, taxol and artemisinin as cancer and malaria medicines, respectively. To short, the application of flavonoids and phenolic compounds are the potential volunteer of bioactive factors in medicinal and pharmaceutical parts to promulgate human health, stop and cure different diseases. Plant steroids are a diverse group of secondary metabolites that can be classified into several groups based on taxonomic considerations and their functions and/or structures. They play important physiological functions within plants and exhibit pharmacological activities beneficial to the mankind. Plants containing steroids are being used in traditional medicines in various countries. On the basis of these traditional medical practices and with the increasing knowledge on the pharmacological activities of their constituent plant steroids, these medicinal plants are used as herbal supplements in many countries. In order to find and progress these alternate elections of applying phytochemical compounds, the study of medicinal plants together with severe profiling research demands to be done. The targeted composites should be employed in pharmaceutical and biomedical research ranging from in vitro, in vivo, and clinical trial step to estimate the security, effect and also the side effects of the tested volunteer compounds. Also, study of the biochemistry of plant natural products has many practical applications. Biotechnological approaches can selectively increase the amounts of defense compounds in crop plants, thereby reducing the need for costly and potentially toxic pesticides. As with pharmaceuticals, flavorings, fragrancery materials, insecticides, fungicides, and other valuable natural products, genetic engineering can increase the productivity of these products as well. In view of their safety and efficacy, these herbs can be useful adjuvants to conventional therapeutic approaches to the management of inflammatory disorders. Although many natural products and their functions have been described in this chapter, the metabolism of natural products in most plant species remains to be elucidated. A great deal of fascinating biochemistry remains to be discovered.

References

[1]   T. Efferth, P. C. H. Li, V. S. B. Konkimalla, B. Kaina. Trends Mol Med. 2007, 13, 353–361.
[2]   T. Goto, N. Takahashi, S. Hirai, T. Kawada. PPAR Res. 2010, 1–9.
[3]   S. Kumar, A. K. Pandey. Sci World J. 2013, 2013, 162750.
[4]   D. Tungmunnithum, A. Thongboonyou, A. Pholboon, A. Yangsabai. Medicines. 2018, 5(93), 1–16.
[5]   D. E. Cane. Comprehensive NaturalProducts Chemistry, Vol. 2, Isoprenoids Including Carotenoids and Steroids. Pergamon/Elsevier, Amsterdam, 1999.
[6]   J. Chappell. Biochemistry and molecular biology of the isoprenoid biosyntheticpathway in plants. Ann Rev Plant Physiol Plant Mol Biol. 1995, 46, 521–547.
[7]   W. Eisenreich, M. Schwarz, A. Cartayrade, D. Arigoin, M. H. Zenk, A. Bacher. Chem Biol. 1998, 5, R221–R233.
[8]   M. Banerjee, D. Parai, S. ChattopadhyayS. K.Mukherjee. Folia Microbiol. 2017, 62(3), 237–244.

[9] W. S. Kim, W. J. Choi, S. W. J. Lee, D. C. Kim, U. D. S. Lee, et al. Korean J Physiol Pharmacol. 2015, 19(1), 21–27.

[10] X. Bi, L. Han, T. Qu, Y. Mu, P. Guan, X. Qu, et al. Molecules. 2017, 22(5), 715.

[11] S. T. S. Hassan, K. Berchová-Bímová, M. Šudomová, M. Malaník, K. Šmejkal, K. R. R. Rengasamy. J Clin Med. 2018, 7(9), 283–15.

[12] F. Wang, D. Wu, F. He, H. Fu, W. Wang. Tumori. 2017, 103, 537–542.

[13] S. T. S. Hassan, R. Masarčíková, K. Berchová. J Pharm Pharmacol. 2015,, 67(10), 1325–1336.

[14] Y. Wenqiang, C. Xu, Y. Li, G. Shaofen, W. Zhen, Y. Xiuling. Nat Prod Commun. 2020, 15(3).

[15] S. Harada, K. Yusa, K. Monde, T. Akaike, Y. Maeda. Biochem Biophys Res Commun. 2005, 329, 480–486.

[16] R. A. Dixon. Isoflavonoids: Biochemistry, molecular biology and biologicalfunctions. In: Barton, D. H. R., Nakanishi, K., Meth-Cohn, O., (Eds.), Comprehensive Natural Products Chemistry, Vol. 1, Polyketides and OtherSecondary Metabolites Including Fatty Acidsand Their Derivatives. inchief Elsevier, Amsterdam, 1999, pp. 773–824.

[17] J. B. Harborne. The Flavonoids: Advances in Research since 1986. Chapman and Hall, London, 1994.

[18] G. Forkmann, W. Heller. Biosynthesisof flavonoids. In: Barton, D. H. R., Nakanishi, K., Meth-Cohn, O., (Eds.), inchief, Comprehensive NaturalProducts Chemistry, Vol. 1, Polyketides andOther Secondary Metabolites Including FattyAcids and Their Derivatives. Elsevier, Amsterdam, 1999, pp. 713–748.

[19] R. Geethalakshmi, S. V. L. Dronamraju. Mol Biol Rep. 2018.

[20] S. K. Hsieh, J. R. Xu, N. H. Lin, Y. C. Li, G. H. Chen, P. C, et al. J Food Drug Anal. 2016, 24, 722–729.

[21] C. Danciu, L. Vlaia, F. Fetea, M. Hancianu, D. E. Coricovac, S. A. Ciurlea, et al. Biol Res. 2015, 48, 1.

[22] M. Hashemzaei, A. D. Far, A. Yari, R. E. Heravi, K. Tabrizian, S. M. Taghdisi, et al. Oncol Rep. 2017, 38, 819–828.

[23] M. Działo, J. Mierziak, U. Korzun, M. Preisner, J. Szopa, A. Kulma. Int J Mol Sci. 2016, 17, 160.

[24] K. E. Heim, A. R. Tagliaferro, D. J. Bobilya. J Nutr Biochem. 2002, 13, 572–584.

[25] T. Oki, M. Masuda, S. Furuta, Y. Nishiba, N. Terahara, A. I. JSuda. Food Sci. 2002, 67, 1752–1756.

[26] M. Mohammadi-Cheraghabadi, S. A. M. Modarres-Sanavy, F. Sefidkon, A. Mokhtassi-Bidgoli, S. Hazrati. Saudi J Biol Sci. 2021a, 28, 7227–7240.

[27] M. Mohammadi-Cheraghabadi, S. A. M. Modarres-Sanavy, F. Sefidkon, A. Mokhtassi-Bidgoli, S. Rashidi-Monfared. Sci Rep. 2021b, 11, 21997.

[28] L. Andreu, N. Nuncio-Jáuregui, Á. A. Carbonell-Barrachina, P. Legua, F. Hernández. J Sci Food Agric. 2018, 98, 1566–1573.

[29] X. H. Meng, C. Liu, R. Fan, L. F. Zhu, S. X. Yang, H. T. Zhu, et al. J Agric Food Chem. 2018, 66, 247–254.

[30] O. Ustun, F. S. Senol, G. Kurkcuoglu, I. E. Orhan, M. Kartal, K. H. C. Baser. Ind Crops Prod. 2012, 38, 115–123.

[31] C. L. Apetrei, C. Tuchilus, A. C. Aprotosoaie, A. Oprea, K. E. Malterud, A. Miron. Molecules. 2011, 16, 7773–7788.

[32] P. J. H. Jones, F. Y. Ntanios, M. R. Raeini-Sarjaz, C. A. Vanstone. Am J Clin Nutr. 1999, 69, 1144–1150.

[33] G. M. Kamal, B. Gunaherath, A. A. Leslie Gunatilaka. Topics in Plant Steroids, V. 31 of Encyclopedia of Analytical Chemistry (Applications, Theory and Instrumentation). R. A. Meyers, (Ed.), Wiley & Sons, Ltd, 2014, pp. 1–26.

[34] Z. D. Yang, D. Z. Duan, W. W. Xue, X. J. Yao, S. Li. Life Sci. 2012, 90, 929–933.

Charu Arora*, Dipti Bharti, Brij Kishore Tiwari, Ashish Kumar,
Dakeshwar Kumar Verma and Bhupender Singh

Chapter 6
Characterization techniques used for the analysis of phytochemical constituents

Abstract: Phytochemicals are chemical constituents isolated from various plants that are used widely for treating several diseases. These plants' botanicals contain various kinds of important bioactive molecules used by various medicinal and pharmaceutical industries globally. In this book chapter, the main focus is to describe various analytical techniques and methods used for the extraction, isolation and characterization of these bioactive phytochemicals. Several types of characterization techniques are used for the identification and characterization of phytoconstituents so there is a need to describe all of these in one place. This chapter mainly focuses on the methods and techniques, used for isolation and characterization of bioactive constituents of the plants, with their applications. The phytochemical screening and analysis of biologically active components present in the extracts of plants involving the applications of common phytochemical investigation such as various chromatographic techniques like high-performance liquid chromatography (HPLC), thin-layer chromatography (TLC), gas chromatography–mass spectrometry (GC-MS), liquid chromatography-mass spectrometry (LC-MS) as well as non-chromatographic techniques such as Fourier Transform Infra-Red (FTIR), UV-visible spectroscopy, various types of nuclear magnetic resonance (NMR) spectroscopic techniques, mass spectrometry have been discussed.

Keywords: Bioactive constituents, high-performance liquid chromatography, thin-layer chromatography, gas chromatography–mass spectrometry, extraction

*Corresponding author: Charu Arora, Department of Chemistry, Guru Ghasidas University, Bilaspur 495009, Chhattisgarh, India, e-mail: charuarora77@gmail.com
Dipti Bharti, Department of Applied Science and Humanities, Darbhanga College of Engineering, Darbhanga 846005, Bihar, India
Brij Kishore Tiwari, Ashish Kumar, Department of Applied Science and Humanities, G. L Bajaj Institute of Technology and Management, Greater Noida 201306, Uttar Pradesh, India
Dakeshwar Kumar Verma, Department of Chemistry, Govt. Digvijay (Autonomous) PG College, Rajnandgaon, Chhattisgarh, India
Bhupender Singh, Institute Instrument Centre, Indian Institute of Technology, Delhi, India

https://doi.org/10.1515/9783110791891-006

6.1 Introduction

The earth is globally fortified with various plants having medicinal properties. These are widely used in the pharmaceutical and food industries and in Ayurveda. Medicinal plants have evolved over some 400 million years, and the richest sources of renewable bioactive phytochemical constituents have effective defense mechanisms during evolution, which secure them in a hostile environment [1]. Plants produce various diverse biologically active important chemical constituents. These phytochemicals get accumulated in fruits and vegetables and protect the plant against free radical damage. Medicinal plants are reported as a rich source of various antioxidants viz. vitamins A, C, E, alkaloids flavonoids, tannins, and lignins. Fruits and vegetables have several medicinal properties and high nutritional value [2].

The total number of medicinal plant phytochemicals may exceed 4,00,000, while secondary metabolites are approximately 10,000. The secondary metabolites protect plants against pests and pathogens. Most of medicinal plant species used for plant protection exhibited feeding deterrent, insect's growth regulator, oviposition deterrents, and insecticidal activity. The medicinal plant families containing most insecticidal and pesticidal medicinal plant species are Meliaceae, Asteraceae, Brassicaceae, Solanaceae, Euphorbiaceae, Apiaceae, and Lamiaceae [3]. The medicinal plant extract or phytochemical compounds isolated from various solvent extracts can act as different medicinal properties. Recent studies in Asian and African countries suggest that medicinal plant extract of locally available plants can be effective as crop protectants, either used alone or in mixing with a synthetic chemical to act as effective antimicrobial agents. These studies may help to understand the knowledge and traditional practices which can make valuable contributions to domestic protection in countries where strict enforcement of pesticide regulations is impractical. Among the well-represented plant pesticides is Azadirachtin compound, an active ingredient found in neem tree, *Azadirachta indica*, is a complex steroid like tetranortriterpenoid that is effectively used against more than 400 species of insects, including many important crop pests and pathogens [1].

The living plants may be considered a biosynthetic laboratory not only for the primary metabolites but also for many secondary products such as glycosides, alkaloids, and flavonoid tannins. Primary metabolites are found in all organisms in nature. These are essential for growth and physiological development due to their basic cell metabolism. These are synthesized by specialized pathways in plant cell metabolism [4]. Secondary metabolites have very limited distribution though these are derived from primary metabolites. They have developed protective chemicals against pests, pathogens, and environmental stress. They are considered to be waste or secretary products of plant metabolism in terms of cellular economy. The chemical structure of secondary metabolites is complex, and a small chemical modification such as hydroxylation and interaction with metal ion methylation leads to a wide spectrum of functionally different substances. Secondary metabolites containing three main groups may be classified into three categories [5].

1. Nitrogen-containing compounds (alkaloids, glycosinolate).
2. Saponins, containing hydrocarbons as major compounds.
3. Phenolics containing a benzene ring.

More than 10,000 secondary products of plants have been identified, and it is estimated that hundreds of thousands of plant chemicals exist. Structures of some secondary metabolites with nitrogen and without nitrogen are reported in Fig. 6.1. There is growing evidence that most of these compounds are involved in the interaction of the plant with other species, primarily for the defense of plants from plant microbes. This represents a large reservoir of chemical structure with biological activity. In terrestrial plants, secondary metabolites are usually present in high structural diversity and often contain more than one functional group, so they exhibit multiple functionalities and are bioactive [6].

Several secondary metabolites in plants play a role in defense against fungi, bacteria, and insects. Plant secondary metabolites are effective against pests and pathogens due to their similarity with certain vital components belonging to the cellular signaling system or their ability to interfere with the vital enzymes to restrict metabolic pathways. Ernst Stanl showed experimentally that secondary metabolites could serve as defensive compounds against snails and other herbivores. Plants were being used to control agricultural pests and pathogens before the discovery of synthetic pesticides. Secondary metabolites showed various biological properties, viz. antifungal, antibacterial, and insecticidal. The chemical structure of secondary metabolites are complex and small chemical modification such as hydroxylation, interaction with metal ions, and methylation leads to wide spectrum of functionally different structure [7, 8].

As per World Health Organization, approximately 21,000 medicinal plants exists around 92 countries. The first premier step to explore the bioactive compounds from the medicinal plant is extraction, isolation and characterization, pharmacological screening, toxicological evaluation, and clinical evaluation to synthesize bioactive molecules. This book chapter summarizes details on extraction, isolation and characterization of biologically active constituents from various important plants, screening assay, thin layer chromatography, immunoassay, and other spectroscopic techniques [9, 10].

6.2 Extraction of plant extract

Plant material is extracted to extract bioactive constituents before isolation and identification. The plant material is washed, dried, and ground before extraction. Several extraction methods including soxhlet, hydro distillation, decoction, maceration, microwave-assisted extraction, ultrasound-assisted extraction, supercritical fluid extraction, and pressurized solvent extraction are used for the extraction of bioactive compounds from plant

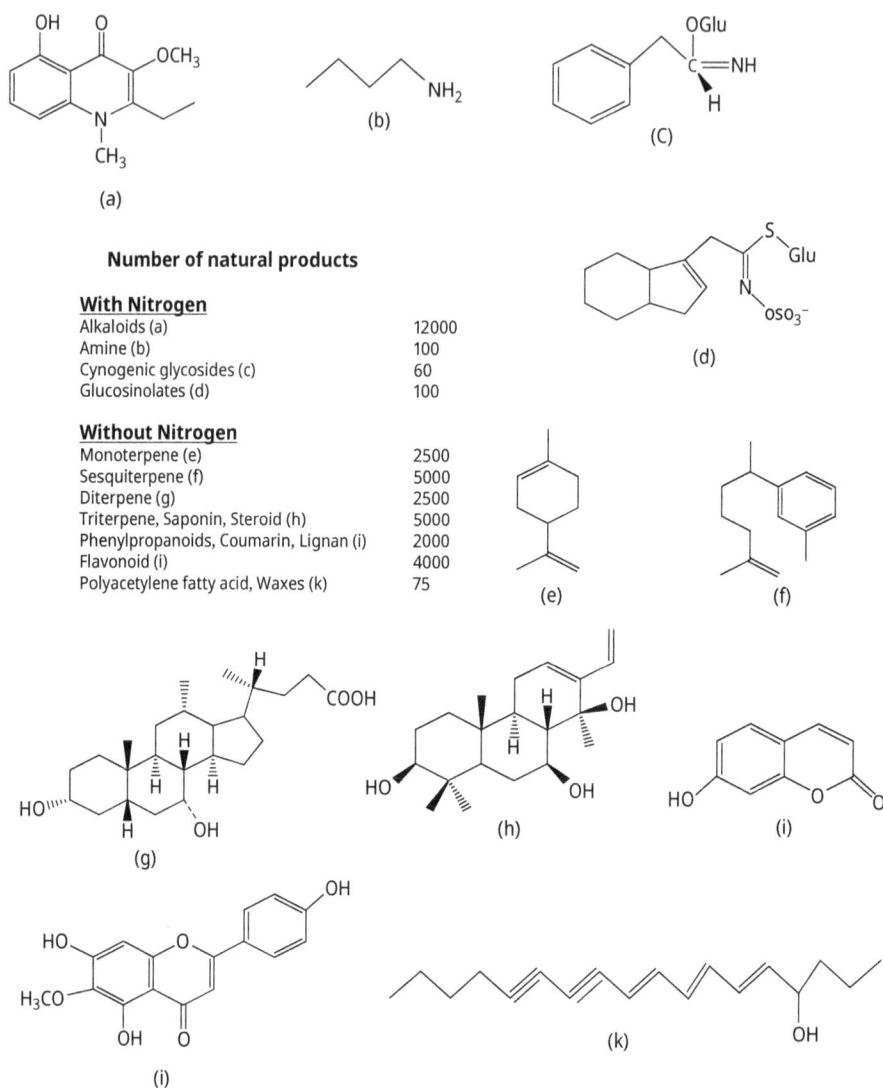

Number of natural products

With Nitrogen

Alkaloids (a)	12000
Amine (b)	100
Cynogenic glycosides (c)	60
Glucosinolates (d)	100

Without Nitrogen

Monoterpene (e)	2500
Sesquiterpene (f)	5000
Diterpene (g)	2500
Triterpene, Saponin, Steroid (h)	5000
Phenylpropanoids, Coumarin, Lignan (i)	2000
Flavonoid (i)	4000
Polyacetylene fatty acid, Waxes (k)	75

Fig. 6.1: Structure of some plant secondary metabolites.

material [11]. Microwave-assisted extraction is useful to reduce time involvement and solvent consumption compared to other conventional methods [12, 13].

The solvent is selected based on the nature of the bioactive compound being isolated. Polar solvents are used to extract hydrophilic compounds. Chlorophyll is removed using hexane for extraction [14].

The extraction method is selected based on the bioactive component being targeted. Different extraction techniques, viz. sonification, heating under reflux, soxhlet extraction and maceration, or percolation of fresh green plants or dried powdered plant material in water and/or organic solvent systems are used for extraction. Various extraction methods and experimental conditions are summarized in Tab. 6.1.

Tab. 6.1: A brief summary of the experimental conditions for various methods of extraction for plants material.

Experimental conditions	Soxhlet extraction	Sonification	Maceration
Common solvents used	Hexane, chloroform, acetone, methanol, ethanol, or mixture of alcohol and water	Methanol, ethanol, or mixture of alcohol and water	Methanol, ethanol, or mixture of alcohol and water
Temperature (°C)	Depending on solvent used	Can be heated	Room temperature
Pressure applied	Not applicable	Not applicable	Not applicable
Time required	3–18 h	1 h	3–4 days
Volume of solvent required (ml)	150–200	50–100	Depending on the sample size
Reference	C. A. Chugh et al. [15]	B. Zygmunt and J. Namiesnik [16]	Phrompittayarata et al. [17]

Solid-phase micro-extraction, supercritical fluid extraction, pressurized liquid extraction, microwave-assisted extraction, solid-phase extraction, and surfactant-mediated techniques are some modern extraction methods having certain benefits, including the reduction in time, solvent consumption, sample degradation, and reduction of additional clean-up of the sample. These methods improve the extraction efficiency, selectivity, and/or kinetics of extraction [13].

6.3 Phytochemical screening assay

Different chemical constituents present in plant species are responsible for the biological activities of plant material [6].

Bioactive ingredients and secondary metabolites are isolated from medicinal plants used for medicinal and agricultural purposes. A phytochemical screening assay is used to detect different types of constituents present in a plant extract [18]. A brief description of the experimental techniques used for screening phytoconstituents and secondary metabolites is summarized in Tab. 6. 2.

Tab. 6.2: Experimental methods for screening of secondary metabolites.

S. no.	Secondary metabolite(s)	Name of test	Methodology	Result(s)	Reference(s)
1	Test for proteins	–	Addition of biuret reagent (2 ml) to the test solution (2 ml)	Development of violet color	G. S. Kumar et al. [19]
2	Alkaloid	Dragendorff's test	Spot a drop of extract on a small piece of precoated TLC plate. Spray the plate with Dragendorff's reagent	Orange spot	C. A. Chugh and D. Bharti [18]
		Wagner test	Add 2 ml filtrate with 1% HCl + steam. Then add 1 ml of the solution with six drops of Wagner's reagent	Brownish-red precipitate	S. V. Chanda et al. [20]
		TLC method 1	Solvent system:chloroform:methanol:25% ammonia (8:2:0.5). Spots can be detected after spraying with Dragendorff reagent	Orange spot	P. B. Mallikharjuna et al. [21]
		TLC method 2	Wet the powdered test samples with a half diluted NH_4OH and lixiviated with EtOAc for 24 h at room temperature. Separate the organic phase from the acidified filtrate and basify with NH_4OH (pH 11–12). Then extract it with chloroform (3×), condense by evaporation and use for chromatography. Separate the alkaloid spots using the solvent mixture chloroform and methanol (15:1). Spray the spots with Dragendorff's reagent	Orange spot	G. S. Kumar et al. [18]

3	Anthraquinone	Borntrager's test	Heat about 50 mg of extract with 1 ml 10% ferric chloride solution and 1 ml of concentrated hydrochloric acid. Cool the extract and filter. Shake the filtrate with equal amount of diethyl ether. Further extract the ether extract with strong ammonia	Pink or deep red coloration of aqueous layer	G. S. Kumar et al. [18]
		Borntrager's test	Add 1 ml of dilute (10%) ammonia to 2 ml of chloroform extract.	A pink-red color in the ammoniacal (lower) layer	D. N. Onwukaeme et al. [22]
4	Cardiac glycosides	Kellar–Kiliani test	Add 2 ml filtrate with 1 ml of glacial acetic acid, 1 ml ferric chloride, and 1 ml concentrated sulphuric acid.	Green-blue coloration of solution	J. Parekh and S. V. Chanda [6]
		Kellar–Kiliani test	Dissolve 50 mg of methanolic extract in 2 ml of chloroform. Add H_2SO_4 to form a layer.	Brown ring at interphase	D. N Onwukaeme et al. [22]
		TLC method	Extract the powdered test samples with 70% EtOH on rotary shaker (180 thaws/min) for 10 h. Add 70% lead acetate to the filtrate and centrifuge at 5,000 rpm/10 min. Further centrifuge the supernatant by adding 6.3% Na_2CO_3 at 10,000 rpm/10 min. Dry the retained supernatant and redissolve in chloroform and use for chromatography. Separate the glycosides using EtOAc-MeOH-H_2O (80:10:10) solvent mixture	The color and hRf values of these spots can be recorded under ultraviolet (UV254 nm) light	P. B. Mallikharjuna et al. [21]

(continued)

Tab. 6.2 (continued)

S. no.	Secondary metabolite(s)	Name of test	Methodology	Result(s)	Reference(s)
5	Flavonoid	Shinoda test	To 2–3 ml of methanolic extract, add a piece of magnesium ribbon and 1 ml of concentrated hydrochloric acid	Pink red or red coloration of the solution	C. A. Chugh et al. [19]
		TLC method	Extract 1 g powdered test samples with 10 ml methanol on water bath (60 °C/5 min). Condense the filtrate by evaporation, and add a mixture of water and EtOAc (10:1 mL), and mix thoroughly. Retain the EtOAc phase and use for chromatography. Separate the flavonoid spots using chloroform and methanol (19:1) solvent mixture	The color and hRf values of these spots can be recorded under ultraviolet (UV254nm) light	P. B. Mallikharjuna et al. [21]
		NaOH test	Treat the extract with dilute NaOH, followed by addition of dilute HCl	A yellow solution with NaOH turns colorless with dilute HCl	D. N. Onwukaeme et al. [22]
6	Phenol	Phenol test	Spot the extract on a filter paper. Add a drop of phosphomolybdic acid reagent and expose to ammonia vapors	Blue coloration of the spot	G. S. Kumar et al. [18]
7	Phlobatannin	–	2 ml extract was boiled with 2 ml of 1% hydrochloric acid HCl	Formation of red precipitates	H. O. Edeoga et al. [2]

8	Pyrrolizidine alkaloid	–	Prepare 1 ml of oxidizing agent, consisting of 0.01 ml hydrogen peroxide (30% w/v) stabilized with tetrasodium pyrophosphate (20 mg/ml) and made up to 20 ml with isoamylacetate, and add to 1 ml of plant extract. Vortex the sample and add 0.25 ml acetic anhydride before heating the sample at 60 °C for 50–70 s. Cool the samples to room temperature. Add 1 ml of Ehrlich reagent and place the test tubes in water bath (60 °C) for 5 min. Measure the absorbance at 562 nm. The method of Holstege et al. [5] should be used to confirm results of the screening method	Peaks were compared with the GC–MS library	G. S. Kumar et al. [18]
9	Reducing sugar	Fehling test	Add 25 ml of diluted sulphuric acid (H_2SO_4) to 5 ml of water extract in a test tube and boil for 15 min. Then cool it and neutralize with 10% sodium hydroxide to pH 7 and 5 ml of Fehling solution	Brick red precipitate	G. S. Kumar et al. [18]
10	Saponin	Frothing test/ Foam test	Add 0.5 ml of filtrate with 5 ml of distilled water and shake well	Persistence of frothing	J. Parekh and S. V. Chanda [6]
		TLC method	Extract 2 g of powdered test samples with 10 ml 70% EtOH by refluxing for 10 min. Condense the filtrate, enrich with saturated *n*-BuOH, and mix thoroughly. Retain the butanol, condense, and use for chromatography. Separate the saponins using chloroform, glacial acetic acid, methanol, and water (64:34:12:8) solvent mixture. Expose the chromatogram to the iodine vapors	The color (yellow) and hRf values of these spots were recorded by exposing chromatogram to the iodine vapors	P. B. Mallikharjuna et al. [21]

(continued)

Tab. 6.2 (continued)

S. no.	Secondary metabolite(s)	Name of test	Methodology	Result(s)	Reference(s)
11	Steroid	Liebermann–Burchardt test	To 1 ml of methanolic extract, add 1 ml of chloroform, 2–3 ml of acetic anhydride, 1 to 2 drops of concentrated sulphuric acid	Dark green coloration	G. S. Kumar et al. [18]
		TLC method	Extract 2 g of powdered test samples with 10 ml methanol in water bath (80 °C/15 min). Use the condensed filtrate for chromatography. The sterols can be separated using chloroform, glacial acetic acid, methanol, and water (64:34:12:8) solvent mixture. The color and hRf values of these spots can be recorded under visible light after spraying the plates with anisaldehyde–sulphuric ac d reagent and heating (100 °C/6 min)	The color (greenish black to Pinkish black) and hRf values of these spots can be recorded under visible light	P. B. Mallikharjuna et al. [21]
12	Tannin	Braemer's test	10% alcoholic ferric chloride will be added to 2–3 ml of methanolic extract (1:1)	Dark blue or greenish grey coloration of the solution	G. S. Kumar et al. [18]; J. Parekh and S. V. Chanda [6]
13	Terpenoid	Liebermann–Burchardt test	To 1 ml of methanolic extract, add 1 ml of chloroform, 2–3 ml of acetic anhydride, 1 to 2 drops of concentrated sulphuric acid	Pink or red coloration	G. S. Kumar et al. [18]
		Salkowski test	5 ml extract was added with 2 ml of chloroform and 3 ml of concentrated sulphuric acid H_2SO_4	Reddish brown color of Interface	H. O. Edeoga et al. [2]
14	Volatile oil	–	Add 2 ml extract with 0.1 ml dilute NaOH and small quantity of dilute HCl. Shake the solution	Formation of white precipitates	D. Dahiru et al. [23]

6.4 Isolation and identification of compounds

Plant extracts possessing bioactive constituents in different types of combinations have different physical properties and chemical polarities. However, separation still remains difficult for the identification and characterization of these bioactive constituents. The isolated compounds have been purified by a chromatographic technique, followed by spectroscopic techniques used to find the structures of the isolated phytoconstituents [9]. Although ultraviolet-visible (UV-Visible) infrared (IR) spectra, mass spectrometry (MS), and nuclear magnetic resonance (NMR) spectroscopy are the main spectroscopic methods for the identification of structures, respectively [10].

6.5 Chromatographic technique

6.5.1 Thin-layer chromatography (TLC) and bio-autographic methods

TLC is an easy, fast, and economical method capable of finding the number of constituents present in the mixture. TLC is useful to identify a constituent present in an extract/mixture by comparing the retention factor (Rf) of a with the Rf of the standard component compared with the standard Rf value. The spray of phytochemical screening reagents is useful for detecting isolated compounds through TLC. Sometimes it is followed by viewing the plate under UV light to detect UV-sensitive phytocomponent. This method is useful to ascertain the purity and identification of isolated constituents.

TLC, in combination with other chromatographic techniques and activity evaluation is helpful for the target-directed isolation of bioactive constituents in various mixtures of compounds [24, 25].

Target-directed separation of biologically active ingredients is termed as bioautography. Solvents used for the chromatographic isolation process are removed before biological detection as these organic solvents can inactivate the enzyme or even cause the death of a living organism under investigation. They are useful for the fast and economical identification of active ingredients present in plant extract. In the present chapter, we have summarized the techniques for the identification of biocidal, antioxidant, and other bioactive constituents by using bioautography [26].

Conventionally in bioautographic methods, researchers investigate growth inhibition of microbes' detection of antibacterial and antifungal components of plant extracts. This method is one of the most powerful assays for detecting antibacterial and antifungal compounds. Bio-autography limits the biocidal efficacy on a chromatogram using the following approaches:

(i) Direct bio-autography: the microbes are grown on the TLC plate.

(ii) Contact bio-autography: the antimicrobial constituents are transferred from the TLC plate to an inoculated agar plate by direct contact.
(iii) Agar overlay bio-autography: TLC plate is loaded with a seeded agar medium.

The growth inhibition zones that occurred on TLC plates during investigations following any of the above methods will be used to ascertain the presence of the bioactive compound with antibacterial and antifungal activity in the TLC fingerprint with reference to Rf values present in the standard table of a known compound. Activity-guided separation by preparatory TLC is used to isolate bioactive components. At last, the isolated compounds are purified and identified by various techniques such as liquid chromatography–mass spectrometry (LC-MS), high-performance liquid chromatography (HPLC), and gas chromatography–mass spectrometry (GC-MS) [27].

6.5.2 Liquid chromatography–mass spectrometry (LC-MS)

The application of LC-MS for toxicology testing has been reviewed [28]. One of the most important uses of LC-MS is for confirmation of drug testing and screening, followed by an immunoassay screen. LC-MS has been used for method development, validation, and implementation.

Various approaches to these applications are reported in the literature and can be referred for the implementation of LC-MS [29].

6.5.3 High-performance liquid chromatography

High-performance liquid chromatography (HPLC) is an important and popular technique for separating and purifying natural products. This technique is helpful in the characterization of bioactive phytoconstituents isolated from plant extract by activity-guided fractionation. The biologically bioactive compound is usually found in very low concentrations in the plant extract, and HPLC is a suitable technique for fast detection of such bioactive components present in trace quantity [30]. Separation takes place due to differences in molecular weights and migration rates of phytoconstituents in a particular column. Usually, isolation and identification of phytoconstituents are carried out using isocratic elution. Gradient mode is useful to study samples consisting more than one constituent to be studied, having significantly different retention times under experimental conditions. The chemical constituent is purified using HPLC by extraction of the compound from other constituents or impurities. Each constituent possesses a specific Rf value for specific chromatographic parameters. Chromatic conditions are optimized based on the requirement of a compound to be isolated and purified, and its relation with other constituents of the sample [31].

The crude source material of plants such as dried leaves, power of dried stems, and fruits should be processed to obtain an appropriate sample for HPLC analysis. The selection of an appropriate mobile phase is important for proper isolation and identification. The dried plant material is extracted with organic solvent followed by maceration. The extract is obtained by the removal of solid material by filtration. The concentrated filtrate is injected into HPLC for further isolation. Guard columns must be used for the separation of crude extract [32]. The presence of components like chlorophyll affects the performance of the analytical column. Guard columns protect the analytical columns and improve their performance [33].

6.5.4 Gas chromatography–mass spectrometry (GC-MS)

It is a hyphenated method coupling the separation efficiency of GC with MS to detect the separated constituent. Volatile components are separated by GC, followed by fragmentation, and identification by MS. GC-MS is an efficient and sensitive technique, which is able to analyze an increased range of analyzable samples at a faster rate [34].

The stationary phase of GC ensures that different chemical compounds are separated and eluted from the column at different times of interval. GC columns act as the stationary phase and separation tool for a gas chromatography analysis at the molecular level. Different types of GC columns equipped with different stationary phases can be used for the separation of different types of constituents.

GC-MS is widely used for the detection of potentially toxic chemicals in foods, quantitation of organic contaminants in water, or for analysis of petroleum products during oil processing [35].

6.6 Purification and characterization of bioactive compounds from plants

Advanced techniques have been developed in recent years for purifying and isolating biologically active constituents from plants. These techniques are more precise for the purification of constituents. Moreover, these are compatible with advanced bioassays. The aim of the research work is to develop/find a suitable method for screening the plant material for bioactivity and isolation, purification, and characterization of bioactive constituents. In vitro methods are preferred over in vivo as they are economical, fast, and free from ethical complications [36]. Different chemical constituents are found in different parts of plant species. Moreover, variation in chemical profiling is observed with seasonal changes and changes in geographic and climatic conditions [37]. The plant species which are relatively disease free are selected based on observation. Ethnobotanical information plays a vital role exploring plant species for bioactive compounds. Different solvents can be used for the isolation and purification of bioactive constituents. Column chromatography followed by preparatory TLC is the conventional technique for the isolation and purification of active ingredients. With the development of HPLC, the process of purification of bioactive compounds has been accelerated. Spectroscopic techniques may be hyphenated with HPLC or used independently for the characterization of isolated compounds [38, 39].

6.7 Structural identification of biologically active molecules

The purified active compounds are characterized using spectroscopic techniques, viz. UV-Visible, FTIR, NMR, and mass spectroscopy. Electromagnetic radiation interacts with molecules to be characterized, and some radiation is absorbed by the molecule leading to excitation followed by de-excitation. By measuring the absorption and emission spectra, we can produce a spectrum that is useful to characterize the compound [10, 40].

6.8 UV-Visible spectroscopy

UV-Vis spectrophotometer is used to record UV-Visible spectra of the purified compound. The samples are dissolved in a suitable solvent for scanning in the UV-Vis range and determining λ_{max} and absorption coefficient. UV-Vis spectra give information about the degree of conjugation in the compound under investigation [41]. It does not play a significant role in characterization [41]. This technique is useful for the qualitative analysis and characterization of some classes of compounds, viz. phenol, aromatic and phenolic compounds, and/or chromophores in biological mixtures or pure form. These techniques are fast and economical but less selective [42].

6.9 Infrared spectroscopy

FTIR spectroscopy provides information about functional groups of chemical constituents. The spectra can be recorded using KBr or dry films, solid, or liquid forms.

Some frequencies are absorbed as infrared radiation passes through the sample, while some pass through the sample without any absorption. Vibrational changes take place in the molecule when it is irradiated with Infrared radiation. Infrared spectroscopy is also known as vibrational spectroscopy [43]. Particular bond has specific vibrational frequencies. We can identify the bond present in the compound on the basis of these frequencies. Resolution power and time consumption are reduced by applying FTIR. Moreover, FTIR is a non-destructive technique for analyzing extracts or products of medicinal plants [44].

6.10 Nuclear magnetic resonance (NMR) spectroscopy

NMR spectroscopy is widely applied for the characterization of purified bioactive compounds. NMR is one of the most useful techniques for the characterization of isolated compounds. The Pulse-Fourier-transform technique has significantly improved the resolution and sensitivity of NMR signal to record the spectra of constituents in trace quantity. This technique is very useful for recording ^{13}C spectra of low-abundance nuclei. The 2D NMR techniques are very useful in resolving any ambiguity in the structure of organic compounds. The sample is irradiated with a single radiofrequency pulse of a few microseconds in the Pulse-Fourier-transform technique. The signal obtained is known as free induction decay (FID). FID produces an NMR signal by Fourier transformation [45].

NMR is correlated to the magnetic properties of nuclei. Proton and carbon 13 are NMR active nuclei present in organic compounds. Researchers can characterize molecules by analyzing the difference of signals of different magnetic nuclei as it can give a clear interpretation of the position of these nuclei in the molecule. It also provides information about atoms attached to neighboring groups. ^{1}H NMR, ^{13}NMR, HMBC, HMQC, and HSQC correlation are helpful in establishing structure of mixture of constituents that are difficult to be isolated by chromatographic techniques [46].

6.10.1 ^{1}H NMR spectroscopy

^{1}H NMR is the primary and basic NMR method for the characterization of organic compounds. The instruments of very high sensitivity (400–800 MHz) have been developed to record the ^{1}H NMR spectrum. These are useful for obtaining high-resolution spectra in a short time, providing the following information:
1. Number of different types of protons is present in the molecule.
2. Information about the chemical shift and electronic environment of protons.
3. Signal intensity and number of protons for each type of signal.
4. Peak splitting gives information about the neighboring environment of a proton [46].

6.10.2 Broad band decoupled ^{13}C NMR spectroscopy

This is a rapid technique using an extremely short relaxation time (<1 s). This method is applicable for trace quantity of samples. It is the preferred technique to study the samples for which quaternary carbon peaks are missed in the J-modulated spectrum. But, all carbons (C, CH, CH$_2$, and CH$_3$) appear on the same side, making it tough to identify different carbon types present in the molecule [47].

6.10.3 J-modulated ^{13}C NMR spectroscopy

It is also termed the attached proton test. This technique is useful for analyzing ^{13}C NMR. In this technique, the resonance from ^{13}C nuclei of methyl and CH appears to be inverted with respect to those from CH$_2$ and quaternary carbons. It can be achieved by using a proton 180° pulse with the carbon 180° pulse, applying a delay time of 1–2 s, and applying broadband decoupling during the acquisition of FID. This technique is advantageous in terms of simplicity for differentiation between C/CH$_2$ and CH/CH$_3$ carbons. Although it has some disadvantages, viz., poor sensitivity in comparison to broadband-decoupled ^{13}C NMR spectroscopy. Moreover, sometimes, signals of quaternary carbons may disappear. However, quaternary carbons can be visualized by increasing the relaxation delay to 6–10 s [48].

6.10.4 Distortion less enhancement by polarization transfer-135

Distortionless enhancement by polarization transfer-135 spectra differentiates CH$_2$ peaks from CH$_3$ and methine peaks. Quaternary carbons do not produce signals in this spectrum, and thus, they differentiate C and CH$_2$ peaks in a J-modulated experiment. It provides useful information for the elucidation of the structure of terpenoids confirming CH$_2$ peaks. The disappearance of quaternary carbon peak in a solvent is beneficial for calibration of the spectrum by using the same solvent reference value or a carbon chemical shift from another ^{13}C (e.g., J-modulated) experiment [49].

6.10.5 Homonuclear correlation spectroscopy

Homonuclear correlation spectroscopy gives information about how the different components of the same nuclei (^{1}H) have a relation with one another through scalar or spatial coupling. ^{1}H–^{1}H correlation spectroscopy (COSY), total correlation spectroscopy (TOCSY), and Nuclear Overhauser Enhancement Spectroscopy (NOESY) are the most commonly used homonuclear correlation spectroscopic techniques for structure elucidation [50].

6.10.6 ^{1}H–^{1}H correlation spectroscopy

^{1}H–^{1}H COSY data gives information about protons that are coupled to each other. In the process of coherence transfer in the COSY technique, the magnetization is transferred between coupled spins. It possesses two 90° pulses separated by a time delay. After the second pulse, the exchange of magnetization owing to scalar coupling (coherence transfer) between spins may take place [50]. Fourier transformation of the FID leads to a COSY spectrum having cross peaks connecting the coupled nuclei. A 45° pulse is used in the COSY-45 technique instead of the second 90° pulse. It results in a simplified spectrum by reducing the cross peaks within multiplets such as CH_2. It simplifies a complex spectrum by identifying the correlations that would otherwise be hidden in the clutter of peaks close to the diagonal. The COSY-lr is another homonuclear 2D NMR technique applying a relatively long delay which enhances the relative intensity of cross peaks from long-range coupling. Generally, the delay time is fixed at 0.3 s. Coupling information is raised among the protons, which are directly coupled to other protons in the COSY technique. Such coupling can take place through two, three, and sometimes four bonds. The diagonal provides the correlations (couplings) of signals among themselves, while identical information is shown on either side of the diagonal. So, interpretation can be made by considering one side of the diagonal [51].

6.10.7 ^{1}H–^{1}H total correlation spectroscopy

The basic principle of TOCSY is similar to that of COSY. But it gives information on the cross peaks for all protons that are part of the same spin system, e.g., the COSY spectra of an abcd system of four protons exhibit cross peaks for "a to b," "b to c," and "c to d," while TOCSY exhibit an additional cross peak for "a to c," "a to d," and "b to d." This observation indicates that the TOCSY spectrum exhibits a total correlation among the ^{1}H values in a particular spin system. TOCSY gives valuable information for the identification of chemical constituents from natural products [52].

6.10.8 ^{1}H–^{1}H Nuclear overhauser enhancement spectroscopy

NOE gives information about the nuclei within the molecule in close spatial. It appears similar to a COSY spectrum, but the cross peaks in NOE correspond to spatial dipolar interaction instead of a scalar coupling through bonds. In a 2D NOE (NOESY) technique, the nuclei get excited, and simultaneously, nonselective pulses are labeled by their processional frequencies. This frequency labeling is equivalent to selectively exciting each of the nuclei and separately measures each interaction pairwise. The NOESY consists of three 90° pulses. It is required to provide a delay time (mixing time) before applying the third 90° pulse. Mixing time is an important parameter in the

NOESY technique. It is dependent on the molecular properties of the compound being investigated. Small molecules need longer mixing time (1–5 s), while larger molecules need much less time (0.1–0.5 s). However, this is not correct for all molecules. It is required to change the mixing time depending on the spectra desired [53].

6.10.9 Heteronuclear correlation spectroscopy

Heteronuclear correlation spectroscopy NMR technique gives information about the correlation of ^{1}H and ^{13}C through scalar coupling. The heteronuclear multiple quantum coherence, heteronuclear single quantum correlation, and HC-COBIDEC exhibit ^{1}J or direct ^{1}H–^{13}C couplings, whereas long-range (^{2}J and ^{3}J) heteronuclear (^{1}H–X–^{13}C–^{13}C) couplings are recorded using heteronuclear multiple bond coherence (HMBC) [54].

6.10.10 Heteronuclear multiple quantum coherence

Heteronuclear multiple quantum coherence (HMQC) is also termed heteronuclear single quantum correlation (HSQC). This technique shows cross peaks that give the direct (^{1}J) correlation between ^{1}H and ^{13}C nuclei. In heteronuclear multiple quantum coherence (HMQC), the ^{1}H spectrum is plotted on the x-axis and ^{13}C on the y-axis. No peak is observed on the diagonal. The ^{1}H NMR reveals a direct correlation to ^{13}C NMR in the HMQC, thereby simplifying structural elucidation. HMQC and HSQC both give the correlation between ^{1}H and ^{13}C nuclei by using a pulse sequence with a delay time set to 1/2 J seconds. This technique suffers from noise artifacts arising from intense protons peaks, viz. signals of methyl groups [55, 56].

6.10.11 Heteronuclear multiple bond coherence

^{1}H and ^{13}C nuclei are correlated through multiple bonds in HMBC. Generally, delay time (variable) (1/ 2 J) is set for approximately 70 ms; it optimizes ^{3}J cross peaks which is prominent compared to those for ^{2}J. It is feasible to turn ^{2}J cross peaks more prominent, and even ^{4}J peaks can be made visible with a change in the value of delay time. Occasionally, direct C–H coupling takes place with a magnitude of JCH being the equal distance between cross peaks (the so-called ^{13}C satellite) in HMBC spectra [57]. HMBC data gives valuable information about the structure that is useful in the characterization of complex compounds. It plays an important role by connecting several fractions of a big molecule through carbon–proton coupling via multiple bonds and then confirms the molecule's structure. HMBC spectra show ^{1}H spectra on the x-axis and ^{13}C on the y-axis, and there are no spectra on the diagonal similar to HMQC [58].

6.11 Mass spectrometry for chemical compounds

Mass spectrometry gives information about the molecular weight of the unknown compound. Several ionization techniques available in MS including atmospheric pressure chemical ionization, electron impact, electrospray ionization, fast atom bombardment, field desorption/field ionization, and matrix-assisted laser desorption ionization. Molecular mass can be precisely measured with an accuracy of 5 ppm using a high-resolution electron impact mass spectrometer. The fragmentation patterns shown in the mass spectra give valuable information about the structural units and structure of the organic compound. Hyphenated techniques, viz. liquid chromatography–mass spectrometry (LC-MS) and gas chromatography mass spectrometry (GC-MS), are very popular for the identification of known compounds [59].

Electrons or laser bombardment is done in mass spectrometry and thereby converted to charged ions, which are highly energetic. The relative abundance of a fragmented ion against the mass/charge ratio of the ions is plotted to exhibit mass spectra. Mass spectrometry gives precise and accurate information on molecular mass (molecular weight) and exact molecular formula, it can be determined with high accuracy, and an exact molecular formula can be determined with knowledge of places where the molecule has been fragmented. In previous work, bioactive molecules from pith were isolated and purified by bioactivity-guided solvent extraction, column chromatography, and HPLC. UV-Visible, IR, NMR, and mass spectroscopy techniques were employed to characterize the structure of the bioactive molecule [60]. Furthermore, molecules may be hydrolyzed and their derivatives characterized. Mass spectrometry provides abundant information for the structural elucidation of the compounds when applied in tandem mass spectrometry (MS). Therefore, the combination of HPLC and MS facilitates rapid and accurate identification of chemical compounds in medicinal herbs, especially when a pure standard is unavailable. Recently, LC/MS has been extensively used to analyze phenolic compounds. Electrospray ionization (ESI) is a preferred source due to its high ionization efficiency for phenolic compounds [61].

6.12 Conclusion

Medicinal plants, either in the form of pure compounds or as standardized extracts, provide various opportunities for the investigation of the new drug because of the presence of unmatched availability of chemical constituent diversity. Plant-based products are a very important part of the human diet and a major source of biologically active substances, such as vitamins, fats, carbohydrates, dietary fiber, antioxidants, and various cholesterol-lowering compounds.

The main focus of this book chapter is on the use of analytical methodologies, which include the extraction, isolation, and characterization of bioactive ingredients in plant botanicals and useful herbal compounds. As plant extraction is the most important and initial step in analyzing chemical constituents present in botanicals and herbal preparations, the strengths and weaknesses of different extraction techniques are discussed in this investigation. Analyses of bioactive compounds present in the plant extracts involving the applications of common phytochemical screening assays, chromatographic techniques such as high-performance liquid chromatography and thin layer chromatography as well as non-chromatographic techniques such as immunoassay and Fourier transform infrared, UV-visible spectroscopy, nuclear magnetic resonance spectroscopy, and mass spectroscopy are discussed. The application of chromatography, mass spectrometry, infrared spectrometry, and NMR allowed quantitative and qualitative measurements of many phytochemical constituents. Besides the importance of the plant itself, such metabolites determine the nutritional quality of food, color, taste, smell, antioxidative, anticarcinogenic, antihypertension, anti-inflammatory, antimicrobial, immunostimulant, and cholesterol-lowering properties. Since bioactive compounds in plant material consist of various bioactive component mixtures, their separation and identification still create problems. Practically most of them have to be purified combining several chromatographic techniques and various other purification and isolation methods to isolate the bioactive compound(s).

References

[1]   R. J. P. Cannell. Natural Products Isolation. Human Press Inc, New Jersey, 1998, pp. 165–208.
[2]   H. O. Edeoga, D. E. Okwu, B. O. Mbaebie. Afr J Biotechnol. 2005, 4, 685–688.
[3]   G. V. Satyavati, M. K. Raina, M. Sharma. Med Plants India, Vol. 1. Indian Council of Medicinal Research, New Delhi, 1976, p. 377.
[4]   C. W. Huie. Anal Bioanal Chem. 2002, 373, 23–30.
[5]   D. M. Holstege, J. N. Seiber, F. D. Galey. J Agric Food Chem. 1995, 43, 691–699.
[6]   J. Parekh, S. V. Chanda. Turk J Biol. 2007, 31, 53–58.
[7]   M. Ikram, H. Inamual. Fitoterapia. 1980, 51, 231.
[8]   D. S. Fabricant, N. R. Farnsworth. Environ Health Perspect. 2001, 109, 69–75.
[9]   J. A. Duke. Handbook of Phytochemical Constituents of GRAS Herbs and Other Economic Plants. CRC Press, London, UK, 1992.
[10] I. Ahmad, A. Z. Beg. J Ethnopharmacol. 2001, 74, 113–123.
[11] A. R. Shahverdi, F. Abdolpour, H. R. Monsef-Esfahani, H. A. Farsam. J Chromatogr B. 2007, 850, 528–530.
[12] P. Raut, D. Bhosle, A. Janghel. J Pharm Technol. 2015, 8(6), 655–666.
[13] O. R. Alara, N. H. Abdurahman, C. I. Ukaegbu. J Appl Res Med Aromat Plants. 2018, 11, 12–17.
[14] S. Sasidharan, Y. Chen, D. Saravanan, K. M. Sundram, L. L. Yoga. Extraction, isolation and characterization of bioactive compounds from plants' extracts. Afr J Tradit Complement Altern Med. 2011, 8(1), 1–10. Epub 2010 Oct 2. PMID: 22238476; PMCID: PMC3218439.
[15] C. Arora Chugh and D. Bharti, Chemical characterization of antifungal constituents of Emblica officinalis, Allelopathy Journal, 2014, 34(1), 155–178.
[16] B. Zygmunt, J. Namiesnik. J Chromatogr Sci. 2003, 41, 109–116.

[17] W. Phrompittayarata, W. Putalunc, H. Tanakad, K. Jetiyanone, S. Wittaya-areekulf, K. Ingkaninana. Naresuan Univ J. 2007, 15(1), 29–34.
[18] G. S. Kumar, K. N. Jayaveera, C. K. A. Kumar, U. P. Sanjay, B. M. V. Swamy, D. V. K. Kumar. Trop J Pharm Res. 2007, 6, 717–723.
[19] C. A. Chugh, D. Bharti. Int J Phytomed. 2012, 4(2), 261–265.
[20] S. V. Chanda, J. Parekh, N. Karathia. J Biomed Res. 2006, 9, 53–56.
[21] P. B. Mallikharjuna, L. N. Rajanna, Y. N. Seetharam, G. K. Sharanabasappa. E-J Chem. 2007, 4, 510–518.
[22] D. N. Onwukaeme, T. B. Ikuegbvweha, C. C. Asonye. Trop J Pharm Res. 2007, 6, 725–730.
[23] D. Dahiru, J. A. Onubiyi, H. A. Umaru. Afr J Trad CAM. 2006, 3, 70–75.
[24] M. O. Hamburger, G. A. A. Cordell. J Nat Prod. 1987, 50, 19–22.
[25] H. B. Li, Y. Jiang, F. Chen. J Chromatogr B. 2004, 812, 277–290.
[26] X. H. Fan, Y. Y. Cheng, Z. L. Ye, R. C. Lin, Z. Z. Qian. Anal Chim Acta. 2006, 555, 217–224.
[27] A. L. Homans, A. Fuchs. J Chromatogr. 1970, 51, 327–329.
[28] M. Beccaria, D. Cabooter. Analys. 2020 Feb 17, 145(4), 1129–1157.
[29] M. S. Lee, E. H. Kerns. Mass Spectrom Rev. 1999 May–Aug, 18(3–4), 187–279.
[30] Y. Wang, X. Xue, Y. Xiao, F. Zhang, X. Qing, X. Liang. J Sep Sci. 2008 Jun, 31(10), 1669–1676.
[31] F. Blum. Br J Hosp Med (Lond). 2014 Feb, 75(2), C18–21.
[32] W. Vine, K. N. Olson, L. D. Bowers. Transplant Proc. 1988 Apr, 20(2 Suppl 2), 354–356.
[33] M. Yang, J. Sun, Z. Lu, G. Chen, S. Guan, X. Liu, B. Jiang, M. Ye, D. A. Guo. J Chromatogr A. 2009, 1216 (11), 2045–2062.
[34] J. Li, L. Ren, G. Sun, H. Huang. Sheng Wu Gong Cheng Xue Bao. 2013 Apr, 29(4), 434–446.
[35] S. Carneiro, R. Pereira, I. Rocha. Methods Mol Biol. 2014, 1152, 197–207.
[36] A. R. Abubakar, H. Mainul. J Pharm Bioallied Sci. 2020, 12(1), 1–10.
[37] D. Bharti, C. Arora, D. Arora . Research Journal of Chemistry and Environment, 2021, 25(10), 114–120
[38] R. Anibogwu, K. Jesus, S. Pradhan, S. Pashikanti, S. Mateen, K. Sharma. Molecules. 2021 Nov 19, 26 (22), 69–95.
[39] J. B. Zygmunt, J. Namiesnik. J Chromatogr Sci. 2003, 41, 109–116.
[40] K. M. Hazra, R. N. Roy, S. K. Sen, S. Laska. Afr J Biotechnol. 2007, 6(12), 1446–1449.
[41] J. Q. Brown, K. Vishwanath, G. M. Palmer, N. Ramanujam. Curr Opin Biotechnol. 2009 Feb, 20(1), 119–131.
[42] J. Tan, R. Li, Z. T. Jiang. Food Chem. 2015 Oct 1, 184, 30–36.
[43] J. Yu, Y. Zhang, S. Y. Pang, J. F. Wang. Guang Pu Xue Yu Guang Pu Fen Xi. 2016 Sep, 36(9), 2807–2811.
[44] Y. Han, S. Ling, Z. Qi, Z. Shao, X. Chen. Phys Chem Chem Phys. 2018 May 3, 20(17), 11643–11648.
[45] H. Duddeck, W. Dietrich. Structure Elucidation by Modern NMR. Springer Verlag, New York, 1989.
[46] R. J. Abraham, J. Fisher, P. Loftus. Introduction to the NMR Spectroscopy. John Willey & Sons Ltd, Chichester, New York, 1993, p. 144.
[47] D. H. Williams, I. Fleming. Spectroscopic Methods in Organic Chemistry, 5th ed. McGraw-Hill Book Company, London, 1995, p. 122.
[48] P. A. Mirau, F. A. Bovey. In: F. A. Bovey (Eds.), Two-Dimensional nuclear magnetic resonance spectroscopy, Nuclear Magnetic Resonance Spectroscopy, 2nd ed. Academic Press, Inc, San Diego, 1988, p. 325.
[49] A. E. Derome. The use of NMR spectroscopy in the structure determination of natural-products-two-dimensional methods. Nat Prod Rep. 1989, 6, 111–141.
[50] A. Bax. Broad-band homonuclear decoupling in heteronuclear shift correlation NMR-spectroscopy. J Mag Reson. 1983, 53, 517e20.
[51] A. Bax, M. F. Summers. J Am Chem Soc. 1986, 108, 2093–2094.
[52] P. Nolis, T. Parella. Magn Reson Chem. 2018 Oct, 56(10), 976–982.

[53] J. SaurI, E. Sistare, R. T. Williamson, G. E. Martin, T. Parella. J Magn Reson. 2015 Mar, 252, 170–175.

[54] T. Parella, J. F. Espinosa. Prog Nucl Magn Reson Spectrosc. 2013 Aug, 73, 17–55.

[55] A. Venkatesh, X. Luan, F. A. Perras, I. Hung, W. Huang, A. J. Rossini. Phys Chem Chem Phys. 2020 Sep 23, 22(36), 20815–20828.

[56] X. M. Kong, K. H. Sze, G. Zhu. J Biomol NMR. 1999 Jun, 14(2), 133–140.

[57] T. X. Lim, G. K. Pierens, P. V. Bernhardt, M. Ahamed, D. C. Reutens. Magn Reson Chem. 2021 Nov, 59 (11), 1154–1159.

[58] P. Sakhaii, B. Haase, W. Bermel. J Magn Reson. 2013 Mar, 228, 125–129.

[59] R. F. Kranenburg, F. A. M. G. van Geenen, G. Berden, J. Oomens, J. Martens, A. C. van Asten. Anal Chem. 2020 May 19, 92(10), 7282–7288.

[60] A. Zhang, H. Sun, X. Wang. Mass Spectrom Rev. 2018 May, 37(3), 307–320.

[61] M. Ye, J. Han, H. Chen, J. Zheng, D. Guo. J Am Soc Mass Spectrom. 2007, 18, 82–91.

Suparna Paul, Subhajit Mukherjee and Priyabrata Banerjee*

Chapter 7
Medicinal bioactivity: anti-cancerous and anti-HIV activity of medicinal plants

Abstract: Cancer is globally recognized to be a major public health disease that severely affects the human population in the developed countries along with the developing countries. It is considered to be the second main reason for morbidity and increasing mortality rate after cardiovascular ailments. Despite much room of advancement toward in-depth understanding of this life-threatening ailment along with the proposition of varying methods for initial stage diagnosis, a growing range of drugs, and treatment modalities, it is anticipated that the subsequent decade will globally give rise to 20 million new cases per year. In this deliberation there arises a surging exigency for new and unprecedented remedial measures aimed at treatment and prevention of this carcinogenic disease. Acquired immune deficiency syndrome (AIDS) is another viral and lethal disease which is primarily triggered by human immunodeficiency virus (HIV). The primary aim of this disease is allied with suppression of immune system, thereby making the body inert to external impetuses. The observatory data obtained from global health unveils that since epidemic over 78 million people were under the clutches of HIV leading to the death of almost 39 million people worldwide. At present to combat against HIV, antiretroviral therapy (ART) has been extensively explored. However, ART is allied with

Acknowledgment: PB is thankful to Department of Higher Education, Science and Technology and Biotechnology, Govt. of West Bengal, India, for providing financial assistance (sanction order no. 78 (Sanc.)/ST/P/S&T/6G-1/2018 dated January 31, 2019, Project No: GAP-2,25612). SP gratefully acknowledges DST INSPIRE for her fellowship [IF1,60302].

Note: Suparna Paul and Subhajit Mukherjee Equally Contributed.

*Corresponding author: Priyabrata Banerjee, Surface Engineering and Tribology Group, CSIR-Central Mechanical Engineering Research Institute (CMERI), Mahatma Gandhi Avenue, Durgapur 713209, West Bengal, India; Academy of Scientific and Innovative Research at CSIR-Central Mechanical Engineering Research Institute (CMERI), Mahatma Gandhi Avenue, Durgapur 713209, West Bengal, India. e-mail: pr_banerjee@cmeri.res.in, priyabrata_banerjee@yahoo.co.in, webpage: www.cmeri.res.in and www.priyabratabanerjee.in
Suparna Paul, Surface Engineering and Tribology Group, CSIR-Central Mechanical Engineering Research Institute (CMERI), Mahatma Gandhi Avenue, Durgapur 713209, West Bengal, India; Academy of Scientific and Innovative Research at CSIR-Central Mechanical Engineering Research Institute (CMERI), Mahatma Gandhi Avenue, Durgapur 713209, West Bengal, India; Department of Chemistry, Seacom Skills University, Bolpur, Kendradangal 731236, Birbhum, West Bengal, India
Subhajit Mukherjee, Department of Chemistry, Seacom Skills University, Bolpur, Kendradangal 731236, Birbhum, West Bengal, India

https://doi.org/10.1515/9783110791891-007

fatal complexities viz. lipodystrophy which curtails its real-day applicability. Owing to the present challenges of anticancer drugs and antiretroviral therapies, scientific and research interest is seeking immense magnetism toward advanced and proficient naturally derived compounds owing to their reduced noxious side effects in comparison to the contemporary treatment methodologies. In the current scenario this has instigated innovative vistas in search of natural anticancer and antiretroviral drugs that are derived from plants since plant-derived products serve to be a reservoir underlying the discovery of new clinical medicines. This chapter systematically consolidates the efficacious nature of the plants or their synthetic derivatives that give rise to naturally producing secondary metabolites, thereby enabling it to display clinical activity. The commercial contemplations and future prospects for treatment of cancer and HIV by employing phytoremediation approach are also well summarized herein.

7.1 Introduction

Cancer can be defined as a combination of diseases having mainly three unique characteristics: uncontrolled cell division, incursion of these defective cells into adjoining tissues, and its distribution to vital body parts through blood and lymph vessels [1]. Thus, the cells of those vital organs in human body continuously divide in an uncontrolled fashion developing malignant tumors with the ability to cause metastatic [2]. Genetic alteration leading to cancer can occur in human body as an outcome of faulty cell division or DNA brake down initiated by certain environmental exposure and carcinogen. The genetic changes that are responsible for cancer mainly affect three categories of genes: viz. proto-oncogene, tumor suppressor gene, and DNA repair gene. Nowadays, cancer becomes one of the major human death reasons, especially in Asia, Africa, and America. Modern medicinal area introduces three ways of treatment for cancer patients such as (a) surgery to remove the affected area from the body, (b) radiation therapy to destroy the malignant cells subjecting to radiation, and (c) chemotherapy to treat the patient by respective doses of medicine. Sometimes combined strategy is also beneficial to get a good result.

7.2 Conventional anticancer drugs and their side effects

Chemotherapy is frequently used to treat cancer where different conventional drugs like Cisplatin, Cyclophosphamide, Cytarabine, Methotrexate, 5-fluorouracil, Doxorubicin, Daunorubicin, and Bleomycinare are applied for different types of cancer treatment. Unfortunately, these conventional chemotherapeutic agents are not devoid of toxicity and they are responsible for savior side effects as they also hamper the growth of the normal or nonmalignant cells and even kill them. Among these adverse effects, bone marrow function inhibition, alopecia, nausea, vomiting, renal impairment, and ototoxicity are commonly found [3, 4]. So there arises a pressing concern to develop alternative medicine in cancer treatment. Recent studies are going on to develop the new anticancer agents with better efficacy with reduced toxicity.

Apart from the cancer-induced life threat, WHO declares HIV (HIV) as a major globally encountered public health concern that have been estimated to infect approximately 75 million people, amongst which 37 million people are till date surviving with this infectious ailment [5, 6]. HIV is a recognized retrovirus which can plausibly integrate its DNA within the host genome. The virus practically makes an entry into the host cell and thereby affects the immune system primarily T lymphocytes, monocytes, and macrophages along with dendritic cells. Its genetic material includes RNA which is composed of nine genes that includes all the requisite instructions for making new viruses. Three of these genes – gag, pol, and env – that deliver the directions to make proteins will ultimately give rise to new viruses. The other six genes rev, nef, vif, vpr, and vpu offer code to make proteins that essentially regulate the potentiality of HIV for infecting a cell thereby producing new replicas of virus or release those viruses from the infected cells. The HIV-1 tends to bind with chemokine receptor 5 or CXC chemokine receptor 4 by interaction with the envelope of protein for gaining an entry within the host cell [7]. The patients who are infected with AIDS have to encounter enormous socioeconomic challenges for obtaining appropriate treatment.

This instigates a surging exigency toward efficacious diagnosis, prevention as well as treatment of this disease such that this antiviral infection becomes a controllable chronic health disorder, thereby empowering the infected people to thrive long and healthy lives [8].

In this relevance global health community has prioritized the unprecedented development of new, innocuous, cost-effective anti-HIV agents for preventing and controlling this viral pandemic which has become the urge of the present scenario [9]. Prompt and appropriate treatment of the diseased people with antiviral medication can alleviate the severe effect, cure the disease, and thereafter minimize the outbreak [10]. Till date conventional antiretroviral therapy (ART) is available that provides a combined drug-induced treatment for combating of HIV infection. These antiretroviral drugs primarily inhibit the various stages of this viral infection that includes viral entry inhibitors (Fostemsavir, i.e., entry inhibitor via gp120, PRO140 (CCR5 monoclonal antibody), reverse transcriptase (RT) inhibitors (abacavir with lamivudine or tenofovir disoproxil fumarate with emtricitabine), protease inhibitors (darunavir or atazanavir), and integrase strand transfer inhibitors (dolutegravir, elvitegravir, or raltegravir) [11].

7.3 Side effects of traditional anti-HIV drugs

Nevertheless, the majority of the contemporary antiviral drugs encounter severe challenges including adverse reactions and toxic side effects like liver diseases, drug resistance along with poor compatibility, and bioavailability which curtails the real-day applicability of antiretroviral drugs [12].

7.4 Advantages of phytomedicinal agents over synthetic drugs

In this consequence the discovery and use of proficient naturally derived compounds as herbal medicines have become an increasingly popular safer alternative to the recently explored treatment methodologies [13] against both HIV and cancer. Natural products from plants give rise to innumerable 2° metabolites namely derivatives of phenol, glycosides, alkaloids, coumarins, terpenoids, and essential oils along with peptides that are endowed with phenomenal biological activities. Amongst these a few exhibits' crucial role in enhancing the immune system, displaying antiviral ability, that includes viral infections allied with the genetic variabilities of HIV that includes Type 1 (HIV-1) and Type 2 (HIV-2). Likewise many natural products like antioxidants and phytochemicals are recently known to have anticancer activity and this is mostly come from their antiproliferative and proapoptotic characteristics [14, 15]. Most importantly these agents originated from the natural sources show minimal or sometimes zero toxicity. Different naturally occurring organic molecules such as flavonoids, terpenes, alkaloids, glycosides, and other secondary metabolites are found to cause selective inhibition to malignant cell proliferation causing ultimate cell death [16, 17].

As a consequence, the medicinal plant-derived extracts and its related products have been extensively explored for the treatment of a wide range of infectious disease including viral infection like HIV as well as disease like cancer caused by genetic malfunction. Most significantly the purified and isolated phytoconstituents serve as an excellent starting precursor for discovery of new unprecedented antiviral drugs [18–19]. Similarly, in continuation for the quest of new anticancer agents, along with the isolated compounds obtained from plant extracts play a crucial role in developing safer and effective chemotherapeutics.

7.5 Plants with anti-cancerous activity

7.5.1 Sugar apple (*Annona squamosa*)

Sugar apple (scientific name: *Annona squamosa*) [Fig. 7.1(a)] is a common plant in India, Bangladesh, South America, and Caribbean islands; however at present it is well cultivated all over the world due to its good taste. The seeds of it contain some toxic alkaloids and thus the corresponding seed oil is being used to treat lice attack in hair [20]. Different acetogenins, terpenes, terpenoids, and alkaloids present in sugar apple have been recognized to have various medicinal impacts over gastrointestinal disorder, improper menstruation, urinary tract infections, and cancer [21, 22]. The fruit, mainly its seed oil, has exhibited antiproliferative activity in combating a broad range of human cancerous cell line. Before testing over human cancer cell line, a good antitumor activity was recorded on hepatocellular carcinoma cell line H22 of mice [23–26]. Anticancer activities of the seed oil are mostly due the presence of annonaceous acetogenins [Fig. 7.1(b)].

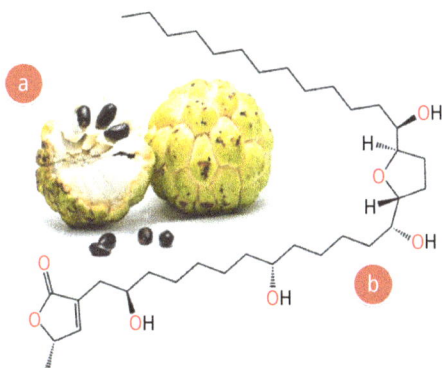

Fig. 7.1: (a) Image of sugar apple and its seeds; (b) chemical structure of acetogenin extracted from its seed oil.

Other part of the plant like pericarp, leaves, and stem bark also contains annonaceous acetogenins as well as various terpenes, and terpenoids, and alkaloids which are also active chemicals against cancer [24–26]. For example, acetogenin squamoxinone-D was discovered to have antiproliferative activity against drug-resistant cell-line SMMC 7,721/T [27]. The study of cell death mechanism revels the cell-cycle arrest and apoptotic approach where the anticancer agents like annonaceous acetogenins activate caspase-3, thereby downregulating the antiapoptotic gens and preventing the fragmentation of DNA [24, 28] of cancer cell lines following the exposure of sugar apple seed oil. This phenomenon of cell death has been accelerated by activating different reactive oxygen species (ROS) and reducing intracellular glutathione species [24, 29]. Additionally, the reduced lipid peroxidation (LPO) level in an animal model of oral carcinoma treated with aqueous-ethanolic extracts of sugar apple stem bark [30] indicates its antiproliferative activity over oral cancer.

7.5.2 Ginger (*Zingiber officinale*)

Ginger (scientific name: *Zingiber officinale*), particularly its rhizome, is widely utilized as traditional kitchen spice all over the globe to induce fragrance and spicy taste in delicious cuisines. The fragrance of the volatile ginger rhizome oil is owing to the existence of gingerol compounds. These isolated compounds being thermally unstable undergo dehydration during cooking to form shogaols (another class of organic compound) which is accountable for pungent odor and taste of ginger [31]. The traditional pharmacological characteristics of ginger mostly come from these two classes of compounds gingerols and shogaols (Fig. 7.2) [32, 33].

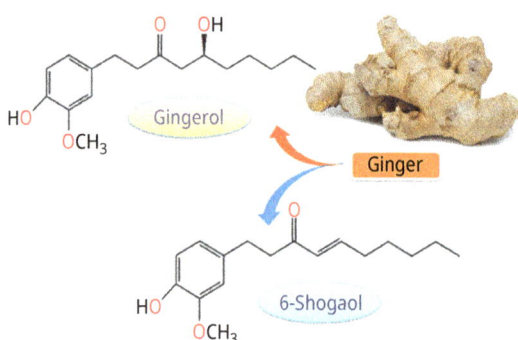

Fig. 7.2: Ginger and the chemical structure of gingerol and 6-shogaol, the two major chemical components extracted from it.

The chemopreventive and anticancer activity of gingerols and shogaols was recognized after the in vitro detection of inhibitory activity against proliferation of cell as well as cell-cycle arrest for unlike human carcinogenic cell lines and in vivo assay over animal tumor cells. This anticancer activity in case both the compounds comes through different mechanistic pathways such as inhibition of attack by the activating role of nuclear receptor peroxisome proliferator-activated receptor-γ (PPAR-γ) [34], downregulating of matrix metalloproteinase 9 transcription [35]; inhibition of tumor angiogenesis [36, 37]; inactivation of elements responsible for regulation of aberrant cell cycle [38, 39]; and interfering activity with microtubule integrity [40, 41].

7.5.3 Garlic (*Allium sativum*)

Garlic, scientifically recognized as *Allium sativum*, is a common plant throughout the world and has a versatile medicinal application like antimicrobial, antithrombotic, hypolipidemic, antiarthritic, and hypoglycemic agents over thousands of years. Apart from these activities, garlic possesses good antitumor activity when different research groups have continued the application of garlic oil, fresh extract of garlic, matured garlic, and a substantial number of organosulfur compounds (OSCs). These OSCs S-allylcysteine and S-allylmercapto-L-cysteine (Fig. 7.3) are mainly responsible for its chemopreventive property. Garlic juice enriched with these organosulfur constituents displayed in vitro activity against many carcinomas including human lung (A549 cell line) [42], breast (MDA-MB-231) [43], colon (SW-480 and HT-29) [44], and leukemia (HL-60) [45]. Sallylcysteine and S-allylmercapto-L-cysteine in aged garlic extracts were recognized to have a very good radical scavenging activity which is an important phenomenon for leading cancer cell death [46, 47]. Further study over different animal models shows that the tumor growth has been prevented by S-allylcysteine, another OSC found in garlic [48]. So, altogether, it is very clear that the regular ingestion of garlic may plausibly offer some sort of defense from developing of malignant tumor in human body.

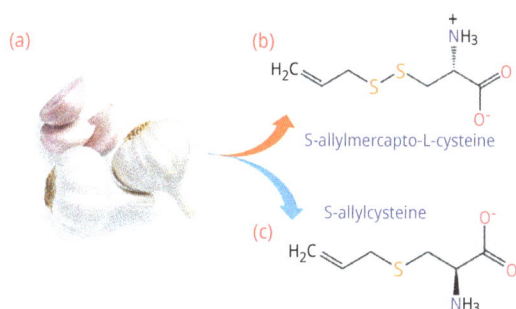

Fig. 7.3: (a) Garlic (*Allium sativum*) and the compounds, (b) S-allylmercapto-L-cysteine, and (c) S-allylcysteine responsible for its anticancer activity.

7.5.4 Turmeric (*Curcuma longa*)

The rhizome of turmeric plat (*Curcuma longa*) has a massive utility in cooking to bring color and taste in food. Anticancer activity of turmeric is mainly generated from curcumin (diferuloylmethane) (Fig. 7.4), a polyphenol compound present in its rhizome [49]. Curcumin was found to be active over different cancer cells due to its potential to undergo downregulation of expression allied with numeral genes, namely NF-kappa B, activator protein 1, epidermal growth receptor 1, cycloxygenase 2, lysyl oxidase, nitric oxide synthase, and matrix metallopeptidase 9 (MMP-9), along with tumor necrosis factor [50–52]. *C. longa* suppresses the expression of different chemokines, cyclins, cell surface adhesive molecules, and various receptors of growth factor like epidermal growth factor receptor (EGFR) as well as human epidermal growth factor receptor 2 (HER2) [50]. Moreover, in vitro analysis showed that turmeric is able to reduce the activity of MMP-2 and HEp2 carcinoma cell inhibition [53]. Curcumin exhibits apoptotic effect on cancer cells which originates from the inhibition of different protein activity [54]. It basically initiates the reduction in membrane potential of mitochondria thereby releasing cytochrome C. Thus the protein caspase 3 and 9 have been activated which leads to the ultimate cell death through apoptotic pathway [55]. Curcumin was found to bring apoptosis over prostate cancer cell-line LNCaP by activating apoptosis-inducing ligand (TRAIL), accelerating the breakdown of pro-caspases 3, 8, and 9 along with the release of cytochrome c [56]. Moreover, besides these in vitro antitumor activities, curcumin has been proved to have enough potential to prevent gastric and colon carcinoma in mice. These results as a whole suggest that turmeric has a good antiproliferative activity on different cancer cells.

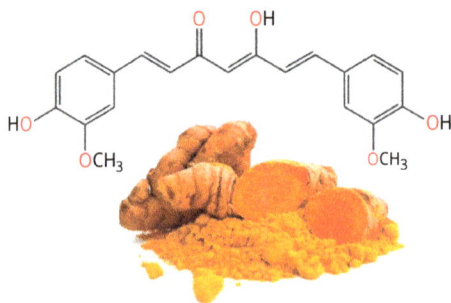

Fig. 7.4: Images of *Curcuma longa* and structural representation of its phytochemical constituent.

7.5.5 Asiatic pennywort (*Centella asiatica*)

Centella asiatica, normally identified as Asiatic pennywort or gotu kola in English [Fig. 7.5(a)], is a therapeutic plant which is native to Southeast Asian countries including India, Bangladesh, Sri Lanka, China, Indonesia, and Malaysia. In Bangladesh, east India especially in West Bengal it is also known as Thankuni. The plants, especially its leaves, are being used as herbal remedies for different diseases since ages. The partly purified extract of *Centella asiatica* has shown its capability of inhibiting the tumor cells development, in particular, Ehrlich ascites along with Dalton's lymphoma ascites tumor cells [57]. Most interestingly, the negative impact of the extract was encountered over normal human lymphocytes [57]. In vivo study over mice demonstrated that the oral ingestion of Asiatic pennywort plant extract increases the survival rate of mice with induced carcinoma. On long term in vitro culture of mouse lung fibroblast cells with the partly purified extract in different concentration also reveals that the extract plant has potential to prevent the cancer cells proliferation. The chemical composition of the extract was found to be the mixture of various components including asiaticoside, hydrocotyline, vallerine, pectic acid, sterol, stigmasterol, flavonoids, thankunosides [58], and ascorbic acid [59]. Among these, asiaticoside [Fig. 7.5(b)] is the major constituent in the extract which is mainly responsible for its antitumor property [58]. *Centella asiatica* extract reduces the LPO production in different human organs including liver, lung, brain, heart, kidney, and spleen [60]. It also showed its free radical

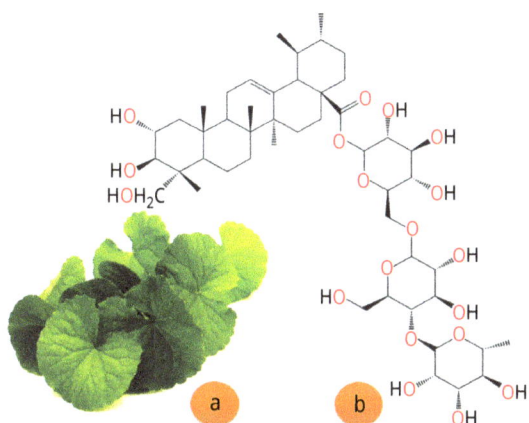

Fig. 7.5: (a) Images of Asiatic pennywort (*Centella asiatica*); (b) structure of Asiaticoside, the major organic component responsible for antitumor properties of Asiatic pennywort.

scavenging ability in different in vitro studies [67]. The mechanistic investigation for *Centella asiatica*-induced cancer cell damage reveals that the proliferative activity comes from the direct inhibition of DNA synthesis [57].

7.5.6 Creat (*Andrographis paniculata*)

Andrographis paniculata, frequently recognized as creat or green chiretta, is a yearly herbaceous plant belonging to the Acanthaceae family and is native to India, Bangladesh, and Sri Lanka. In India and Bangladesh, it is well acclaimed as Kalmegh by the common people. The roots and leaves are well known to have the versatile medicinal utility for the treatment of jaundice and cholestasis and also serve as an antidote for hepatotoxins which mainly comes from its major component diterpene andrographolide (Fig. 7.6) [61]. This diterpene lactone andrographolide crystallizes as a colorless solid and possesses a ring-like structure and extensively bitter in taste. The maximum accumulation of andrographolide (~2.25%) is present in the leaves of *Andrographis paniculata* [61]. Andrographolide exhibited activity against a wide range of carcinoma, such as KB human epidermoid cancer cells, P388 lymphocytic leukemia cells, MCF-7 breast cancer cells, HCT-116, and HT-29 colon cancer cells [62]. It is also found to accelerate the rapid development and multiplication of human peripheral blood lymphocytes and exerts prodifferentiative impact on the myeloid leukemia M1 cell line of mouse [63, 64]. The application of andrographolide over laboratory mice reveals that it stimulates the immune system of the mice under investigation and thus activates immune responses in both antigen-specific and nonspecific way [65]. This makes *A. paniculata* potent chemoprotective agent. Parallel to this, extract of *A. paniculata* in alcohol was found to bring a substantial enhancement in the activity of glutathione-*S*-transferase (GST), DT-diaphorase

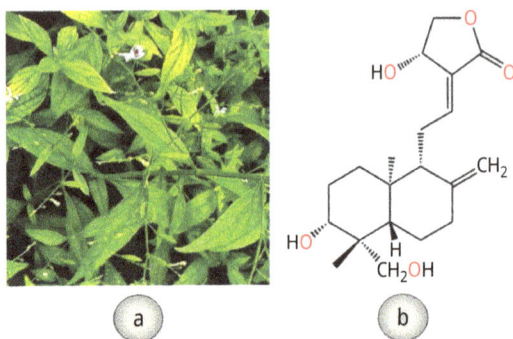

Fig. 7.6: (a) Creat (*Andrographis paniculata*) and (b) the anticancer agentandrographolide extracted from it.

(DTD), superoxide dismutase (SOD), and catalase, differentially within the vital organs including lung, liver, kidney, and forestomach [66]. The in vivo treatment of andrographolide over immunocompetent Swiss albino mice showed that the isolated compound appreciably prevents the cancer cell proliferation with almost zero toxicity even at relatively high doses in this mice model [67].

7.5.7 Cat's claw (*Uncaria tomentosa*)

Uncaria tomentosa is frequently acclaimed as cat's claw and is a native plant in South and Central America such as Paraguay, Brazil, Bolivia, Peru, Ecuador, Colombia, Venezuela, and the Guyanas. The leaves and stem bark extracts are being used traditionally for the treatment of osteoarthritis and rheumatoid arthritis [68–70]. This medicinal activity comes from oxindole which is a pentacyclic alkaloid compound [70]. The anticancer activity of oxindole, extracted from *U. tomentosa* on different human cancer cell line and a mouse model, had been detected. This pentacyclic alkaloid induces cancer cell killing through apoptotic pathway [71]. On clinical trial, more interesting results were encountered in the patients with advanced solid tumors treated with the dried extract of this cat's claw plant which showed better social functioning, increase in body weight, and complete advancement in their life quality [72]. But in this study, no considerable biochemical and inflammatory responses were encountered. In a separate clinical trial, it showed lower toxicity in comparison to the anticancer drug 5-fluorouracil-doxorubicin-cyclophosphamide combination used in fighting human breast cancer [73]. This low toxicity of oxindole originates from its reduced chances of causing neutropenia compared to this chemotherapeutic combination. In another human trial an oral ingestion of ethanolic extract of stem back in the tablet form caused almost no adverse effect while a massive side effect was encountered for 5-fluorouracil-oxaliplatin administration in case of colorectal cancer patients [74]. All these results focus on the ability of

Uncaria tomentosa extract to have anticancer activity with reduced toxicity which is the main aim for cancer research to develop a drug with reduced adverse effects.

7.5.8 Apple (*Malus* sp., Rosaceae)

Apple (*Malus* sp., Rosaceae) is a common fruit consumed by human worldwide. The high food value of apple is on account of the existence of different polyphenols and dietary phytochemicals like various flavonoids and other phytochemicals [75]. Various advantageous impacts of apple polyphenols on human health such as antiproliferative, apoptotic, and antioxidative effects have been revealed through different modern studies. Parallel to its antioxidant behavior, polyphenols present in apple affect significantly to the signaling pathways, controlling cell survival, growth, and proliferation [76, 77]. Treatment of phloretin, a polyphenol indicated dose-dependent growth inhibition on different lung cancer cell lines (A549, Calu-1, H838, and H520) and induced cell death through apoptosis. The therapeutic efficacy of phloretin (10 mg/kg body weight) was proved when it had been jointly administered with some clinically approved anticancer drug namely paclitaxel (1 mg/kg body weight) in the body of SCID mouse model [76].

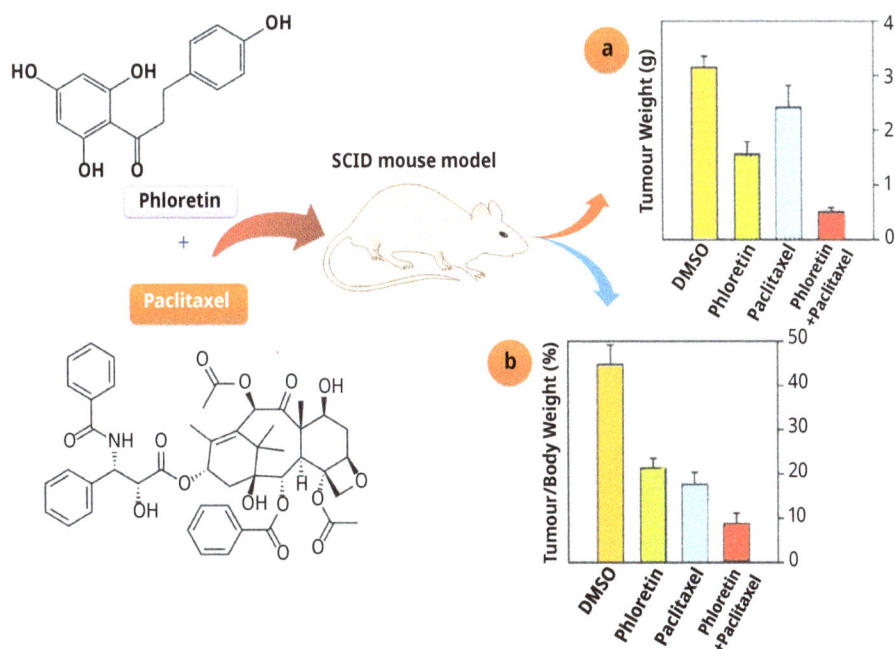

Fig. 7.7: Phloretin treatment in association with the anticancer drug paclitaxel in a human Hep G2-xenografted tumor-bearing SCID mouse: (a) the average tumor weight in gram and (b) tumor/body weight ratio in % after the treatment with negative control DMSO, phloretin, paclitaxel, and (phloretin + paclitaxel) (Ref [76]).

Another study [78] showed that phloretin has antiproliferation effect on human colorectal cancer and liver cancer cells via impeding of GLUT2. The in vitro treatment of Pelingo apple juice over MCF-7 and MDA-MB-231 indicates the antiproliferative efficacy on these human breast cancer cells [79]. 3-β-*trans*-cin-namoyloxy-2α-hydroxy-urs-12-en-28-oic acid, present in apple peels showcased in vivo antitumor activity against mammary tumor on mice. Interestingly, no loss of body weight and mortality was encountered for the treated mice [80]. Moreover, present report says that ingestion of minimum an apple a day will curtail the chances of colorectal cancer in human.

7.5.9 Indian gooseberry (*Phyllanthus emblica*)

Phyllanthus emblica, frequently known as Indian gooseberry, is a well-recognized Indian medicinal plant whose fruit is a good vitamin C source and often used as potential antioxidant and found to be beneficial over heart health [81]. Apart from these, both in vitro as well as in vivo experimentations revealed anticancer property of this fruit extract against different carcinoma like breast, colon, ovarian, and prostate [82, 83]. The activity mostly comes from different polyphenols (most importantly tannins and flavonoids) present in the extract. The aqueous fruit extract (50–100 µg/mL) brought significant cell death over the breast cancer cell lines whilst induced no significant changes in normal breast cancer cells (Fig. 7.8) [81]. The mechanistic study of cell death induced by the fruit extract showed apoptotic death pathway for these human breast carcinomas. *Phyllanthus emblica* aqueous extract was also found to be active over cervical and ovarian carcinoma [84, 85]. In vivo investigation over ovarian cancer xenograft model in mice [86] reveals that the oral intake of Indian gooseberry by the mice dissipates the ovarian tumor growth. Thereby *Phyllanthus emblica* or its aqueous extract can be used as therapeutic agents over these different types of cancer with zero or very less nontransformed or normal cell toxicity.

7.5.10 *Ziziphus jujube* (Jajube)

Ziziphus jujuba or simply jujube known in English is a plant which is a member of Rhamnaceae family. It is mostly abundant in subtropical regions of Asia and America [87]. Jujube fruit contains different bioactive species such as triterpenic acids, flavonoids, cerebrosides, phenolic acids, α-tocopherol, and β-carotene as well as polysaccharides which altogether make its food value high and beneficial for human health [88]. Guo et al. in the year 2009 [89] have identified 10 varieties of triterpenic acids in the dried jujube fruit among which three triterpenic acids, namely betulinic acid (BA), oleanolic acid (OA), and ursolic acid (UA), are observed to have antiproliferative activity over different malignant cells. The structural representation of these compounds

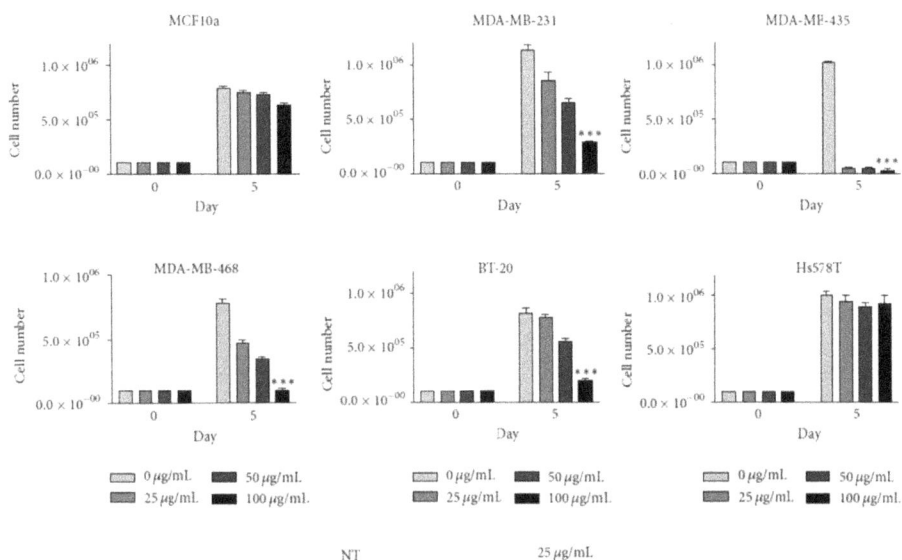

(a)

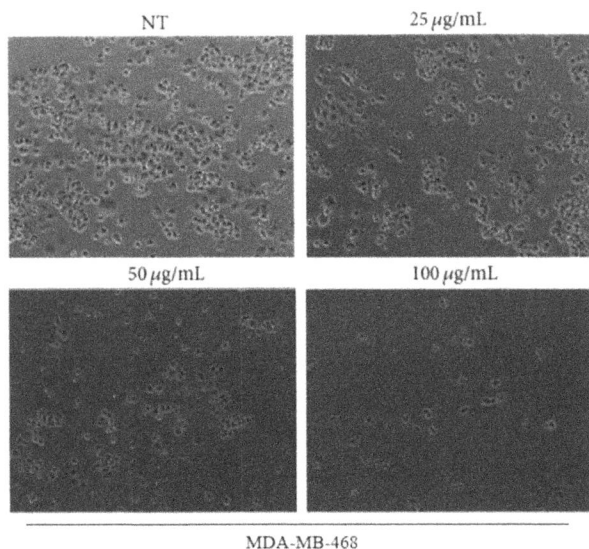

(b)

Fig. 7.8: Therapeutic effect of *Phyllanthus emblica* (Indian Gooseberry) extract against untransformed mammary epithelial cells and triple-negative breast cancer cells upon five-day exposure period: (a) cell count for the gooseberry extract treated (0–100 µg/mL) untransformed mammary epithelial cells (MCF10A) and the other breast cancer cell lines (MDA-MB-231, MDA-MB-468, and MDA-MB-435); (b) five days viability of MDA-MB-468 cells posttreatment with (0–100 µg/mL) gooseberry extract (Ref. [81]).

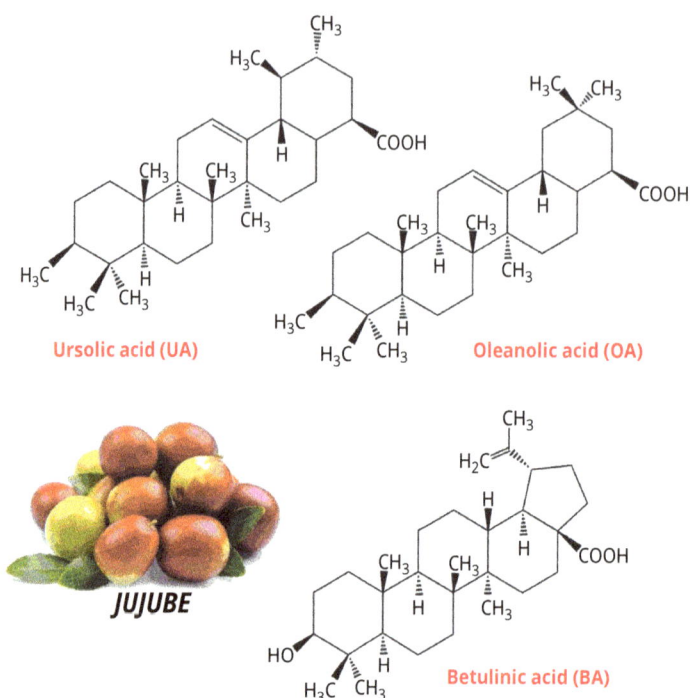

Fig. 7.9: Image of jujube (*Ziziphus jujube*) and chemical structure of its constituents, ursolic acid (UA), oleanolic acid (OA), and betulinic acid (BA).

is shown in Fig 10. An in vitro study over Hep G2 cell line indicates antiproliferative activity of UA on this liver carcinoma [90].

The in vitro as well as in vivo investigation on female ovariectomized C57BL/6 mice had also been carried out separately where the dose-dependent and time-dependent treatment of UA were investigated. UA inhibited WA4 cell proliferation at concentration of 25 and 50 μm that showed its antitumor activity on mouse model of postmenopausal breast cancer [91]. OA is also an isomer of UA where a methyl group has its different position. An investigation over colon carcinoma cell line HC-T15 revealed notable antitumor effects of these two triterpene acids UA and OA [92]. BA was also found to show its anticancer activity on human melanoma cell lines while very less cytotoxic effect of it had been encountered for normal human melanocytes [93]. BA is also active over a long range of cancer such as lung, leukemia, cervical, colon, prostate, head, and neck [94]. BA induces apoptosis via mitochondrial pathway leading to cancer cell death.

Apart from these phytochemicals, a list of other medicinal bioplants with proven anticancer activity has tabulated in Tab. 7.1.

Tab. 7.1: List of plants and the phytoconstituents traditionally used as anticancer agents.

Sl. no.	Botanical name of the plant	Common name (English)	Family	Part of the plant	Active constituents	Reference(s)
1.	*Allium cepa*	Onion	Alliaceae	Bulb	Allicin alliin, allyldisulphide, quarcetin, flavonoids, and vitamin C and E	[95]
2.	*Aloe ferox, Aloe barbadenis*	Indian aloe	Liliaceae	Leaves	Aloe-emodin, emodin, aloin, acemannan	[96]
3.	*Aphanamixis polystachya*	Rohituka tree	Meliaceae	Seeds, bark, leaves	Amooranin	[95]
4.	*Butea monosperma*	Palash	Fabaceae	Bark	Butin, isobutin, coreoopsin, butrin, and palasitrin	[97–100]
5.	*Calotropis gigantean*	Madar	Apocynaceae	Whole plant	Calotropnapthalene, calotropises fuiterpinol	[101]
6.	*Cassia auriculata*	Tarwar	Caesalpinaceae	Root, bark, and leaves	Avaraol, avaraoside	[102]
7.	*Cleistanthus collinus*	Karra	Euphorbiaceae	Leaves	Cleistanthin, collinusin	[103]
8.	*Colchicum luteum*	Suranjan	Liliaceae	Leaves	Colchicines, demecolcine	[104]
9.	*Dysoxylum binectariferum*	Rosewoods	Meliaceae	Barks	Rohitukine	[105]
10.	*Glycine max*	Soyabean	Leguminosae	Oil, seeds, and flower	Genistein and daidzein	[106]

7.6 Plants with anti-HIV activity

7.6.1 Bitter leaf (*Vernonia amygdalina*)

Vernonia amygdalina is a bush which primarily grows in tropical Africa and its neighboring places. It is a traditionally used phytomedicine for treating HIV/AIDS. It belongs to the Asteraceae family and its trivial name is derived from its bitter taste. Literature reports that bitter leaf possesses nutritive, health-improving properties, and most significantly, it demonstrated superior antioxidant activity against HIV in comparison with market accessible tablets like Immunace™. The extract of bitter leaf is testified to display anti-HIV activity on HIV-affected patients undergoing antiretroviral treatment. An investigation was performed by using aqueous leaf extracts of this shrub in association with ART for assessing its effect on the cell count of CD4$^+$ cells for an extended time span of four months. It is to be noted that the cell count of CD4 is reliant on WBC count. In preliminary phase of HIV, cell count of CD4 is tremendously reduced, such that the patient is incapable of competing with contagions. Interestingly the outcome of this analysis revealed a phenomenal growth in cell count of WBCs in case of patients using bitter leaf extract or vitamin supplements or both. Impressively, skin rashes of patients were also recovered. Owing to the immunological effect of this shrub on HIV-affected patients, it is extensively used in HIV management. Its fresh leaves are also useful for treating chronic fever, headache, and pain in the joints particularly in AIDS patients. Additionally, it serves as a nutrition-containing supplement [107, 108].

7.6.2 Stone apple (*Aegle marmelos*)

The *Aegle marmelos* plant, generally known as "bilwa" or "bael," is naturalized in various countries including Burma, Pakistan, Bangladesh, Sri Lanka, Thailand, and South-eastern Asian parts wherein it is recognized for its significant and indigenous system of Homeopathic and Ayurvedic medicinal properties [109]. Most importantly all parts of this plant, viz., leaves, root, bark, seeds, and fruits, are eatable and explored for several Ayurvedic preparations. The methanol extract of *A. marmelos* fruits primarily comprises coumarins and furanocoumarins (imperatorin (1) and xanthotoxol) as well marmeline alkaloid. Imperatorin is a major furanocoumarin that demonstrated 60% anti-HIV efficacy at barely 8 µg/mL concentration. Whereas other structurally analogous furanocoumarin-based compounds xanthotoxol (2) and xanthotoxin (3) displayed feeble activity at elevated concentration of 20 µg/mL owing to the absence of prenyl group within their molecular scaffold. Other coumarin compounds isolated from *A. marmelos*, umbelliferone (4) did not exhibit any promising inhibition efficiency even at concentration of 20 µg/mL. Whilst, aurapten (5) comprising geranyl group presented feeble anti-HIV efficacy (Fig. 7.10). These observations clearly validate that the presence of prenyl moiety and furanocou-marin backbone is a prerequisite for showcasing of anti-HIV property. Further in-vestigation revealed that imperatorin (1) could readily hinder the replication of HIV in H9 lymphocytes possessing EC50 values lower than 0.10 μg/mL and approximately greater than 1,000 as therapeutic index (TI) [110]. The mechanistic pathway of activ-ity of imperatorin disclosed that (1) neither hindered the phenomenon of reverse transcription (RT) nor the steps involved in integration within the viral cell cycle. Nevertheless, 1 intensely obstructed the cyclin D1 expression and underwent G1 phase cell-cycle arrest, that is, 1 impeded replication of HIV-1 via Sp1-reliant path-way [111]. Alkaloid and marmeline (6) did not exhibit anti-HIV activity even at its concentration of noncytotoxicity.

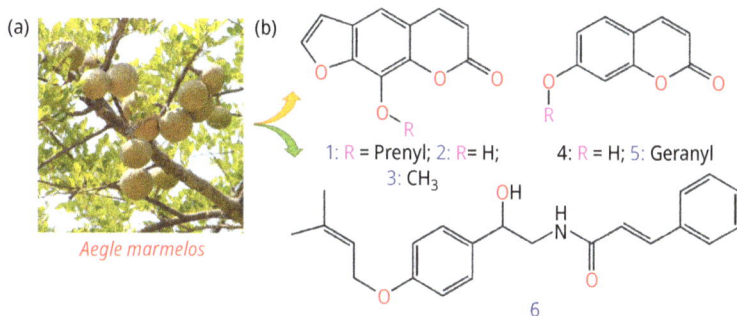

Fig. 7.10: (a) Image of stone apple (*Aegle marmelos*); chemical structure of coumarins and furanocoumarins extracted from *A. marmelos*; (b) 1: imperatorin, 2: xanthotoxol, 3: xanthotoxin, 4: umbelliferone, 5: auraptene, 6: marmeline.

7.6.3 Common madder or Indian madder (*Rubia cordifolia*)

Rubia cordifolia is a flowering plant belongs to the coffee family, Rubiaceae and is normally originated in India and China. In particular its roots and rhizomes are ex-tensively utilized as a natural remedy. The extracts of *R. cordifolia* have demonstrated diverse biological activities including antiviral property. The plant extracts primarily comprise several naphthoquinones and anthraquinones [112]. Literature reports that ethyl acetate extract (15 μg/mL) of its roots were found to be active against HIV-1. Xan-thopurpurin (20) is an organic compound that could be isolated from the above plant extract and interestingly it displayed 42% efficiency of HIV inhibition at 15 μg/mL. Sev-eral other anthraquinones could also be obtained including purpurin (21), alizarin (22), rubiadin (23), 1,4-dihydroxy-2-methyl-5-methoxyanthraquinone (24), and triter-pene rubifolic acid (25). However, these aromatic keto compounds were perceived to be inactive against HIV-1. Whilst, among the naphthoquinones, 1,4-naphthoquinone, juglone, and plumbagin displayed effective anti-HIV efficacy [113].

7.6.4 Common marigold (*Calendula officinalis*)

Common marigold is a flowering plant of Indian origin that belongs to the class of daisy family Asteraceae. The ointments extracted from this plant are extensively utilized for treating wounds, ulcers, skin damage, and purification of blood. The leaves of it are used for preparing infusions that is conventionally used for curing of varicose veins. It is reported in literature that an extract of *C. officinalis* in a dichloromethane-methanol solvent demonstrated effective anti-HIV activity in vitro (3-(4,5-dimethylthiazolyl-2)-2,5-diphenyltetrazolium bromide) tetrazolium-derived test. The therapeutic efficacy was rationalized toward suppression of HIV1-RT and HIV instigated fusion at 1,000 µg/mL and 500 µg/mL concentration, respectively [114]. Both of organo-aqueous extracts of its dried flowers were surveyed on account of their capability to impede the replication of HIV-l. Although both of the extracts were comparatively nontoxic to human lymphocytic Molt-4 cells, however, the extract in organic medium exhibited effective in vitro anti-HIV activity in MTT ketrazolium-functionalized assay. Additionally, in existence of 500 µg/mL of organic extract, the healthy Molt-4 cells were totally safeguarded for a day from fusion and succeeding decease owing to cocultivation along with insistently affected U-937/HIV-1 cells. One step ahead it has been observed that the organoextract obtained from the flowers triggered a substantial time and dosage-dependent diminution of HIV-l RT effectivity. Around 85% of RT suppression was attained posttreatment of partly decontaminated enzyme for half an hour in particularly a cell-free structure. The outcome of this analysis proposed that the organic extract of this flowering plant is endowed with anti-HIV properties that are of immense therapeutic interest [115].

7.6.5 Mexican poppy (*Argemone mexicana*)

In literature Mexican poppy has been investigated that the methanolic extract of this flowering plant displayed good anti-HIV activity at 5 µg/mL concentration. Whilst the tBuOH extract demonstrated moderate activity, interestingly two different alkaloids namely protopine (1) and allocryptopine (2) could be extracted from *A. mexicana* [Fig. 7.11 (a) and (b)]. Compound 1 exhibited moderate anti-HIV property owing to the presence of methylenedioxy moiety at both the terminals whereas Compound 2 remained practically inert due to the existence of methylenedioxy moiety at one terminal and two methyl groups at the other terminal. Most significantly the alkaloid isolated from tBuOH extract was observed to display maximum inhibition efficacy (66.66%) at a concentration of 0.1 lg/mL. Existence of berberine (3) in *A. mexicana* [Figure 11(c)] clearly authenticates the adequate anti-HIV property of tBuOH extract of *A. mexicana*. However, the superior anti-HIV activity of MeOH extract over tBuOH may plausibly be due to some unknown principles of activity [116]. Interestingly, berberine revealed an anti-HIV RT inhibitory action with IC50 of 164 µM [117]. Previously, from this flowering plant, alkaloid isolated namely benzo[c]phenanthridine, 6-acetonyldihydro-chelerythrine demonstrated potential anti-HIV activity in H9 lymphocytes possessing EC50 of 1.77 lg/mL and TI of 14.6 [118].

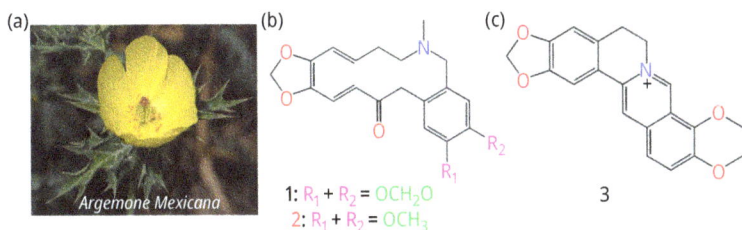

Fig. 7.11: (a) Image of Mexican poppy (*Argemone Mexicana*); chemical structure of anti-HIV compounds extracted from *A. Mexicana*; (b) alkaloids; 1: protopine, 2: allocryptopine; (c) 3: berberine.

7.6.6 Flame-of-the-forest (*Butea monosperma*)

Butea monosperma is a popularly known conventional medicinal plant which is widespread as moderate-sized deciduous tree in India, Ceylon, and Burma. It is a part of the Leguminosae family and is typically recognized as "Palash" or "Dhak." In particular the flowers, seeds, barks, fruits, and leaves are known for exhibiting pharmacological activities. Its phytoconstituents primarily include alkaloids, flavonoids, phenolic compounds, amino acids, glycosides, and steroids which are predominantly responsible for demonstrating various medicinal activities. An in-depth investigation reveals that the methanolic extract of flowers of this plant at a concentration of 600 and 800 mg/kg revealed concentration-reliant anti-inflammatory efficacy. It was observed to restrict the paw edema and granuloma in the carrageenan instigated paw edema and cotton pellet granuloma model in the rats [119]. This plausibly may be owing to the existence of several polyphenols butrin, isobutrin, isocoreopsin, and butein (Fig 7.12) [120]. In addition to the methanolic extract, it was unveiled that an oral administration of the seed extracts of this plant displayed substantial anti-inflammatory activity owing to the existence of fixed oil, fatty acids along with unsaponified portion present in the extract [121]. Likewise, diverse extracts of its leaves also presented anti-inflammatory activity in human red blood cells via stabilization of membrane methodology [122–123].

7.6.7 *Asparagus racemosus (A. racemosus)*

It is commonly known as Satawar, Satamuli, and Satavari, belongs to the Liliaceae family, and is present all over India especially at lower altitudes. The dried roots of this plant particularly possess phytomedicinal properties. The extract of its roots is taken in polar solvents; EtOAc, BuOH as well as H_2O demonstrated effectivity toward combating HIV. Several steroidal saponin glycosides were isolated from its extracts [124]. The

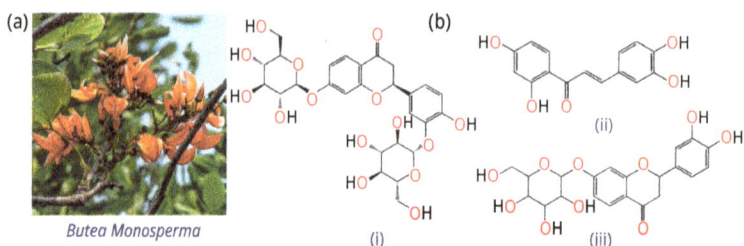

Fig. 7.12: Image of flame-of-the-forest (*Butea monosperma*); chemical structure of several polyphenols extracted from *Butea monosperma*: (i) butrin; (ii) butein; and (iii) isocoreopsin.

glycoside comprising one sugar (1) was obtained from EtOAc extract, whilst the BuOH extract produced glycosides with multiple sugar units (two, three, and four) (2–6) (Fig. 7.13).

Amongst the various saponin glycosides; 1 demonstrated the maximum anti-HIV activity of 53.3% inhibition at 6 µg/mL. Whereas the glycosidic 2 comprising two sugars displayed feeble activity at 6 µg/mL. The saponins bearing three units of sugar (3–6) showcased feeble to zero efficacy. From the linking array of these isolated sugars, it was speculated that the presence of deoxy sugar Rhamnose (Rha) is essential at the C-4 terminal of Glucose (Glu) appended to the sapogenin. It was noted that the Rha or Glu existing at the C-2 terminal of Glu being appended to sapogenin was unable to induce any substantial change in anti-HIV property. Furthermore, the four units of sugar bearing compound 6, attached to the nucleus of sapogenin, demonstrated feeble to modest activity in combating HIV-1. The compound isolated from EtOAc extract primarily accountable for the perceived activity whereas the other compounds (3–5) were investigated for inducing toxicity.

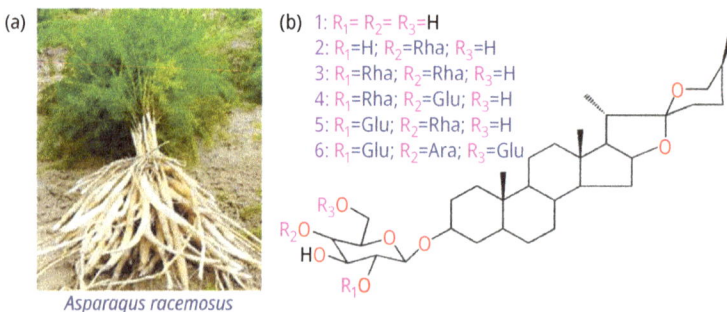

(b) 1: $R_1 = R_2 = R_3 = H$
     2: $R_1 = H$; $R_2 = Rha$; $R_3 = H$
     3: $R_1 = Rha$; $R_2 = Rha$; $R_3 = H$
     4: $R_1 = Rha$; $R_2 = Glu$; $R_3 = H$
     5: $R_1 = Glu$; $R_2 = Rha$; $R_3 = H$
     6: $R_1 = Glu$; $R_2 = Ara$; $R_3 = Glu$

Fig. 7.13: (a) Image of Satawar (*Asparagus racemosus*); (b) chemical structure of several saponin glycosides extracted from *Asparagus racemosus*.

7.6.8 *Coleus forskohlii* (*C. forskohlii*)

This plant is commonly known as Makandi or kaffir potato that belongs to the family of Labiatae. Singh and coworkers were the first to unveil the anti-HIV property of isolated constituents of this flowering plant [125]. Varying extracts of *Coleus forskohlii* investigated at their noncytotoxic concentration of 20 µg/mL in chloroform, ethyl acetate along within-butanol displayed 45.6%, 66.5%, and 37.7% against HIV-1NL4-3-infected CEM-GFP cells. In particular, four diterpenes namely 1-deoxyforskolin (1), 1,9-dideoxyforskolin (2), forskolin (3), and isoforskolin (4) could be separated from CHCl3 extract and subsequently investigated toward combating the virus. Amongst the varying isolated compounds that were under investigation, 5 µg/mL of 1 and 3 demonstrated highest efficiency of inhibition against HIV-1NL4-3. One step ahead, for verifying the effect of structural changes, a series of semisynthetically derived compounds (5–10) were successfully produced. All the compounds remained inactive except compound 7, which at its noncytotoxic concentration of 40 µg/mL displayed moderate activity of inhibition (22.6%) against HIV-1NL4-3. An in-depth investigation underlying the structural prerequisite toward showcasing anti-HIV activity proposed that the existence of –OH at C-9 terminal, –OAc at C-7 terminal, along with –OH at C-6, C = O at C-11 as well as a double bond at C-14 might plausibly be crucial functionalities for the desired action. Absence or elimination of any of these groups, that is, –OH from C-9 (2) or –OAc from C-7 (4, 5) caused a depletion in anti-HIV activity. Additionally, C = O reduction at C-11 (8–10) and double bond at C-14 (6) also reduced the inhibition efficacy (Fig 14). Moreover, the –OH oxidation at C-1 and C-6 may also contribute toward loss in activity (7).

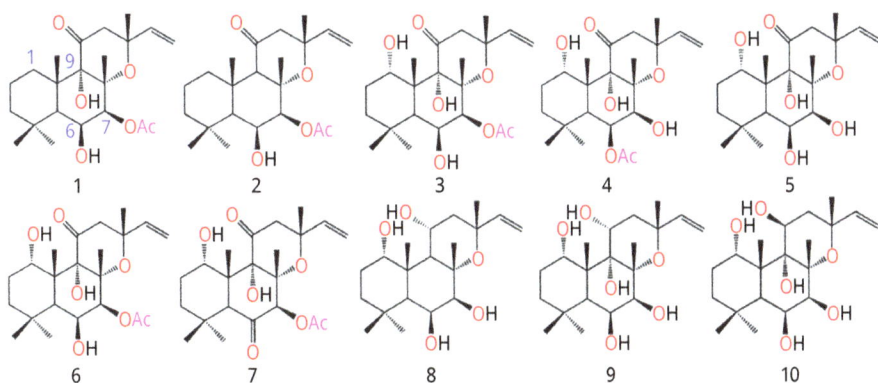

Fig. 7.14: Chemical structures of several diterpenes and semisynthetically derived compounds from *Coleus forskohlii*.

7.6.9 Banaba plant (*Lagerstroemia speciosa* L.)

This plant is frequently acknowledged as the Pride of India and is native to Philippines, Southeast Asia. Interestingly the tea produced from its leaves is conventionally utilized in many countries toward treatment of diabetes and obesity [126–128]. Additionally, the organo-aqueous extracts as-prepared from its leaves as well as the stem portion are known to exhibit remarkable anti-HIV property. The components yielded from the ethanolic extract of leaves included gallic acid and ellagic acid (Fig. 7.15) that demonstrated the novel inhibition activity. The several extracts displayed dose-reliant inhibition of anti-HIV activity in TZM-bl and CEMGFP cell lines with maximum activity obtained from ethanolic extract of its leaves (IC50 = 1–25 µg/mL). These findings were affirmed by the evaluation of virus load (p24) in the CEM-GFP cells infected with the virus upon being treated with the extracts. An in-depth investigation of in vitro reporter gene-based cell-assay systems delineate the underlying mechanism wherein it reveals that these acids may plausibly act as postentry inhibitors via inhibition of RT/HIV activity of discrete protease/integrase. In particular, gallic acid exhibited RT inhibition whilst ellagic acid inhibited the HIV-1 protease activity. This instigated the extensive use of banaba plant as potential contender toward plant-derived microbicide development for combating HIV infection [129].

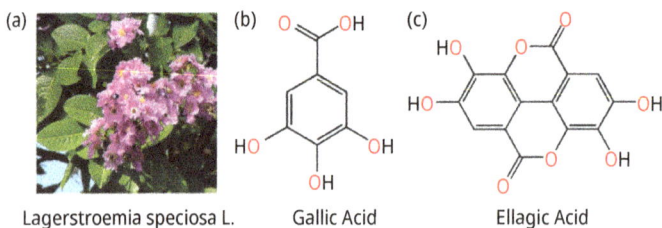

(a) Lagerstroemia speciosa L.   (b) Gallic Acid   (c) Ellagic Acid

Fig. 7.15: (a) Image of banaba (*Lagerstroemia speciosa* L.); chemical structure of (b) gallic acid and (c) ellagic acid extracted from banaba plant.

7.6.10 Garlic (*Allium sativum*)

This is a commonly ingested herb which is utilized worldwide in the form of functional edible materials as well as conventional remedies toward the prevention of various infectious ailments. Uddin and coworkers have investigated that garlic along with its noninnocent OSCs demonstrated potential antiviral activity and alleviate viral infections toward fighting of diverse viruses affecting humans, animals, and plant pathogens (Fig. 7.16) [130]. The preclinical diagnosis demonstrated the hindering the viral entrance within the host cells followed by constraining of viral RNA polymerase, RT, biomolecule like DNA synthesis and in succession transcription of an immediate early gene 1. Furthermore, clinical analysis displayed a prophylactic effect of *Allium sativum* in preventive

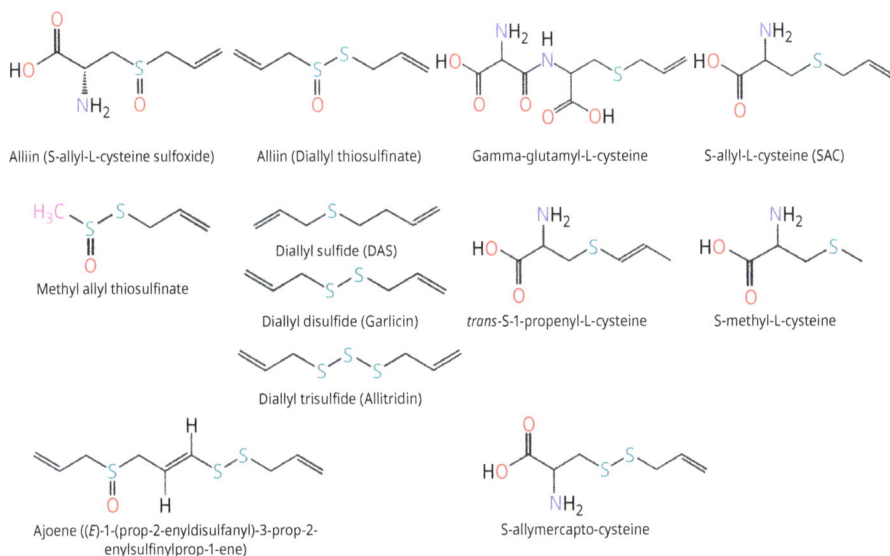

Fig. 7.16: Structure of the major organosulfur compounds (OSCs) obtained from garlic.

action of extensive viral infections prevailing in human body via improvement of their immunological response.

In addition to the above-described medicinal bio plants, other biologically active plants with proven anti-HIV activity have been tabulated in Tab. 7.2.

Tab. 7.2: List of plants and the phytoconstituents traditionally used to exhibit anti-HIV activity.

S. no	Botanical name of the plant	Common name (English)	Family	Part of the plant	Active constituents	Assay	References
1.	*Adhatoda vasica*	Malabar nut	Acanthaceae	Leaves	Alkaloids	p24 antigen assay	[131]
2.	*Alstonia scholaris*	Chatim tree	Apocynaceae	Stem, bark, and leaves	Alkaloids and terpenoids	p24 antigen assay	[131]
3.	*Aconitum kusnezoffi*	Indian Aconite	Ranunculaceae	Aerial	Diterpene alkaloids, flavonoids, and polysaccharides	MT-4 cell assay	[132]

Tab. 7.2 (continued)

S. no	Botanical name of the plant	Common name (English)	Family	Part of the plant	Active constituents	Assay	References
4.	*Azadirachta indica*	Neem tree	Meliaceae	Leaves	Azadirachtin, quercetin and β-sitosterol, and polyphenolic flavonoids	Syncytium reduction assay, ELISA, anti-HIV-1RT inhibitory activity	[133]
5.	*Garcinia indica*	Kokum butter tree	Clusiaceae	Leaves	Garcinol (camboginol) and isogarcinol (cambogin)	MT-4 cell assay	[134]
6.	*Gossampinus malabarica*	Red silk-cotton tree	Bombacaeae	Flower	Alkaloids, tannins, glycosides, reducing sugar, flavones (saponarin), phlobatanins, and terpenoids	MT-4 cell assay	[135]
7.	*Madhuca indica*	Mahuwa	Sapotaceae	Bark	Glycosides, flavonoids, terpenes, and saponins	p24 antigen assay	[136]
8.	*Moringa oleifera*	Drumstick tree	Moringaceae	Leaves	Phenols, flavonoids	Vector-based antiviral assay	[137]
9.	*Ocimum sanctum*	Tulsi	Lamiaceae	Leaves	Essential oil components viz eugenol, 1,8-cineole, β-bisabolen, etc.	RT inhibition assay, Gp120-binding inhibition assay	[138]
10.	*Phyllanthus emblica*	Indian gooseberry	Phyllanthaceae	Fruit	Polyphenols, tannins, and flavonoids	p24 production assay	[139]

7.7 Conclusion

In summary, this chapter solely consolidates a comprehensive outline describing a few Indian medicinal plants with proven anticancer and anti-HIV activity. Scientists and researchers are gradually being lured toward isolation of naturally derived plant-based compounds and subsequent exploration of their underlying properties that makes them potential agents for combating ghastly disorders like cancer and HIV. The medicinal plants produce naturally abundant secondary metabolites which are known to possess less poisonous side effects in comparison to the contemporary methodologies of treatment and clinically available anticancer and anti-HIV drugs. Therefore, the anticancer and anti-HIV activity of the phytoextracts are being exhaustively explored for sustainable development of new clinical drugs with reduced toxicity and superior efficacy. Keeping in mind the high demand of plant-derived agents as efficacious inhibitors of cancer and HIV, a detailed discussion of phytoremediation approach for treatment of cancer and HIV has been summarized herein.

References

[1]   H. A. Idikio. Human cancer classification: A systems biology- based model integrating morphology, cancer stem cells, proteomics, and genomics. J Cancer. 2011, 2, 107–115.

[2]   I. D. O. Ochwang', C. N. Kimwele, J. A. Oduma, P. K. Gathumbi, J. M. Mbaria, S. G. Kiama. Medicinal plants used in treatment and management of cancer in Kakamega County, Kenya. J Ethnopharmacol. 2014, 151, 1040–1055.

[3]   K. Sak. Chemotherapy and Dietary Phytochemical Agents. Chemother Res Pract. 2012, 2012, 1–11.

[4]   R. Baskar, J. Dai, N. Wenlong, R. Yeo, K.-W. Yeoh. Biological response of cancer cells to radiation treatment. Front Mol Biosci. 2014, 1, 24.

[5]   S. G. Deeks, J. Overbaugh, A. Phillips, S. Buchbinder. HIV infection. Nat Rev Dis Prim. 2015, 1, 15035.

[6]   World Health Organization (WHO). 2017. Available online: http://www.who.int/hiv/data/epi_plhiv_2016_regions.png?ua=1 (accessed on 1 December 2017).

[7]   B. Salehi, V. Nanjangud, A. K., B. Şener, M. Sharifi-Rad, M. Kılıç, B. M. Gail, S. Vlaisavljevic, M. Iriti, F. Kobarfard, W. N. Setzer, S. A. Ayatollahi, A. Ata, J. Sharifi-Rad. Medicinal Plants Used in the Treatment of Human Immuno deficiency Virus. Int J Mol Sci. 2018, 19(5), 1459.

[8]   R. Raghavi, S. K. Deborah, J. Joseph, W. Aruni. Evaluation of Anti-Hiv Activity of Selected Medicinal Plants: A Short Review. Biosc Biotech Res Comm. 2020, 13(2), 401–409.

[9]   S. Sabde, H. S. Bodiwala, A. Karmase, P. J. Deshpande, A. Kaur, N. Ahmed, S. K. Chauthe, K. G. Brahmbhatt, R. U. Phadke, D. Mitra, K. K. Bhutani, I. P. Singh. Anti-HIV activity of Indian medicinal plants. J Nat Med. 2011, 65, 662–669.

[10] L. M. Koonin, A. Patel. Timely antiviral administration during an influenza pandemic: Key components. Am J Public Health. 2018, 108(S3), S215–S220.

[11] H. F. Günthard, J. A. Aberg, J. J. Eron, J. F. Hoy, A. Telenti, C. A. Benson, D. M. Burger, P. Cahn, J. E. Gallant, M. J. Glesby. Antiretroviral treatment of adult HIV infection: Recommendations of theInternational Antiviral Society-USA panel. J Am Med Assoc. 2014, 312, 410–425.

[12] H. J. Field, M. A. Wainberg. Antiviral drug development. Future Virol. 2011, 6(5), 545–547.

[13]   N. E. Thomford, D. A. Senthebane, A. Rowe, D. Munro, P. Seele, A. Maroyi, K. Dzobo. Natural Products for Drug Discovery in the twenty-first Century: Innovations for Novel Drug Discovery. Int J Mol Sci. 2018, 19(6), 1578.

[14]   S. Chikara, L. D. Nagaprashantha, J. Singhal, D. Horne, S. Awasthi, S. S. Singhal. Oxidative stress and dietary phytochemicals: Role in cancer chemoprevention and treatment. Cancer Lett. 2018, 413, 122–134.

[15]   S. Singh, B. Sharma, S. S. Kanwar, A. Kumar. Lead Phytochemicals for Anticancer Drug Development. Front Plant Sci. 2016, 7, 8973.

[16]   J. Iqbal, B. H. Abbasi, R. Batool, T. Mahmood, B. Ali, A. T. Khalil, S. Kanwal, A. Shah, R. Ahmad. Potential phytocompounds for developing breast cancer therapeutics: Nature's healing touch. Eur J Pharmacol. 2018, 827, 125–148.

[17]   D. Avtanski, L. Poretsky. Phyto-polyphenols as potential inhibitors of breast cancer metastasis. Mol Med. 2018, 24, 29.

[18]   R. K. Ganjhu, P. P. Mudgal, H. Maity, D. Dowarha, S. Devadiga, S. Nag, G. Arunkumar. Herbal plants and plant preparations as remedial approach for viral diseases. Virus Disease. 2015, 26(4), 225–236.

[19]   L. T. Lin, W. C. Hsu, C. C. Lin. Antiviral natural products and herbal medicines. J Tradit Complement Med. 2014, 4(1), 24–35.

[20]   M. Zahid, M. Mujahid, P. K. Singh, S. Farooqui, K. Singh, S. Parveen, A. M. *Annona squamosa* Linn. (custard apple): An aromatic medicinal plant fruit with immense nutraceutical and therapeutic potentials. Int J Pharm Sci Res. 2018, 9, 1745–1759.

[21]   R. Saha. Pharmacognosy and pharmacology of *Annona squamosa*: A review, Int. J Pharm Life Sci. 2011, 2,1183–1189.

[22]   W. M. Oo, M. M. Khine. Pharmacological activities of *Annona squamosa*: Updated review. Int J Med Chem. 2017, 3, 86–93.

[23]   Y. Chen, S. S. Xu, J. W. Chen, Y. Wang, H. Q. Xu, N. B. Fan, X. Li. Anti-tumor activity of *Annona squamosa* seeds extract containing annonaceous acetogenin compounds. J Ethnopharmacol. 2012, 142, 462–466.

[24]   C. Ma, Q. Wang, Y. Shi, Y. Li, X. Wang, X. Li, Y. Chen, J. Chen. Three new anti-tumor annonaceous acetogenins from the seeds of *Annona squamosa*. Nat Prod Res. 2017, 31, 2085–2090.

[25]   Y. Y. Chen, Y. Z. Cao, F. Q. Li, Z. Xl, C. X. Peng, J. H. Lu, J. W. Chen, X. Li, Y. Chen. Studies on anti-hepatoma activity of *Annona squamosa* L. pericarp extract. Bioorg Med Chem Lett. 2017, 27, 1907–1910.

[26]   Y. Y. Chen, C. X. Peng, Y. Hu, C. Bu, S. C. Guo, X. Li, Y. Chen, J. W. Chen. Studies on chemical constituents and anti-hepatoma effects of essential oil from *Annona squamosa* L. pericarp. Nat Prod Res. 2017, 31, 1308–1308.

[27]   N. S. Vilanova, S. M. Morais, M. J. C. Facao, L. K. A. Machado, C. M. L. Becilaqua, I. R. S. Costa, N. V. G. P. S. Brasil, H. F. A. Júnior. Leishmanicidal activity and cytotoxicity of compounds from two Annonacea species cultivated in Northeastern Brazil. Rev Soc Bras Med Trop. 2011, 44, 567–571.

[28]   Y. Chen, J. W. Chen, X. Li. Cytotoxic bistetrahydrofuran annonaceous acetogenins from the seeds of *Annona squamosa*. J Nat Prod. 2011, 74, 2477–2481.

[29]   B. V. Pardhasaradhi, M. Reddy, A. M. Kumari, A. L. Ali, A. Khar. Differential cytotoxic effects of *Annona squamosa* seed extracts on human tumor cell lines: Role of reactive oxygen species and glutathione. J Biosci. 2005, 30, 237–244.

[30]   K. Suresh, S. Manoharn, D. Blessy. Protective role of *Annona squamosa* Linn bark extracts in DMBA induced genotoxicity. Kathmandu Univ Med J. 2008, 6, 364–369.

[31]   K. An, D. Zhao, Z. Wang, J. Wu, Y. Xu, G. Xiao. Comparison of different drying methods on Chinese ginger (*Zingiber officinale* Roscoe): Changes in volatiles, chemical profile, antioxidant properties, and microstructure. Food Chem. 2016, 197(Part B), 1292–1300.

[32] A. H. Rahmani, F. M. Shabrmi, S. M. Aly. Active ingredients of ginger as potential candidates in the prevention and treatment of diseases via modulation of biological activities. Int J Physiol Pathophysiol Pharmacol. 2014, 6, 125–136.

[33] R. Gupta, P. K. Singh, R. Singh, R. L. Singh. Pharmacological activities of *Zingiber officinale* (ginger) and its active ingredients: A review. Int J Of Innov And Sci res. 2016, 4, 1–18.

[34] B. S. Tan, O. Kang, C. W. Mai, K. H. Tiong, A. S. Khoo, M. R. Pichika, T. D. Bradshaw, C. O. Leong. 6-Shogaol inhibits breast and colon cancer cell proliferation through activation of peroxisomal proliferator activated receptor gamma (PPARgamma). Cancer Lett. 2013, 336,127-139.

[35] H. Ling, H. Yang, S. H. Tan, W. K. Chui, E. H. Chew. 6-Shogaol, an active constituent of ginger, inhibits breast cancer cell invasion by reducing matrix metalloproteinase-9 expression via blockade of nuclear factor-kappa B activation. Br J Pharmacol. 2010, 161, 1763–1777.

[36] J. Rhode, S. Fogoros, S. Zick, H. Wahl, K. A. Griffith, J. Huang, J. R. Liu. Ginger inhibits cell growth and modulates angiogenic factors in ovarian cancer cells, BMC Complement. Altern Med. 2007, 7, 44.

[37] A. C. Brown, C. Shah, J. Liu, J. T. Pham, J. G. Zhang, M. R. Jadus. Ginger's (*Zingiber officinale* Roscoe) inhibition of rat colonic adenocarcinoma cells proliferation and angiogenesis in vitro. Phytother Res. 2009, 23, 640–645.

[38] Y. J. Park, J. Wen, S. Bang, S. W. Park, S. Y. Song. [6]-Gingerol induces cell cycle arrest and cell death of mutant p53-expressing pancreatic cancer cells. Yonsei Medical Journal. 2006, 47, 688–697.

[39] E. Lee, K. K. Park, J. M. Lee, K. S. Chun, J. Y. Kang, S. S. Lee, Y. J. Surh. Suppression of mouse skin tumor promotion and induction of apoptosis in HL-60 cells by *Alpina oxyphylla* Miquel (Zingiberaceae. Carcinogenesis. 1998, 19, 1337–1381.

[40] K. Ishiguro, T. Ando, O. Maeda, N. Ohmiya, K. Niwa, K. Kadomatsu, G. Hidemi. Ginger ingredients reduce viability of gastric cancer cells via distinct mechanisms. Biochem Biophys Res Commun. 2007, 362, 218–223.

[41] F. F. Gan, A. A. Nagle, X. Ang, O. H. Ho, S. H. Tan, H. Yang, W. K. Chui, E. H. Chew. Shogaols at proapoptotic concentrations induce G(2)/M arrest and aberrant mitotic cell death associated with tubulin aggregation. Apoptosis. 2011, 16, 856–867.

[42] K. Sakamoto, L. D. Lawson, J. A. Milner. Allyl Sulfides from Garlic Suppress the in vitro Proliferation of Human A549 Lung Tumor Cells. Nutr Cancer. 1997, 29, 152–156.

[43] H. Nakagawa, K. Tsuta, K. Kiuchi, H. Senzaki, K. Tanaka, K. Hioki, A. Tsubura. Growth Inhibitory Effects of Diallyl Disulfide on Human Breast Cancer Cell Lines. Carcinogenesis. 2001, 22, 891–897.

[44] H. Shirin, J. T. Pinto, Y. Kawabata, J. W. Soh, T. Delohery, S. F. Moss, V. Murty, R. S. Rivlin, P. R. Holt, I. B. Weinstein. Antiproliferative Effects of SAllylmercaptocysteine on Colon Cancer Cells when Tested Alone or in Combination with Sulindac Sulfide. Cancer Res. 2001, 61, 725–731.

[45] V. M. Dirsch, A. L. Gerbes, A. M. Vollmar. Ajoene, a Compound of Garlic, Induces Apoptosis in Human Promyeloleukemic Cells, Accompanied by Generation of Reactive Oxygen Species and Activation of Nuclear Factor κB. Molec Pharmacol. 1998, 53, 402–407.

[46] M. Thomson, M. Ali. Garlic (*Allium sativum*): A review of its potential use as an anti- cancer agent. Curr Cancer Drug Targets. 2003, 3(1), 67–81.

[47] S. Adaki, R. Adaki, K. Shah, A. Karagir. Garlic: Review of literature. Indian J Cancer. 2014, 51(4), 577–581.

[48] V. Petrovic, A. Nepal, C. Olaisen, S. Bachke, J. Hira, C. K. Søgaard. Anticancer potential of home made fresh garlic extract is related to increased endoplasmic reticulum stress. Nutrients. 2018, 10(4), 450.

[49] I. Chattopadhyay, K. Biswas, U. Bandyopadhyay, R. K. Banerjee. Turmeric and curcumin: Biological actions and medicinal applications. Curr Sci. 2004, 87(1), 44–53.

[50] B. B. Aggarwal, A. Kumar, A. C. Bharti. Anticancer potential of curcumin: Preclinical and clinical studies. Anticancer Res. 2003, 23(1A), 363–398.

[51]   Z. M. Shao, Z. Z. Shen, C. H. Liu, M. R. Sartippour, V. L. Go, D. Heber, M. Nguyen. Curcumin exerts multiple suppressive effects on human breast carcinoma cells. Int J Cancer. 2002, 98(2), 234–240.

[52]   A. J. Smith, J. Oertle, D. Prato. Multiple Actions of Curcumin Including Anticancer, Anti-Inflammatory, Antimicrobial and Enhancement via Cyclodextrin. J Cancer Ther. 2015, 6(3), 257–272.

[53]   A. Mitra, J. Chakrabarti, A. Banerji, A. Chatterjee, B. Das. Curcumin, a potential inhibitor of MMP-2 in human laryngeal squamous carcinoma cells HEp2. J Environ Pathol Toxicol Oncol. 2006, 25(4), 679–690.

[54]   H. W. Chen, H. C. Huang. Effect of curcumin on cell cycle progression and apoptosis in vascular smooth muscle cells, Br. J Pharmacol. 1998, 124(6), 1029–1040.

[55]   N. R. Jana, P. Dikshit, A. Goswami, N. Nukina. Inhibition of proteasomal function by curcumin induces apoptosis through mitochondrial pathway. J Biol Chem. 2004, 279(12), 11680–11685.

[56]   D. Deeb, Y. X. Xu, H. Jiang, X. Gao, N. Janakiraman, R. A. Chapman, S. C. Gautam. Curcumin (diferuloyl-methane) enhances tumor necrosis factor-related apoptosis-inducing ligand-induced apoptosis in LNCaP prostate cancer cells. Mol Cancer Ther. 2003, 2(1), 95–103.

[57]   T. D. Babu, G. Kuttan, J. Padikkala. Cytotoxic and anti-tumour properties of certain taxa of Umbelliferae with special reference to *Centella asiatica* (L.) Urban. J Ethnopharmacol. 1995, 48(1), 53–57.

[58]   R. Srivastava, Y. N. Shukla, S. Kumar. Chemistry and pharmacology of *Centella asiatica*: A review. Ind J Med Arom Plants. 1997, 19, 1049–1056.

[59]   R. Sharma, J. Sharma. Modification of gamma ray induced changes in the mouse hepatocytes by *Centella asiatica* extract: In vivo studies. Phytother Res. 2005, 19(7), 605–611.

[61]   P. Siripong, K. Preechanukool, P. Picha, K. Tunsuwan, W. C. Taylor. Cytotoxic Diterpenoid Constituents From *Andrographis paniculata* Nees Leaves. ScienceAsia. 1992, 18, 187–194.

[62]   S. R. Jada, G. S. Subur, C. Matthews, A. S. Hamzah, N. H. Lajis, M. S. Saad, M. F. Stevens, J. Stanslas. Semisynthesis and in vitro anticancer activities of andrographolide analogues. Phytochemistry. 2007, 68(6), 904–912.

[63]   R. A. Kumar, K. Sridevi, N. V. Kumar, S. Nanduri, S. Rajagopal. Anticancer and immunostimulatory compounds from *Andrographis paniculata*. J Ethnopharmacol. 2004, 92(2–3), 291–295.

[64]   T. Matsuda, M. Kuroyanagi, S. Sugiyama, K. Umehara, A. Ueno, K. Nishi. Cell differentiation-inducing diterpenes from *Andrographis paniculata* Nees. Chem Pharm Bull (Tokyo). 1994, 42(6), 1216–1225.

[65]   A. Puri, R. Saxena, R. Saxena, K. C. Saxena, V. Srivastava, J. S. Tandon. Immunostimulant agents from *Andrographis paniculata*. J Nat Prod. 1993, 56(7), 995–999.

[66]   R. P. Singh, S. Bannerjee, A. Rao. Modulatory influence of *Andrographis paniculata* on mouse hepatic and extrahepatic carcinogen metabolizing enzymes and antioxidant status. Phytother Res. 2001, 15, 382–390.

[66]   R. P. Singh, S. Bannerjee, A. Rao. Modulatory influence of *Andrographis paniculata* on mouse hepatic and extrahepatic carcinogen metabolizing enzymes and antioxidant status. Phytother Res. 2001, 15, 382–390.

[68]   R. A. DeFilipps, S. L. Maina, J. Crepin. Medicinal Plants of the Guianas (Guyana, Surinam, French Guiana). DC, Smithsonian Institution, Washington, 2004.

[69]   T. R. Van Andel, S. Ruysschaert. Medicinale En Rituele Planten van Suriname (Medicinal and Ritual Plants of Suriname). KIT Publishers, Amsterdam, 2011.

[70]   M. E. Heitzman, C. C. Neto, E. Winiarz, A. J. Vaisberg, G. B. Hammond. Ethnobotany, phytochemistry and pharmacology of Uncaria (Rubiaceae. Phytochemistry. 2005, 66, 5–29.

[70]   M. E. Heitzman, C. C. Neto, E. Winiarz, A. J. Vaisberg, G. B. Hammond. Ethnobotany, phytochemistry and pharmacology of Uncaria (Rubiaceae). Phytochemistry. 2005, 66, 5–29.

[72] L. C. De Paula, F. Fonseca, F. Perazzo, F. M. Cruz, D. Cubero, D. C. Trufelli, S. P. D. S. Martins, P. X. Santi, E. A. da Silva, A. D. Giglio. *Uncaria tomentosa* (cat's claw) improves quality of life in patients with advanced solid tumors. J Altern Complement Med. 2015, 21, 22–30.

[73] M. C. S. Araújo, I. L. G. Farias, J. Gutierres, S. L. Dalmora, N. Flores, J. Farias, I. de Cruz, J. Chiesa, V. M. Morsch, M. R. C. Schetinger. *Uncaria tomentosa*-Adjuvant treatment for breast cancer: Clinical trial. Evid Based Complement Alternat Med. 2012, 676984.

[74] I. L. G. Farias, M. C. S. Araújo, J. G. Farias, L. V. Rossato, L. I. Elsenbach, S. L. Dalmora, N. M. P. Flores, M. Durigon, I. B. M. Cruz, V. M. Morsch, Schetinger. M.R.C.;*Uncaria tomentosa* for reducing side effects caused by chemotherapy in CRC patients: Clinical trial. J Evid Based Complementary Altern Med. 2012, 892182.

[74] I. L. G. Farias, M. C. S. Araújo, J. G. Farias, L. V. Rossato, L. I. Elsenbach, S. L. Dalmora, N. M. P. Flores, M. Durigon, I. B. M. Cruz, V. M. Morsch, M. R. C. Schetinger. *Uncaria tomentosa* for reducing side effects caused by chemotherapy in CRC patients: Clinical trial. J Evid Based Complementary Altern Med. 2012, 892182.

[75] D. A. Hyson. A comprehensive review of apples and apple components and their relationship to human health. Adv Nutr. 2011, 2, 408–20.

[76] K. C. Yang, C. Y. Tsai, Y. J. Wang, P. L. Wei, C. H. Lee, J. H. Chen, C. H. Wu, Y. S. Ho. Apple polyphenol phloretin potentiates the anticancer actions of paclitaxel through induction of apoptosis in humanhep G2 cells. Mol Carcinog. 2009, 48, 420–431.

[77] S. T. Lin, S. H. Tu, P. S. Yang, S. P. Hsu, W. H. Lee, C. T. Ho, C. H. Wu, Y. H. Lai, M. Y. Chen, L. C. Chen. Apple polyphenol phloretin inhibits colorectal cancer cell growth via inhibition of the type 2 glucose transporter and activation of p53-mediated signalling. J Agric Food Chem. 2016, 64, 6826–6837.

[78] S. M. Hong, C. W. Park, S. W. Kim, Y. J. Nam, J. H. Yu, J. H. Shin, C. H. Yun, S. H. Im, K. T. Kim, Y. C. Sung, K. Y. Choi. NAMPT suppresses glucose deprivation-induced oxidative stress by increasing NADPH levels in breast cancer. Oncogene. 2016, 35, 3544–3554.

[79] G. F. Schiavano, G. De Santi, M. Brandi, A. Fanelli, A. Bucchini, L. Giamperi, G. Giomaro. Inhibition of breast cancer cell proliferation and in vitro tumorigenesis by a new red applecultivar. PLoS One. 2015, 10, e0135840.

[80] A. Qiao, Y. Wang, L. Xiang, C. Wang, X. He. A novel triterpenoid isolated from apple functions as an anti-mammary tumor agent via a mitochondrial and caspase-independent apoptosis pathway. J Agric Food Chem. 2015, 63, 185–191.

[81] T. Zhao, Q. Sun, M. Marques, M. Witcher. Anticancer Properties of *Phyllanthus emblica* (Indian Gooseberry. Oxid Med Cell Longev. 2015, 950890.

[82] C. Spagnuolo, M. Russo, S. Bilotto, I. Tedesco, B. Laratta, G. L. Russo. Dietary polyphenols in cancer prevention: The example of the flavonoid quercetin in leukemia. Ann NY Acad Sci. 2012, 1259(1), 95–103.

[83] J. L. Steiner, J. M. Davis, J. L. McClellan, R. T. Enos, J. A. Carson, R. Fayad, M. Negratti, P. S. Nagarkatti, D. Altomare, K. E. Creek, E. A. Murphy. Dose-dependent benefits of quercetin on tumorigenesis in the C3(1)/SV40Tag transgenic mouse model of breast cancer. Cancer Biol Ther. 2014, 15(11), 1456–1467.

[84] A. De, C. Papasian, S. Hentges, S. Banerjee, I. Haque, S. K. Banerjee. Emblica officinalis extract induces autophagy and inhibits human ovarian cancer cell proliferation, angiogenesis, growth of mouse xenograft tumors. PLoS ONE. 2013, 8(8), e72748.

[85] X. Zhu, J. Wang, Y. Ou, W. Han, H. Li. Polyphenol extract of *Phyllanthus emblica* (PEEP) induces inhibition of cell proliferation and triggers apoptosis in cervical cancer cells. Eur J Med Res. 2013, 18(1), 46.

[86] N. V. Rajeshkumar, M. R. Pillai, R. Kuttan. Induction of apoptosis in mouse and human carcinoma cell lines by *Emblica officinalis* polyphenols and its effect on chemical carcinogenesis. J Exp Clin Cancer Res. 2003, 22(2), 201–212.

[87] P. Plastina, D. Bonofiglio, D. Vizza, A. Fazio, D. Rovito, C. Giordano, I. Barone, S. Catalano, B. Gabriele. Identification of bioactive constituents of *Ziziphus jujube* fruit extracts exerting antiproliferative and apoptotic effects in human breast cancer cells. J Ethnopharmacol. 2012, 140, 325–332.

[88] Q. H. Gao, C. S. Wu, M. Wang. The jujube (*Ziziphus jujuba* Mill.) fruit: A review of current knowledge of fruit composition and health benefits. J Agric Food Chem. 2013, 61, 3351–3363.

[89] S. Guo, J. A. Duan, Y. Tang, S. Su, E. Shang, S. Ni, D. Qian. High-performance liquid chromatography–Two wavelength detection of triterpenoid acids from the fruits of *Ziziphus jujuba* containing various cultivars in different regions and classification using chemometric analysis. J Pharm Biomed Anal. 2009, 49, 1296–1302.

[90] D. K. Kim, J. H. Baek, C. M. Kang, M. A. Yoo, J. W. Sung, H. Y. Chung, N. D. Kim, Y. H. Choi, S. H. Lee, K. W. Kim. Apoptotic activity of ursolic acid may correlate with the inhibition of initiation of DNA replication. Int J Cancer. 2000, 87, 629–636.

[91] R. E. De Angel, S. M. Smith, R. D. Glickman, S. N. Perkins, S. D. Hursting. Antitumor effects of ursolic acid in a mouse model of postmenopausal breast cancer. Nutr Cancer. 2010, 62, 1074–1086.

[92] J. Li, W. J. Guo, Q. Y. Yang. Effects of ursolic acid and oleanolic acid on human colon carcinoma cell line HCT15. World J Gastroenterol. 2002, 8, 493–495.

[93] E. Selzer, E. Pimentel, V. Wacheck, W. Schlegel, H. Pehamberger, B. Jansen, R. Kodym. Effects of betulinic acid alone and in combination with irradiation in human melanoma cells. J Invest Dermatol. 2000, 114, 935–940.

[94] Y. X. Zhang, C. Z. Kong, H. Q. Wang, L. H. Wang, C. L. Xu, Y. H. Sun. Phosphorylation of Bcl-2 and activation of caspase-3 via the c-Jun N-terminal kinase pathway in ursolic acid-induced DU145 cells apoptosis. Biochimie. 2009, 91, 1173–1179.

[95] P. Govind. Some important anticancer herbs: A review. Int Res J Pharm. 2011, 2(7), 45–52.

[96] L. Wasserman, S. Avigad, E. Beery, J. Nordenberg, E. Fenig. The effect of aloe-emodin on the proliferation of a new merkel carcinoma cell line. Am J Dermatopathol. 2002, 24(1), 17–22.

[97] B. M. R. Bandara, N. S. Kumar, K. M. S. Samaranayake. An antifungal constituent from the stem bark of *Butea monosperma*. J Ethnopharmacol. 1989, 25, 73–75.

[98] M. Sumitra, P. Manikanand, L. Suguna. Efficacy of *Butea monosperma* on dermal wound healing in rats. Int J Biochem Cell Biol. 2005, 37, 566–573.

[99] H. Wagner, B. Geyer, M. Fiebig, Y. Kiso, H. Hikino. Isobutrin and butrin, the antihepatotoxic principles of *Butea monosperma* flowers. Planta Med. 1986, 2, 77–79.

[100] R. Wright, H. D. Colby, P. R. Miles. Cytosolic factors that affect microsomal lipid peroxidation in lung and liver. Arch Biochem Biophys. 1981, 206, 296–304.

[101] G. Roja, P. S. Rao. Anticancer compounds from tissue cultures of medicinal plants. J Herbs Spices Med Plants. 2000, 7(2), 71–102.

[102] M. Umadevi, K. P. Sampath, D. Bhowmik, S. Duraivel. Traditionally Used Anticancer Herbs In India. J Med Plants Stud J Med Plants Stud. 2013, 1(3), 56–74.

[103] S. Shital, M. G. Chavan, P. B. Damale, D. P. Shamkuwar, Pawar. Traditional medicinal plants for anticancer activity. Int J Curr Pharm Res. 2007, 5(4), 201–202.

[104] J. Bruneton. Pharmacognosy, Phytochemistry Medicinal Plants. Lavoisier publisher, France, 2003, pp. 151–832.

[105] G. M. Cragg, D. Newman. J.;Plants as a source of anti-cancer agents. J Ethnopharmacol. 2005, 100, 72–79.

[106] V. C. Dennis. Awang.;Tyler's Herbs of Choice. The Therapeutic Use of Phytomedicinals. Haworth press, New York, 1994, pp. 32–33.

[107] U. Laila, M. Akram, M. A. Shariati, A. M. Hashmi, N. Akhtar, I. M. Tahir, A. O. Ghauri, N. Munir, M. Riaz, N. Akhter, G. Shaheen, Q. Ullah, R. Zahid, S. Ahmad. Role of medicinal plants in HIV/AIDS therapy. Clin Exp Pharmacol Physiol. 2019, 46, 1063–1073.

[108] M. Momoh, U. Muhamed, A. Agboke, E. Akpabio, U. E. Osonwa. Immunological effect of aqueous extract of *Vernonia amygdalina* and a known immune booster called immunace® and their admixtures on HIV/AIDS clients: A comparative study. Asian Pac J Trop Biomed. 2012, 2, 181–184.

[109] S. K. Dash, S. J. Padhy. Review on ethnomedicines for diarrhoea diseases from Orissa: Prevalence versus culture. Human Ecol. 2006, 20, 59–64.

[110] P. Zhou, Y. Takaishi, H. Duan, B. Chen, G. Honda, M. Itoh, Y. Takeda, O. K. Kodzhimatov, K. H. Lee. Coumarins and bicoumarin from *Ferula sumbul*: Anti-HIV activity and inhibition of cytokine release. Phytochemistry. 2000, 53, 689–697.

[111] R. Sancho, N. Marquez, G. M. Gomez, M. A. Calzado, G. Bettoni, M. T. Coiras, J. Alcami, C. M. Lopez, G. Appendino, E. Munoz. Imperatorin inhibits HIV-1 replication through an Sp1-dependent pathway. J Biol Chem. 2004, 279, 37349–37359.

[112] L. C. Chang, D. Chavez, J. J. Gills, H. S. Fong, J. M. Pezzuto, A. D. Kinghorn, A.-C. Rubiasins. new anthracene derivatives from the roots and stems of *Rubia cordifolia*. Tetrahedron Lett. 2000, 41, 7157–7162.

[113] B. S. Min, H. Miyashiro, M. Hattori. Inhibitory effects of quinones on RNase H activity associated with HIV-1 reverse transcriptase. Phytother Res. 2002, 16, S57-S62.

[114] Z. Kalvatchev, R. Walder, D. Garzaro. Anti-HIV activity of extracts from *Calendula officinalis* flowers. Biomed Pharmacother. 1997, 51, 176–180.

[116] S. Sabde, H. S. Bodiwala, A. Karmase, P. J. Deshpande, A. Kaur, N. Ahmed, S. K. Chauthe, K. G. Brahmbhatt, R. U. Phadke, D. Mitra, K. K. Bhutani, I. P. Singh. Anti-HIV activity of Indian medicinal plants. J Nat Med. 2011, 65, 662–669.

[117] G. T. Tan, J. M. Pezzuto, A. D. Kinghorn, S. H. Hughes. Evaluation of natural products as inhibitors of human immunodeficiency virus type 1 (HIV-1) reverse transcriptase. J Nat Prod. 1991, 54, 143–154.

[118] Y. C. Chang, P. W. Hsieh, F. R. Chang, R. R. Wu, C. C. Liaw, K. H. Lee, Y. C. Wu. Two new protopines argemexicaines A and B and the anti-HIV alkaloid 6-acetonyldihydrochelerythrine from Formosan Argemone Mexicana. Planta Med. 2003, 69, 148–152.

[119] V. M. Shahavi, S. K. Desai. Anti-inflammatory activity of *Butea monosperma* flowers. Fitoterapia. 2008, 79(2), 82–85.

[120] Z. Rasheed, N. Akhtar, A. Khan, K. A. Khan, T. M. Haqqi. Butrin, isobutrin, and butein from medicinal plant *Butea monosperma* selectively inhibit nuclear factor-κB in activated human mast cells: Suppression of tumor necrosis factor-α, interleukin (IL)-6, and IL-8. J Pharmacol Exp T. 2010, 333(2), 354–363.

[121] A. Gunakunru. Chemical investigations and anti-inflammatory activity of fixed oil of *Butea monosperma* seeds. Nat Prod Sci. 2004, 10(2), 55–58.

[122] V. S. Borkar. Evaluation of in vitro anti-inflammatory activity of leaves of *Butea monosperma*. Indian Drugs. 2010, 47(6), 62–63.

[123] V. S. Borkar. In vitro evaluation of *Butea monosperma* Lam. for antioxidant activity. Orient J Chem. 2008, 24, 753–755.

[124] A. N. Jadhav, K. K. Bhutani. Steroidal saponins from the roots of *Asparagus adscendens* Roxb and *Asparagus racemosus* Willd. Indian J Chem Sect B. 2006, 1515–1524.

[125] H. S. Bodiwala, S. Sabde, M. Debashis, K. K. Bhutani, I. P. Singha. Anti-HIV Diterpenes from *Coleus forskohlii*. Nat Prod Commun. 2009, 4(9), 1173–1175.

[126] F. Garcia. On the hypoglycemic effect of decoction of *Lagerstroemia speciosa* (banaba) administered orally. J Phil Med Assoc. 1940, 20, 395–402.

[127] E. Quisumbing. Medicinal Plants of the Philippines. Katha Publishing, Quezon, Philippines, 1978, pp. 640–642.

[128] F. Matsuyama. Composition for inhibiting increase in blood sugar level or lowering blood sugar level. United States Patent Application No. 09/730, 2008, 74.

[129] Nutan, M. Modi, T. Goel, T. Das, S. Malik, S. Suri, A. K. S. Rawat, S. K. Srivastava, R. Tuli, S. Malhotra, S. K. Gupta. Ellagic acid & gallic acid from *Lagerstroemia speciosa* L. inhibit HIV-1 infection through inhibition of HIV-1 protease & reverse transcriptase activity. Indian J Med Res. 2013, 137, 540–548.

[130] R. Rouf, S. J. Uddin, D. K. Sarker, M. T. Islam, E. S. Ali, J. A. Shilpi, L. Nahar, E. Tiralongo, S. D. Sarker. Antiviral potential of garlic (*Allium sativum*) and its organosulfur compounds: A systematic update of pre-clinical and clinical data. Trends Food Sci Technol. 2020, 104, 219–234.

[131] S. Sudeep, B. Hardik, K. Aniket. Anti-HIV activity of Indian medicinal plants. J Nat Med. 2011, 65, 662–669.

[132] L. M. Bedoya, S. S. Palomino, M. J. Abad, P. Bermejo, J. Alcami. Anti-HIV activity of medicinal plant extracts. J Ethnopharmacol. 2001, 77(1), 113–116.

[133] P. E. David, S. G. L. Benjamín, E. Delia, Á. E. A., N. H. M. Patricia, V. M. M. Del Carmen. HIV-1 infection inhibition by neem (*Azadirachta indica* A. Juss.) leaf extracts and Azadirachtin. Indian J Tradit Knowl. 2017, 16(3), 437–441.

[134] S. Padhye, A. Ahmad, N. Oswal, F. H. Sarkar. Emerging role of Garcinol, the antioxidant chalcone from *Garcinia indica* Choisy and its synthetic analogs. J Hematol Oncol. 2009, 2, 38.

[135] J. A. Wu, A. S. Attele, L. Zhang, C. S. Yuan. Anti-HIV activity of medicinal herbs: usage and potential development. Am J Chinese Med. 2001, 29(1), 69–81.

[136] R. P. Gujjeti, E. Mamidala. In Vitro Hiv-1 Rt Inhibitory Activity of *Madhuca Indica* Inner Bark Extracts. Biolife. 2014, 2(3).

[137] C. S. Nworu, E. Okoye, E. GO, C. O. Esimone. Extracts of *Moringa oleifera* Lam. showing inhibitory activity against early steps in the infectivity of HIV-1 lentiviral particles in a viral vector-based screening. Afr J Biotechnol. 2013, 12(30), 4866–4873.

[138] K. Silprasit, S. Seetaha, P. Pongsanarakul, P. Pongsanaraku, K. Choowongkomon. Anti-HIV-1 reverse transcriptase activities of hexane extracts from some Asian medicinal plants. J Med Plant Res. 2011, 5(19), 4899–4906.

[139] M. Estari, L. Venkanna, D. Sripriya, R. Lalitha. Human immunodeficiency virus (HIV-1) reverse transcriptase inhibitory activity of *phyllanthus emblica* plant extract. Biol Med. 2012, 4(4), 178–182.

[139] M. Estari, L. Venkanna, D. Sripriya, R. Lalitha. Human immunodeficiency virus (HIV-1) reverse transcriptase inhibitory activity of *phyllanthus emblica* plant extract. Biol Med. 2012, 4(4), 178–182.

Hilal Ahmed, Tousief Irshad Ahmed, Roli Jain, Jyoti Rathore,
Shanthi Natarajan, Reena Rawat and Bhawana Jain*

Chapter 8
Anti-cancerous and anti-HIV activity of medicinal plants

Abstract: Cancer, as numerous diseases, can start from each organ including the development of abnormal and uncontrolled cells in human body. On the other hand, human immunoviros (HIV) is caused by virus and makes humans susceptible to infection which is then acquired to immune deficiency syndrome (AIDS). Every year many people are diagnosed and around 1 million people die due to both diseases. Thus, for treatment, natural products especially from plants were produced from related medicinal compounds used for the preparation of ideal and medicinal anticancer and anti-HIV agents like Taxol, Vinblastine, Podophyllotoxin, Betulinic acid, Camptothecin, and Vincristine. For the research of drugs from medicinal plants, biological, botanical, phytochemical, and molecular techniques were involved. Medicinal plants provide new and important medicinal eminent for various pharmacological treatment of cancer and HIV/AIDS.

8.1 Introduction

Over the past decades, humans always look up to nature for their basic needs such as foodstuffs, houses, clothing, fertilizers, spices, flavors, and fragrances including medicines. In many countries, for thousand years, plants used traditional medicine systems [1, 2]. These medicinal plants are used as an essential part in health care system. For example, antimalarial drug, quinine, was plant-derived and obtained from the bark of *Cinchona officinalis* plant; analgesics, codeine, and morphine were obtained from *Papaver somniferum*; antihypertensive reserpine was obtained from *Rauwolfi aserpentina*; and cardiac glycoside and digoxin were obtained from *Digitalis purpurea* [3].

*Corresponding author: Bhawana Jain, Siddhachalam Laboratory, Raipur 493221, Chhattisgarh, India
Hilal Ahmed, Department of Zoology, University of Jammu, Jammu, 180006, Jammu and Kashmir, India
Tousief Irshad Ahmed, Department of Clinical Biochemistry, SKIMS, Soura, Srinagar, Jammu and Kashmir, India
Roli Jain, Department of Chemistry, Dr. Hari Singh Gour University, Sagar, Madhya Pradesh, India
Jyoti Rathore, Govt. Engineer Vishwesarraiya Post Graduate College, Korba 495677, India
Reena Rawat, Department of Chemistry, Echelon Institute of Technology, Faridabad 121101, Haryana, India
Shanthi Natarajan, Department of Botany, Pachaiyappa's College, Affiliated to University of Madras, Chennai 600030, Tamil Nadu, India

https://doi.org/10.1515/9783110791891-008

Division and uncontrolled growth of cells are due to mutations called cancer. Clinically, cancer is a large group of cells that fluctuate from starting of case, pace development, division of cell, perceptibility, intrusiveness, metastatic potential, and treatment. Because of cancer cells, incursion, damage of adjoining tissues, and organs occur [4]. According to WHO, worldwide, the second-most killer disease is cancer [5]. In females, breast, lung, stomach, and cervix cancer are the most common cancers and for men liver, lung, colorectal, prostate, and stomach cancer are the most common ones. The general methods for treatment of cancer were immunotherapy, radiotherapy, and chemotherapy, but these methods showed harmful effects on healthy cells and organs. Thus, for the treatment and preventing harmful actions from healthy tissues and organs, it encourages finding and developing new and safe methods for the treatment of cancer like natural bioactive compounds which are used as anticancer drugs that can be extracted from plants [6]. There are many other bioactive compounds that can be used in clinical and preclinical treatment [7].

On the other side, human immunodeficiency virus (HIV) is a virus and the infection of it causes acquired immune deficiency syndrome (AIDS) in human. First, around 75 million people were infected with this virus and almost half of them died [8]. Plant-derived natural products work as a reservoir and used for the recognition of new medicines with anti-HIV agents. Thus, new, less toxic, but more effective agents and drugs are needed for treatment. Natural products, which are derived from plants, are used for the discovery of new medicines for anti-HIV agents. These days, around 30 anti-HIV agents were developed for the treatment of AIDS [9].

8.2 Anti-cancerous medicinal plants

8.2.1 History

Since ancient times, humans treated their diseases by natural products like using plants, roots, stems, or fruits. First, medicinal plants were used as instinctive and treated only a few diseases [10]. In India, approximately 5,000 years ago, with written proof, medicinal plants were used to treat diseases. It described 12 methods for the preparation of numerous medicines which used more than 250 varieties of plants [10].

At 2,500 BC, a Chinese emperor Shen Nung wrote a book "Pen T'Sao" regarding the parts of plants (leaves and stems), which contained 365 different medicines. It also included few recently used products like cinnamon bark, ephedra, and ginseng [10].

At 1,550 BC in Egypt, El Papiro de Ebers wrote 800 prescriptions and described 700 plant species with drugs use for treatment of diseases, for example, pomegranate, castor oil, aloe, senna, garlic, onion, fig, willow, and cilantro use for treatment [10].

At 371–267 BC, in Athens, Greece, Theophrastus wrote books regarding botanical science "De Causis Plantarium" and "De Historia Plantarium." Both books described the classification of 500 plus medicinal plants, for example, cinnamon, iris rhizome, false hellebore, mint, pomegranate, cardamom, fragrant hellebore, and aconite plants [10].

At 77 BC, "father of pharmacognosy," the Greek Dioscorides wrote the most document of famous medicinal plant "De Materia Medica." It described 944 medications, in which 657 plants were derived from medicinal plants. It also contained the collection, synthesis, and their therapeutic effect. From 500 years, this book is used for medical treatment in different systems [10].

These days, people are interested to isolate the chemical compounds from plants. However, conventional medicines are divided into two types: first one is derived from plants and second one is manually synthesized in laboratories. Some diseases, like cancer, did not show results with conventional medicine; thus medicinal plants were used as alternative therapy [10, 11].

8.2.2 Plant-based anticancer agents

Some medicinal plants are used in cancer treatment [12]. It played important and special role as source-effective anticancer agents, and importantly over 60% of current anticancer agents were derived from natural products like plants, marine organisms, and micro-organisms [13, 14]. From early 1950s, development of anticancer agents from plant started with the development of vinca alkaloids, vinblastine, and vincristine and also isolated cytotoxic podophyllotoxins. The discovery of various chemotypes showed selective cytotoxic activities [15] with taxanes and camptothecins.

8.3 Anticancer agents derived from plant

8.3.1 For clinical use

The first agent for clinical treatment was vinca alkaloids, vinblastine (VLB), and vincristine (VCR) (Fig. 8.1(A)) which were isolated from Madagascar periwinkle, *Catharanthusroseus G.* Don. (Apocynaceae) and use for diabetes treatment in human [16]. First, drugs were discovered during the experiment of plants as oral hypoglycaemic agents; later, their interest attributed with the observation of nonmedicinal plants [17]. Semisynthetic analogs of these agents were vinorelbine and vindesine [14]. Another semisynthetic derivatives of natural product epipodophyllotoxin were etoposide and teniposide (Fig. 8.1(C) and (D)), which were used for the treatment of cancer. Epipodophyllotoxin was an isomer of podophyllotoxin (Fig. 8.1(B)), isolated as active antitumor agent from the roots of numerous species of *Podophyllum* which is used for the treatment of skin cancers and warts [17, 18].

Taxanes and camptothecins were plant-derived chemotherapeutic agents. Paclitaxel (Fig. 8.1(E)) [19] was isolated from the bark of *Taxus brevifolia* plant [20]. Initially, native American tribes use *T. brevifolia* and other *Taxus* species (e.g. *canadensis, baccata*) for the treatment of few noncancerous other diseases (Hartwell, 1982). The leaves of *T. baccata* are used as authenticate Ayurvedic medicine (Kapoor, 1990) and also used in cancer treatment [21]. Paclitaxel, along with baccatins (a precursor), is present in the leaves of *Taxus* species and plants and also takes part in semisynthetic conversion of abundant baccatins to paclitaxel. Also active paclitaxel belongs to docetaxel [22] (Fig. 8.1(F)) which provided renewable natural resources for drugs. For example, clinically active agent, topotecan (hycamptamine) (Fig. 8.1(G)), irinotecan (CPT-11;CAMPTOSAR) (Fig. 8.1(H)), and 9-aminocamptothecin were semisynthetically derived from camptothecin (Fig. 8.1(I)), which isolated from Chinese tree, *Camptotheca acuminata* [23] an ornamental tree. Topotecan is used for the treatment of ovarian and lung cancers, where Irinotecan is used for colorectal cancer treatment.

Other examples of agents which are isolated from Chinese tree are homo harringtonine (Fig. 8.1(J)), *Cephalotaxus harringtonia* var. *drupacea* (Sieb and Zucc.), and ellaptinium (Fig. 8.1(K)) were derivative of ellipticine which isolated from several species of *Bleekeriavitensis*, belongs to the family of *Apocynaceae*. *Bleekeriavitensis* was a Fijian medicinal plant that also had important anticancer properties [18]. Homo harringtonine showed efficacy for leukemias, while elliptinium was used for breast cancer treatment. The synthetic flavone, flavopiridol (Fig. 8.1(L)), due to its ideal structure which related to natural product, rohitukine, is derived from *Dysoxylum binectariferum*. Now, it comes in Phase II clinical trials of tumors [24]. While only flavopiridol probably was not a viable treatment, it conjugated with the other agents like paclitaxel and cisplatin and led to partial and complete remissions of Phase I and II experiments in patients led with paclitaxel-resistant tumors [25]. Similarly, related trials also currently underway in various cancer institutes.

The combretastatins isolate from *Combretum caffrum*, related to stilbenes family and used as medicinal antiangiogenicagents, is the cause of vascular shutdown in tumors and results in tumor necrosis [26]. A water-soluble, combretastatin A-4 phosphate [Fig. 8.1(M)] showed early clinical trials. Various plant-derived agents come under the process of clinical trials and terminate because of the absence of toxicity, for example, ronycine, bruceant [Fig. 8.1(N)], maytansine [Fig. 8.1(O)], and thalicarpine [20]. Rather, botulin in different stages of preclinical development exhibited selective activity for melanoma cell lines [27] and thapsigargin (Fig. 8.1(P)) analogs, which show activity *in vivo*, after conjugation with hexapeptide targeting prostate cancers [28].

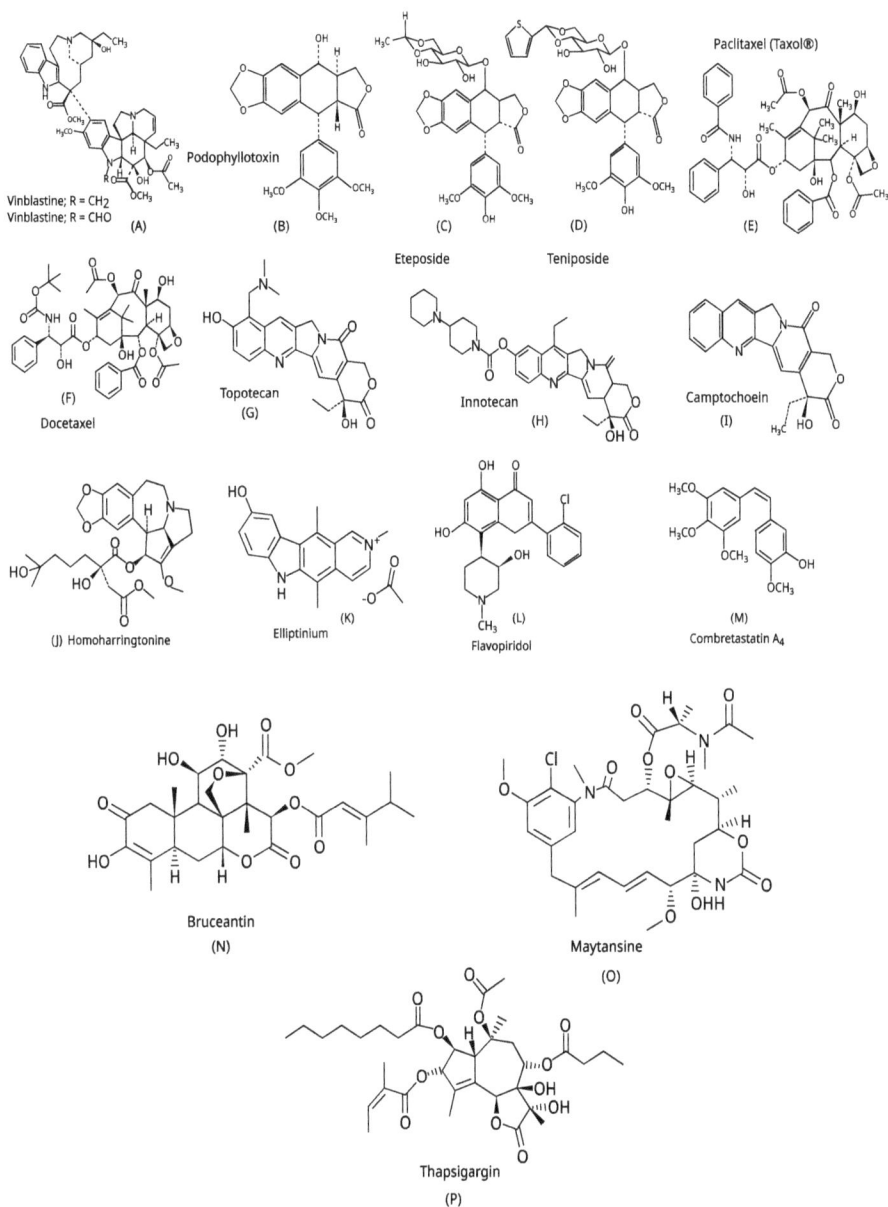

Fig. 8.1: Anticancer agents from plant derived for clinical use.

8.3.2 For cell-cycle target inhibitors

Since 1990s, research of antitumor agents from plants was specially based on cyto-toxic activity which examine against *in vivo* and *in vitro* models of cancer cell-line growth. Various natural specially plant-derived anticancer agents recognized by such system showed to exert their cytotoxic activity through interaction with tubulin. Thus, vinblastine, vincristine, colchicine, combretastatin, and maytansine agents influenced tubulin depolymerization [29]. And for taxanes, microtubules were bun-dled due to the result of stabilization against depolymerization [30]. Because of search of more effective tubulin interactive agents prepared taxol mimics which had micro-bial metabolites, epothilones, and marine invertebrate metabolites, discodermolide, eleutherobin, sarcodictyins, and laulimalides [31]. Other specific examples were deriv-atives of camptothecin, topotecan, and irinotecan which exhibit their cytotoxic actions via topoisomerase I inhibition which was specific enzyme complex and used in "wind-ing and unwinding" double-stranded DNA.

Cyclin-dependent kinases (CDKs) played an important role to regulate the pro-gression and inhibition activity of cells at special level of the cell cycle. Bis-indole, in-dirubin (Fig. 8.2 (A)) isolate from authenticate Chinese medicine, first CDK inhibitor which used by human for treatment of leukemia. Now, indirubin recognized [32] as CDK2 inhibitor which bind with ATP and showed weak affinity for CDK1. It attached with each stage of cell cycle. On the other hand, CDK1 works at G2/M interface. The other substituted indorubicins specially 3'-monooxime (Fig. 8.2(B)) and the 5-sulfonic acid (Fig. 8.2(C)) are also known as inhibitors of CDK1.

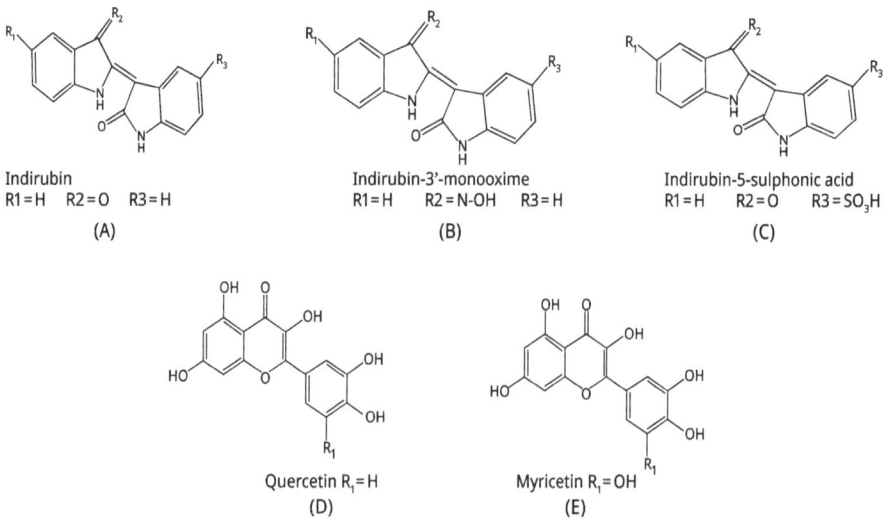

Indirubin
R1 = H   R2 = O   R3 = H
(A)

Indirubin-3'-monooxime
R1 = H   R2 = N-OH   R3 = H
(B)

Indirubin-5-sulphonic acid
R1 = H   R2 = O   R3 = SO$_3$H
(C)

Quercetin R$_1$ = H
(D)

Myricetin R$_1$ = OH
(E)

Fig. 8.2: Anticancer agents derived from plants for cell-cycle target inhibitors.

Ranelletti et al. [33] discover the CDK inhibitors, quercetin (Fig. 8.2(D)) which had antitumor property. This flavanoid could be ATP-mimic, where planar bicyclic ring of chromone was adenine isostere. Quercetin exerts their antitumor property by blocking of progression of cell cycle at $G0/G1$ interface and CDK inhibition consistent. However, myricetin (Fig. 8.2(E)) had IC_{50} which were 10 µM for CDK2 [34]. Flavopiridol (Fig. 8.1(L)) showed 100-folds and were more selective for CDKs [35].

8.3.3 For clinical development

Flavopiridol was synthetic but according to flavonoid structure, it is shown as a natural product rohitukine which derived from *Dysoxylum binectariferum* for activity of anti-inflammatory and immune modulatory. Hook. f. (Meliaceae), was an Ayurvedic plant, *Dysoxylum malabaricum* Bedd., which is applied for rheumatoid arthritis. Flavopiridol (Fig. 8.1(L)) was synthesized and used in tyrosine kinase with growth inhibitory activity for breast and lung carcinoma cell lines [36]. It showed their effect *in vivo*, which led for preclinical and clinical experiments. Recently, it is under clinical trial of phase I and II, which combine with other anticancer agents for large area of tumors like leukemias, lymphomas, and solid tumors too.

Another synthetic agent was roscovitine (Fig. 8.3(A)) which was isolated from olomucine. It also derived from cotyledons of radish, *Raphanus sativus* L. (Brassicaceae) [37]. Olomucine (Fig. 8.3(B)) inhibited CDKs proteins (Cdk proteins) and played an important role in the progression of cell cycle. Resulted, chemically modified and more influenced inhibitor, roscovitine, persist their clinical trial Phase II. Further synthesis, focused in combinatorial chemistry which led to purvalanols and more influenced, also comes under preclinical development [38].

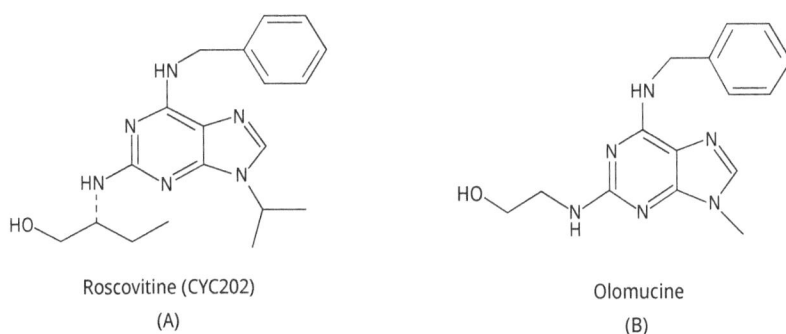

Roscovitine (CYC202)

(A)

Olomucine

(B)

Fig. 8.3: Anticancer agents from plant-derived, use for clinical development.

8.3.4 Target to natural products

Natural products had specific importance as anticancer agents; especially for cancer chemotherapy, it had low solubility in aqueous with lesser therapeutic indices. These factors demised various pure natural products like plant-derived agents, bruceantin and maytansine, but it is an optional method which utilized those agents to investigate their potential as "warheads" which attached monoclonal antibodies on tumors [39].

At early 1970s, maytansine (Fig. 8.1(O)) was derived from Ethiopian plant, *Maytenus serrata* (Hochst. Ex A. Rich.) Wilczek (Celastraceae) had less yield (2×10^{-5}% dry plant dry weight) which had high potency for experiment on cancer cell lines which used for preclinical and clinical development. Related compounds, ansamitocins, were derived from actinomycete (*Actinosynnema pretiosum*). The derivative of maytansine, DM1 conjugate with monoclonal antibody (mAb) which target small cell, lung cancer cells which developed huN901-DM1 and use for small-cell lung cancer treatment. Another conjugate of DM1 with J591, a mAb target the prostate-specific membrane antigen, were under clinical trials of prostate cancer. Another conjugation, known as SB408075 or huC242-DM1 (*Cantuzumab mertansine*), developed through combination of DM1 with huC242, a mAb directed through the *muc1* epitope of cancers, where pancreatric, biliary, colorectal, and gastric cancers come under Phase I clinical trials in the United States.

Thapsigargin (TG) (Fig. 8.1(P)) was derived from umbelliferous plant, *Thapsiagarganica* L. (Apiaceae) [40]. It induced cell death in quiescent and proliferating prostate cancer cells. It conjugated with small peptide, carried, and produced a water-soluble prodrug which specially activated through prostate-specific antigen, protease at metastatic prostate cancer. It is used for the treatment of prostate cancer xenograft tumor which generally occurs in animal and inhibits the tumor growth. This prodrug stable in plasma of human is also used in human prostate cancer treatment.

8.3.5 Anti-HIV medicinal plants

From the last few decades, more than 75 million people in world were infected by HIV virus and around 37 million people are still alive and live with this viral infection [41, 42]. It was estimated that around ~26 million HIV patients are present in Africa; ~3.3 million in America; ~3.5 million in Southeast Asia; ~2.4 million in Europe; ~360,000 in eastern Mediterranean; and 1.5 million in western Pacific [42].

Since 1980s, HIV was recognized by the world and still there was no treatment or any useful vaccine available for HIV infection, but there were few special advance treatment, control, and prevention available [43]. The anti-HIV agents and highly active antiviral therapy discovered in 1996, which ideally kill the virus or decrease the mortality of HIV/AIDS. Now, antiviral therapy is highly prescribed for adult HIV patients. The reduction of mortality of HIV virus changed the disease from fatal to chronic disease which comes under manageable condition (42 [44–46]). Antiretroviral agents attract their interest for HIV treatment and prevention [47]. These days, HIV-infected patients are treated with antiretroviral therapy (ART) for preventing disease progression which improve the treatment and reduce AIDS-associated eminent, HIV eminent and all-cause mortality, and also reduced transmission of virus [47]. Agents of antiretroviral suppress HIV and also prevent from new HIV infections [47].

ART did not use without serious adverse case, especially the patient undergoing long-term treatment. Those therapies were controlled by multidrug resistance [48]. And new drugs with their targets had to overcome from HIV reservoirs in body with complete eradication of HIV and AIDS. For the last 10 years, HIV latency had influenced to discover various drugs which was able to reactivate the selective proviruses without activation of polyclonal T cell [49]. Nowadays, significant potential for naturally derived products or authenticate medicines form medicinal plants for preventing HIV infections and symptoms *in vivo* too.

8.3.6 Anti-HIV agents that derived from plant

From 1987 to 1996, more than 30,000 plant extracts are used *in vitro* cell for anti-HIV screening and treatment which identify replication of HIV-1 from infected lymphoblastic cells on the bases of treated, uninfected control cells.

Michellamine B (Fig. 8.4(A)) were isolated from leaves of liana (*Ancistrocladus korupensis*) as main *in vitro* active agent. This plant occurred Korup region of southwest Cameroon [50]. According to experiments in dogs, concentration of anti-HIV effective *in vivo* was achieved only close to toxic level. Thus, *in vitro* activity of HIV-1 and HIV-2 strains, where the difference between both strains, has high toxicity level and the anticipated level required small effective antiviral activity. However, the discovery of antimalarial agents, korupensamines, from same species [51] proves to be more prominent.

The extract of leaves and twigs of *Calophyllum lanigerum*tree which collected from Sarawak, Malaysia in 1987, yielded (+)-calanolide A (Fig. 8.4(B)) and had identical anti-HIV activity [52]. According to the detailed discovery of *C.lanigerum* with species, resultant extract of latex of *Calophyllumteysmanii* had special activity of anti-HIV agents. The active eminent was (−)-calanolide B (Fig. 8.4(C)), which isolated in 20–30% yield, where (−)-calanolide B was less active than (+)-calanolide A. These days, drugs were under Phase II clinical trials and (−)-calanolide B in preclinical development.

From the wood of *Homalanthus nutans* tree, prostratin (Fig. 8.4(D)) was isolated as the HIV active eminent [53], which is discovered by Dr Paul Cox and applied it in treatment of yellow fever (hepatitis). Prostratina HIV activator, expressed in infected T-cell linesn [54], and its potential value lie as viral activator rather than an anti-HIV agent.

Extracts of "smokebush" isolated from *Conospermum incurvum* were collected from Western Australia, and resultant conocurvone was used as HIV active agents [55].

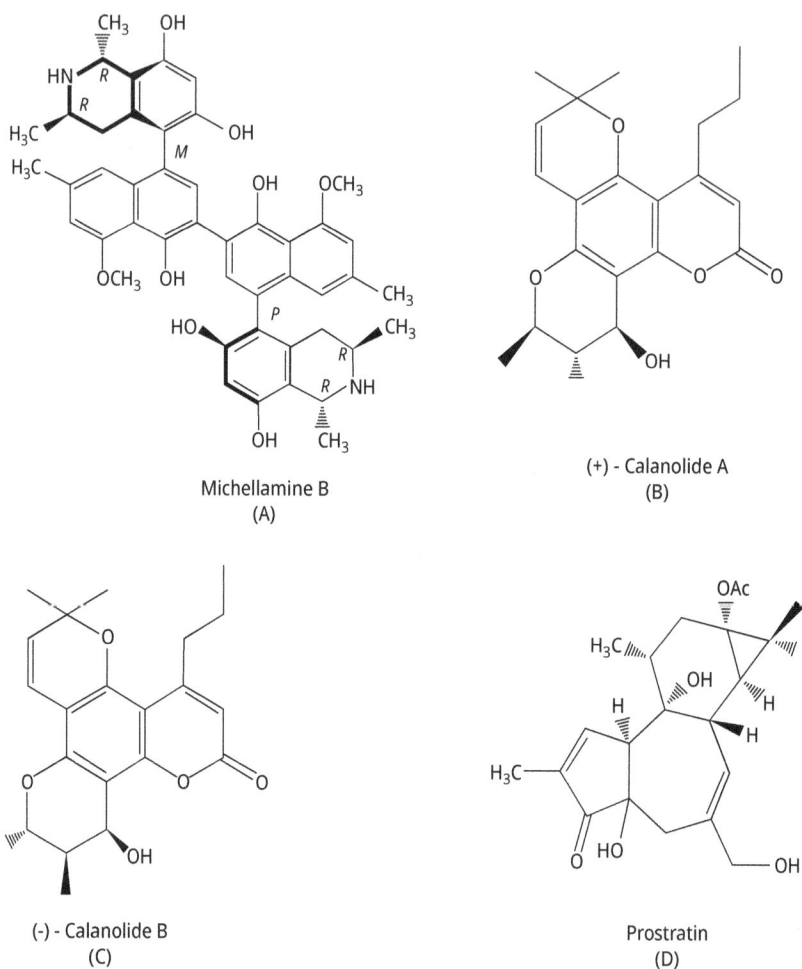

Fig. 8.4: Plant-derived anti-HIV agents.

8.4 Extracts of plant and their metabolites for anti-HIV activity

8.4.1 *Artemisia annua* L. (Asteraceae)

From *Artemisia annua* L. (*Asteraceae*), a Chinese medicinal plant prepared anti-HIV active tea infusion with cellular systems. It was highly active with IC_{50} (2.0 µg/mL). In generally, at 25 µg/mL, artemisinin was inactive. But in their related species *Artemisia afra*, where artemisinin was absent, also showed similar activity effect [56].

8.4.2 *Astragalus membranaceus* Bunge (Fabaceae)

Astragalus membranaceus Bunge (*Fabaceae*) was a Chinese traditional medicine for immune-stimulation. Immune-suppressed and immune-competent human patients used *Astragalus* extracts for the demonstration of restoration and augmentation of local graft. *Astragalus* extracts had better symptomology for HIV patients. Thus, it was safe and mutagenecity is still in the examination process [57].

8.4.3 *Calendula officinalis* (Asteraceae)

Calendula officinalis (*Asteraceae*) was as Indian medicinal plant. Their flower was used as ointments for wounds treatment, herpes, ulcers, frostbite, skin damage, scars, blood purification, etc. The infusion from leaves could be used for treating varicose veins. The *Calendula officinalis* flower extract with dichloromethane-methanol (1:1) exhibited potent anti-HIV activity, (3-(4,5-dimethylthiazolyl-2)-2,5-diphenyltetrazolium bromide) (MTT)/tetrazolium based assay *in vitro*. At 1,000 µg/mL concentration of HIV1-RT, this activity was inhibited and the suppression of HIV mediated fusion attributed at 500 µg/mL [58]. The aqueous and organic extracts of dried flowers of *Calendula officinalis* were tested for the inhibition of HIV-l replication. Both flower extracts were nontoxic for lymphocytic Molt-4 cells in human, but organic extract exhibited, MTT ketrazolium-based assay showed anti-HIV activity *in vitro*. In the presence of 500 µg/mL organic extract, the uninfected Molt-4 cells were completely protected up to 24 h from fusion and subsequent death because of cultivation with persistently infected U-937/HIV-1 cells. The organic extract of *Calendula* was used as significant and time-dependent reduction of HIV-l reverse transcription (RT) activity. After 30 min treatment of partially purified enzyme, 85% RT inhibition was achieved in cell-free system. It proved that the organic extract of *Calendula oflicinalis* flower had anti-HIV properties [52].

8.4.4 *Cassia abbreviate* (Fabaceae)

Cassia abbreviate (*Fabaceae*) occurs in Botswana and also used as authenticate medicine for HIV/AIDS patients. It examined the inhibitory effects on HIV replication against cytopathic effect of HIV-1c (MJ4), which was protected by viral p24 antigen in infected PBMCs. *Cassia sieberiana* and *Cassia abbreviata* extraction showed remarkable inhibition of HIV-1c (MJ4) replication. The concentration-dependent, root and bark extracts of *Cassia sieberiana* and *Cassia abbreviate* had effective concentration (EC$_{50}$) on 65.1, 85.3, and 102.8 µg/mL, respectively [59].

8.4.5 *Chelidonium majus* L. (Papaveraceae)

The antiviral active and freshly synthesized crude extract of *Chelidonium majus*
L. (*Papaveraceae*) was characterized and low-sulfated poly-glycosaminoglymoiety
(molecular weight of ~3,800 Da) was isolated from it [173]. The extract of *Chelidonium*,
prevented infection of HIV-1 at concentration of 25 µg/mL in human CD4$^+$ T-cell lines
AA2 and H9. It also determined reverse transcriptase (RT) activity of cell-to-cell virus
spreading in H9 cells which continuously infected with HIV-1 and culture p24 contents
in cell [60].

Tab. 8.1: List of following plant species with activation of different HIV inhibition.

Family	Plant	Plant part	HIV-RT	HIV-PR	HIV-IN	Anti-HIV
Asteraceae	*Artemisia annua* L.	Aerial part				Crude [56]
Fabaceae	*Astragalus membranaceus* Bunge	Aerial part				Crude [57]
Asteraceae	*Calendula officinalis* L.	Leaf	Crude [57]			
Fabaceae	*Cassia abbreviate*	Bark		Crude [52]		
Papaveraceae	*Chelidonium majus* L.	Seed				Crude [59]
Combretaceae	*Combretum molle* R. Br. ex G. Don	Root				Crude [60]
Ebenaceae	*Diospyros lotus* L.	Stem	Crude [61]			
Asteraceae	*Dittrichia viscosa*		Crude [62]			
Amaryllidaceae	*Galanthus nivalis* L.					Crude [63]
Clusiaceae	*Garcinia edulis* Exell	Stem bark				Crude [64]
Asteraceae	*Helichrysum populifolium* DC. A	Aerial part				Crude [65]
Hypericaceae	*Hypericum perforatum* L.					Crude [66]
Lamiaceae	*Hyssopus officinalis* L.	Leaf	Crude [67]			
Cucurbitaceae	*Momordica charantia* L.	Seed, Fruit				Crude [68]

Tab. 8.1 (continued)

Family	Plant	Plant part	HIV-RT	HIV-PR	HIV-IN	Anti-HIV
Polyporaceae	*Pachyma hoelen Rumph*	Whole plant				Crude [69]
Euphorbiaceae	*Phyllanthus pulcher*		Crude [70]			
Anacardiaceae	*Rhus chinensis* Mill.	Leaf, Root, Stem, Bark, Fruit				Read phyto [71]
Smilaceae	*Smilax corbularia* Kunth	Plant				Crude [72]
Combretaceae	*Terminalia paniculate*					Crude [73]

8.4.6 *Combretum mole* (R. Br. ex. G. Don.) Engl and Diels (Combretaceae)

Anti-HIV activity of *Combretum mole* extract bark used *in vitro* is largely applied in Ethiopian authenticate medicine for the treatment of liver diseases, malaria, and tuberculosis. It is also used for HIV virus type 1 (HIV-1) and type 2 (HIV-2). These extracts are synthesized by percolation with petroleum ether, chloroform, acetone, and methanol extract which obtained through successive hot extraction by soxhlet. Selective inhibition of viral growth measured by simultaneous determination of cytotoxicity of extracts verses MT-4 cells *in vitro* [61]. The obtained results showed acetone fraction which had high inhibition of HIV-1 replication. Phytochemical experiments of the acetone fraction showed isolation of two tannins and two oleanane-type pentacyclic triterpene glycosides. One of the tannins was punicalagin (an ellagitannin), and the other molecule (CM-A) did not fully elucidate. Both punicalagin and CM-A had selective inhibition of HIV-1 replication with selective indices.

8.4.7 *Diospyros lotus* L. (Ebenaceae)

Methanol extract of *Diospyros lotus* L. (*Ebenaceae*) fruits was anti-HIV-1 active. Gallic acid, highly active compound against HIV-1 with the value of therapeutic index (TI) > 32.84 but other compounds had lesser active. *Diospyros lotus* fruits provided a chemical reservoir of anti-HIV agents. All compounds examined for their cytotoxicity and anti-HIV-1 activities. Gallic acid inhibited HIV-1 replication and EC_{50} was 6.09 μg/mL including the value of TI > 32.84 higher than the other compounds [62].

8.4.8 *Dittrichia viscosa* (L.) *Greuter* (Asteraceae)

For therapeutic interest, the extract from *Dittrichia viscosa* (L.) *Greuter* (*Asteraceae*) was examined to inhibit the HIV replication. MT-2 cells are used in experiments of determination of rapid antiviral effect and sensitive assay for the identification of significant antiviral drugs which are effective on AIDS. The *Dittrichia viscosa* extract had inhibitory effects against HIV-1 which induced infections in 25–400 µg/mL concentrated MT-2 cells [63].

8.4.9 *Galanthus nivalis* L. (Amaryllidaceae)

Agglutinin extracts from *Galanthus nivalis* L. (*Amaryllidaceae*) (GNA), member of monocotyledonous plants, also had a large area for biological activities like antitumor, antiviral, and antifungal [64]. The molecular structure of GNA exerting antiviral activities by stop the entry of virus into target cells, prevent transmission of virus, and also force virus to remove glycan from packet of protein and stimulate neutralized antibody. These results showed new perspective of GNA-related lectins and ideal drug for virus protection and treatment of HIV in the future.

8.4.10 *Garcinia edulis* Exell (Clusiaceae)

The derivative of iso-prenylated xanthone is determined as 1,4,6-trihydroxy-3-methoxy-2-(3-methyl-2-butenyl)-5-(1,1-dimethyl-prop-2-enyl) xanthone, which isolated from root bark ethanol extract of *Garcinia edulis* Exell (*Clusiaceae*) plant. It exhibited anti-HIV-1 protease activity with 11.3 µg/mL IC_{50} *in vitro* and acetyl pepstatin was used as anti-HIV-1 PR activity with 2.2 µg/mL IC_{50} [65]. This compound also showed potent lethality with LC50 (2.36 µg/mL) versus brine shrimp larvae *in vitro*.

8.4.11 *Helichrysum populifolium* (Asteraceae)

Helichrysum populifolium (*Asteraceae*) occurs in South Africa and methanol: water (1:1) mixture extract of aerial parts of *Helichrysum* examine for anti-HIV test by HeLa-SXR5 expressed by CD4 receptor, tCXCR4/CCR5 chemokine receptors, and extract was active (IC50 value of 12 µg/mL) [66]. Anti-HIV compounds from *Helichrysum* had three dicaffeoylquinic acid derivatives: 3,4-dicaffeoylquinic acid, 3,5-dicaffeoylquinic acid, and 4,5-dicaffeoylquinicacid with two tricaffeoylquinic acid derivatives: 1,3,5-tricaffeoylquinic acid and 5-malonyl-1,3,4-tricaffeoylquinic or 3-malonyl-1,4,5-tricaffeoylquinic acid.

8.4.12 *Hypericum perforatum* L. (Hypericaceae)

Hypericum perforatum L. (*Hypericaceae*), also called St. John's Wort, was used as medicinal drug and wound healing. It is also applied in the treatment of AIDS [67]. For clinical trial, hypericin and pseudo hypericin were isolated from it and examined the antiviral activity in HIV-infected patients [74].

8.4.13 *Hyssopus officinalis* L. (Lamiaceae)

Hyssopus officinalis L. (*Lamiaceae*) was used as herbal medicine. Due to the presence polysaccharide-type compounds in extract of this species, it demonstrated strong activity against HIV-1 [68]. The 50% hydro-alcoholic extract of *Hysoppus officinalis* was examined for the inhibition of HIV replication. Among the numerous evaluating antiviral experiments, HIV infectious MT-2 cells were used as rapid and sensitive assay system for the identification of antiviral drugs which is effective for AIDS. This extract had inhibitory effects at 50–100 µg/mL concentration against HIV-1-induced infections in MT-2 cells.

8.4.14 *Momordica charantia* L. (Cucurbitacae)

Momordica charantia L. (*Cucurbitacae*) also called bitter melon and widely used in folkloric medicine, which showed inhibition of HIV-1 RT because of MRK29-coded protein [75]. The molecular structure of bitter gourd-induced anti-diabetic, anti-HIV, and antitumor activities used 20 components which were already known. Thus, bitter gourd also had clinical application in near future. The extract of *Momordica* fruit pulp is commonly applied in northern part of Nigeria for its antiviral efficacy in poultry and also had potent inhibitor of HIV-1 replication [76].

8.4.15 *Pachyma hoelen* Rumph (Polyporaceae)

Pachyma hoelen Rumph (*Polyporaceae*) occurs in Korea and the hexane extract was used in folk medicine for anti-HIV-1 activity. This extract had 37.3 µg/mL (EC_{50}) concentration on the p24 antigen assay and had high value, 36.8% on RT activity test (200 µg/mL). It showed 58.2% highest protective effects on infected MT-4 cells. The 50% cytotoxic concentration (CC_{50}) of hexane extract had 100.6 µg/mL concentration value [69].

8.4.16 *Phyllanthus pulcher* (Euphorbiaceae)

Phyllanthus pulcher (*Euphorbiaceae*) species are found in Malaysia and the methanol extract is measured for anti-HIV-1 RT activity with HIV-RT assay by the inhibition of HIV-1 RT enzyme-based IC_{50} values. Azido-deoxythymidine-triphosphate (AZT151TP) applies for controlling the infection of virus. The inhibition of HIV-RT for *P. pulcher* had 5.9 μg/mL IC_{50} [70].

8.4.17 *Rhus chinensis Mill* (Anacardiaceae)

Rhus chinensis Mill (*Anacardiaceae*) occurs in China and Japan, and their petroleum ether, ethyl acetate, butanol, and aqueous extracts are used for anti-HIV-1 activities. It is also called Chinese Sumac. The petroleum ether extract had significantly suppressed the HIV-1 activity *in vitro* and inhibit syncytium formation and HIV-1 p24 antigen at noncytotoxic concentrations, the EC50 were 0.71 and 0.93 μg/mL, respectively. The petroleum ether extract did not show any activity on inhibiting HIV-1 into host cell cycle. *R. chinensis* was highly important and ideal medicinal plant for chemotherapy against HIV-1 infection. The petroleum ether extract was also able to target the new sites of HIV-1 replication [71].

8.4.18 *Smilax corbularia Kunth* (Smilaceae)

Ethanolic and aqueous extracts of *Smilax corbularia Kunth* (Smilaceae) were examined for the inhibition effects against HIV-1 protease (HIV-PR) and HIV-1 integrase (HIV-1 IN). The obtained results showed that ethanolic extract of *S. corbularia* exhibited anti-HIV-1 IN activity and had 1.9 μg/mL of IC_{50} value. The ICF_{50} value of water extract was 5.4 μg/mL, respectively [72].

8.4.19 *Terminalia paniculate* (Combretaceae)

The anti-HIV-1 active, acetone, and methanol extracts were synthesized from *Terminalia paniculate* (*Combretaceae*) fruit. The EC_{50} value was ≤10.3 μg/mL. The enzymatic assays determined by the mechanism of action significantly showed anti-HIV1 activity because of the inhibition of RT enzymes (≥77.7% inhibition) and protease (≥69.9% inhibition) [73, 77].

8.5 Conclusion

Plants had ideal source of highly effective conventional drugs for the treatment of many dangerous forms of cancer and HIV but actual compounds isolated from those plant frequently but itself did not use as drugs. Those plants influenced the development of ideal agents. Some agents failed earlier clinical trials and now stimulate interest in cancer. The ability of attached agents to carry molecules directed to specific tumors and targeted highly cytotoxic natural products for tumors which had nontoxic effects on normal healthy tissues. With rapid recognition of new proteins have regulatory effects on tumor cell-cycle progression and their conversion into targets for high throughput screening, molecules isolated from plants, and other natural organisms proved as important source of novel inhibitors and also had potential for development of selective anticancer agents.

In HIV chemotherapy, anti-HIV drugs are always needed, and medicinal plants are used as an ideal role in discovery of anti-HIV drugs. Various plant species showed significant anti-HIV activity, especially *Artemisia annua, Calendula officinalis* L., *Galanthus nivalis* L., *Phyllanthus pulcher, Rhus chinensis, Smilax corbularia, Terminalia paniculate*, etc. These plant species are worthy for the further development of new anti-HIV for chemotherapeutic options. Specially, *in vivo* experiments and clinical trials on humans are necessary for carrying out this research on phytochemical isolation from plants with regular recognition of medicinal plants for anti-HIV and anticancer activities.

References

[1]   H.-M. Chang, P. P. But, S.-C. Yao. World Scientific. 1986.
[2]   L. Kapoor. CRC Handbook of Ayurvedic Medicinal Plants. CRC Press, Boca Raton, 1990, pp. 86.
[3]   A. D. Kinghorn. Discovery Novel Nat Prod Ther Potential. 1994, 81–108.
[4]   S.-J. Li, X. Zhang, X.-H. Wang, C.-Q. Zhao. Eur J Med Chem. 2018, 156, 316–343.
[5]   M. Wink, A. W. Alfermann, R. Franke, B. Wetterauer, M. Distl, J. Windhövel, et al. J Plant Genetic Resour. 2005, 3(2), 90–100.
[6]   A. Roy, N. Jauhari, N. Bharadvaja. Anticancer Plants: Natural Products and Biotechnological Implements. Springer, 2018, pp. 109–139.
[7]   A. Majumder, S. Jha. J Biol Sci. 2009, 1(1), 46–69.
[8]   X. Xin-Ya, W. Dong-Ying, K. Chuen-Fai, Z. Yang, H. Cheng, L. Kang-Lun, et al. Chin J Nat Med. 2019, 17 (12), 945–952.
[9]   X. Zuo, Z. Huo, D. Kang, G. Wu, Z. Zhou, X. Liu, et al. Expert Opin Ther Pat. 2018, 28(4), 299–316.
[10]   B. B. J. Petrovska. Pharmacognosy Rev. 2012, 6(11), 1.
[11]   M. Fridlender, Y. Kapulnik, H. J. Koltai. Front Plant Sci. 2015, 6, 799.
[12]   J. L. Hartwell. Quarterman Publications, 1982.
[13]   G. M. Cragg, D. G. Kingston, D. J. Newman. CRC press. 2005.
[14]   D. J. Newman, G. M. Cragg, K. M. Snader. J Nat Prod. 2003, 66(7), 1022–1037.
[15]   J. M. Cassady, J. D. Douros. Academic press. 1980.

[16] F. Gueritte, J. Fahy. Brunner-Routledge Psychology Press. Taylor & Francis Group, Boca Raton, FL, 2005, pp. 123–136.

[17] M. Snader, T. G. McCloud. Ethnobot Search New Drugs. 1994, 185, 178.

[18] G. M. Cragg, M. R. Boyd, J. H. Cardellina 2nd, D. J. Newman, K. M. Snader, T. G. McCloud. Ciba Found Symp. 1994, 185, 178–190. discussion 90-6.

[19] D. G. J. C. C. Kingston. Chem Commun. 2001, 10, 867–880.

[20] G. M. Cragg, S. A. Schepartz, M. Suffness, M. R. J. Grever. J Nat Prod. 1993, 56(10), 1657–1668.

[21] J. J. Q. P. Hartwell. Quarterman Publication, 1982, 438–439.

[22] J. E. Cortes, R. Pazdur. Journal of clinical oncology: Official journal of the American Society of Clinical Oncology. 1995, 13(10), 2643–2655.

[23] M. Potmesil, H. M. Pinedo. CRC press. 1994.

[24] M. C. Christian, J. M. Pluda, P. Ho, S. G. Arbuck, A. J. Murgo, E. A. Sausville. editors. Seminars in Oncology. 1997.

[25] A. Kaubisch, G. K. J. C. J. Schwartz. Cancer J (Sudbury, Mass). 2000, 6(4), 192–212.

[26] S. Holwell, P. Cooper, K. Grosios, J. Lippert 3rd, G. Pettit, S. Shnyder, et al. Anticancer Res. 2002, 22 (2A), 707–711.

[27] E. Pisha, H. Chai, I. Lee, T. Chagwedera. Nat Med. 1995, 1(10), 1046–1051.

[28] C. M. Jakobsen, S. R. Denmeade, J. T. Isaacs, A. Gady, C. E. Olsen, S. B. Christensen. J Med Chem. 2001, 44(26), 4696–4703.

[29] E. K. D. R. C. Rowinsky. Cancer Chemotherapy and Biotherapy. 1996, pp. 263–296.

[30] M. Rowinsky, K. Eric. Annu Rev Med. 1997, 48(1), 353–374.

[31] L. He, G. A. Orr, S. B. Horwitz. Drug Discovery Today. 2001, 6(22), 1153–1164.

[32] R. Hoessel, S. Leclerc, J. A. Endicott, M. E. Nobel, A. Lawrie, P. Tunnah, et al. Nature Cell Biology. 1999, 1(1), 60–67.

[33] F. O. Ranelletti, R. Ricci, L. M. Larocca, N. Maggiano, A. Capelli, G. Scambia, et al. Int J Cancer. 1992, 50(3), 486–492.

[34] D. Walker. Cyclin Dependent Kinase (CDK) Inhibitors, 1998, 149–165.

[35] T. M. Sielecki, J. F. Boylan, P. A. Benfield, G. L. J. Trainor. J Med Chem. 2000, 43(1), 1–18.

[36] E. A. Sausville, D. Zaharevitz, R. Gussio, L. Meijer, M. Louarn-Leost, C. Kunick, et al. Pharmacology Therapeutics. 1999, 82(2-3), 285–292.

[37] L. Meijer, E. J. Raymond. Acc Chem Res. 2003, 36(6), 417–425.

[38] Y.-T. Chang, N. S. Gray, G. R. Rosania, D. P. Sutherlin, S. Kwon, T. C. Norman, et al. Chem Biol. 1999, 6(6), 361–375.

[39] release]. San Diego,: Academic Press, 1997.

[40] S. R. Denmeade, C. M. Jakobsen, S. Janssen, S. R. Khan, E. S. Garrett, H. Lilja, et al. J Natl Cancer Inst. 2003, 95(13), 990–1000.

[41] S. G. Deeks, J. Overbaugh, A. Phillips, S. J. Buchbinder. Nat Rev Dis Primers. 2015, 1(1), 1–22.

[42] (WHO) WHO. 2017 [Available from: http://www.who.int/hiv/data/epi_plhiv_2016_regions.png?ua=1

[43] P. Piot, S. S. A. Karim, R. Hecht, H. Legido-Quigley, K. Buse, J. Stover, et al. The Lancet. 2015, 386 (9989), 171–218.

[44] A. B. Kharsany, Q. A. Karim. The Open AIDS J. 2016, 10, 34.

[45] B. Auvert, D. Taljaard, E. Lagarde, J. Sobngwi-Tambekou, R. Sitta, A. J. Puren. PLoS Med. 2005, 2(11), e298.

[46] R. C. Bailey, S. Moses, C. B. Parker, K. Agot, I. Maclean, J. N. Krieger, et al. Lancet. 2007, 369(9562), 643–656.

[47] H. F. Günthard, M. S. Saag, C. A. Benson, C. Del Rio, J. J. Eron, J. E. Gallant, et al. Jama. 2016, 316(2), 191–210.

[48] J. Sharifi-Rad. Cell Mol Biol (Noisy-le-grand). 2016, 62(9), 1–2.

[49] M. B. Lucera, C. A. Tilton, H. Mao, C. Dobrowolski, C. O. Tabler, A. A. Haqqani, et al. J Virology. 2014, 88(18), 10803–10812.

[50] M. R. Boyd, Y. F. Hallock, J. H. Cardellina, K. P. Manfredi, J. W. Blunt, J. B. McMahon, et al. J Med Chem. 1994, 37(12), 1740–1745.

[51] Y. F. Hallock, K. P. Manfredi, J. W. Blunt, J. H. Cardellina, M. Schaeffer, K.-P. Gulden, et al. J Org Chem. 1994, 59(21), 6349–6355.

[52] Y. Kashman, K. R. Gustafson, R. Fuller, J. Cardellina 2nd, J. McMahon, M. Currens, et al. J Med Chem. 1992, 35(15), 2735–2743.

[53] K. R. Gustafson, J. H. Cardellina, J. B. McMahon, R. J. Gulakowski, J. Ishitoya, Z. Szallasi, et al. J Med Chem. 1992, 35(11), 1978–1986.

[54] R. J. Gulakowski, J. B. McMahon, R. W. Jr Buckheit, K. R. Gustafson, M. R. J. Boyd. Antiviral Res. 1997, 33(2), 87–97.

[55] L. A. Decosterd, I. Parsons, K. Gustafson, J. Cardellina, J. McMahon, G. Cragg, et al. J Am Chem Soc. 1993, 115(15), 6673–6679.

[56] J. H. Burack, M. R. Cohen, J. A. Hahn, D. I. Abrams. J Acquired Immune Deficiency Syndromes. 1996, 12(4), 386–393.

[57] B. Muley, S. Khadabadi, N. Banarase. Trop J Pharm Res. 2009, 8, 5.

[58] Z. Kalvatchev, R. Walder, D. J. B. Garzaro. Pharmacotherapy Biomed Pharmacother. 1997, 51(4), 176–180.

[59] M. Gerenčer, P. L. Turecek, O. Kistner, A. Mitterer, H. Savidis-Dacho, N. P. Barrett. Antiviral Res. 2006, 72(2), 153–156.

[60] K. Asres, F. J. Bucar. Ethiopian Med J. 2005, 43(1), 15–20.

[61] K. Rashed, X.-J. Zhang, M.-T. Luo, Y.-T. J. Zheng. Phytopharmacology. 2012, 3(2), 199–207.

[62] L. Bedoya, S. S. Palomino, M. Abad, P. Bermejo, J. Alcami. Phytothe Res: Int J Devoted Pharmacol Toxicol Eval Na Prod Derivatives. 2002, 16(6), 550–554.

[63] L. Wu, J.-K. Bao. Glycoconjugate J. 2013, 30(3), 269–279.

[64] J. J. J. Magadula. Fitoterapia. 2010, 81(5), 420–423.

[65] J. Hudson, L. Harris, G. J. Towers. Antiviral Res. 1993, 20(2), 173–178.

[66] H. M. Heyman, F. Senejoux, I. Seibert, T. Klimkait, V. J. Maharaj, J. J. M. Meyer. Fitoterapia. 2015, 103, 155–164.

[67] R. M. Gulick, V. McAuliffe, J. Holden-Wiltse, C. Crumpacker, L. Liebes, D. S. Stein, et al. Ann Internal Med. 1999, 130(6), 510–514.

[68] S. Miraj, N. Azizi, S. J. D. P. L. Kiani. Der Pharmacia Lett. 2016, 8(6), 229–237.

[69] S.-A. Lee, S.-K. Hong, C.-I. Suh, M.-H. Oh, J.-H. Park, B.-W. Choi, et al. J Microbio. 2010, 48(2), 249–252.

[70] -R.-R. Wang, Q. Gu, L.-M. Yang, -J.-J. Chen, S.-Y. Li. Zheng Y-T. J Ethnopharmacol. 2006, 105(1-2), 269–273.

[71] A. Durge, P. Jadaun, A. Wadhwani, A. A. Chinchansure, M. Said, H. Thulasiram, et al. Natural Product Research. 2017, 31(12), 1468–1471.

[72] S. Tewtrakul, S. Subhadhirasakul, S. J. S. J. S. T. Kummee. Songklanakarin J Sci Technol. 2006, 28(4), 785–790.

[73] W. Jian, Y. Feng-Zhen, Z. Min, Z. Yun-Hui, Z. Yong-Xiang, L. Ying, et al. Chin J Integr Med. 2006, 12(1), 6–11.

[74] H.-J. Zhang, E. Rumschlag-Booms, Y.-F. Guan, D.-Y. Wang, K.-L. Liu, W.-F. Li, et al. J Nat Prod. 2017, 80 (6), 1798–1807.

[75] Y. Bot, L. Mgbojikwe, C. Nwosu, A. Abimiku, J. Dadik, D. J. Damshak. Afr J Biotechnol. 2007, 6, 1.

[76] I. Eldeen, E. Seow, R. Abdullah, S. Sulaiman. S Afr J Botany. 2011, 77(1), 75–79.

[77] B. Salehi, N. V. Kumar, B. Şener, M. Sharifi-Rad, M. Kılıç, G. B. Mahady, S. Vlaisavljevic, M. Iriti, F. Kobarfard, W. N. Setzer, S. A. Ayatollahi. Medicinal plants used in the treatment of human immunodeficiency virus. Int J Mol Sci. 2018, 19(5), 1459.

Mayur Mausoom Phukan, Pranay Punj Pankaj, Samson Rosly Sangma,
Ramzan Ahmed, Kumar Manoj, Jayabrata Saha, Manjit Kumar Ray,
Amenuo Susan Kulnu, Rupesh Kumar, Kalpana Sagar,
Pranjal Pratim Das and Plaban Bora

Chapter 9
Antimicrobial, anti-inflammatory, and wound-healing activities of medicinal plants

Abstract: Plants and the compounds derived from them have a long history of usage in the treatment and management of a wide variety of microbial diseases, wounds, and inflammation. This is likely due to the decreased risk of side effects associated with their use. The hunt for substitute methods or drug therapy, particularly those derived from medicinal plants, can be utilized to develop an economical, sustainable, stable, and effective delivery system for the treatment and allopathic management. The article aims to provide an overview to the readers, which will enable them in having a better comprehensive understanding of the functions that plant-based components play in the treatment and management of microbial disease, wounds, and inflammation.

Mayur Mausoom Phukan, Samson Rosly Sangma, Department of Forestry, School of Science, Nagaland University, Lumami 798627, Nagaland, India

Pranay Punj Pankaj, Department of Zoology, School of Science, Nagaland University, Lumami 798627, Nagaland, India

Ramzan Ahmed, Jayabrata Saha, Manjit Kumar Ray, Department of Applied Biology, School of Biological Sciences, University of Science and Technology, Meghalaya, Baridua, Ribhoi 793101, Meghalaya, India

Kumar Manoj, Department of Botany, Marwari College, Tilka Manjhi Bhagalpur University, Bhagalpur 812007, Bihar, India

Amenuo Susan Kulnu, Department of Environmental Science, School of Science, Nagaland University, Lumami 798627, Nagaland, India

Rupesh Kumar, Department of Biotechnology, The Assam Royal Global University, Betkuchi 781035, Guwahati, Assam, India

Kalpana Sagar, Department of Botany and Microbiology, Faculty of Life Science, Gurukula Kangri University, Haridwar, 249404, Uttarakhand, India

Pranjal Pratim Das, Department of Biotechnology, Guwahati University, Tezpur 784001, Assam, India

Plaban Bora, Department of Energy Engineering, Assam Science and Technology University, Guwahati 781013, Assam, India

https://doi.org/10.1515/9783110791891-009

9.1 Introduction

Medicinal plants and herbs have been in use even much before the dawn of human civilization. Medicinal plants form the basis of sophisticated traditional medicine systems that have been in existence since millennia and still providing mankind with new and advanced remedies [1]. Plant-based conventional medications have been extensively used throughout the long annals of medical history, Ayurveda, Unani, Siddha, and Traditional Chinese Medicine being the most sought after. Indian saints like Charaka and Sushruta had long documented the therapeutic uses of medicinal plants for various diseases in their masterpieces viz. the *Charak Samhita* and the *Sushruta Samhita.*

It is a well-established fact that a significant fraction of the global population in underdeveloped nations rely on traditional plant-based medicines (sometimes in conjugation with modern medicines) to treat various ailments. Recently, there has been a renewed interest in these traditional plant-based medicines in contrast to synthetic drugs, the latter mainly being credited for their side effects and drug resistance. With this renewed interest in plant-based remedies and the burgeoning need to develop novel effective drugs, conventionally used medicinal plants have received the attention of pharmaceutical and scientific societies [2].

Medicinal plants exhibit healing properties due to phytochemicals, which have recently been found to act as antimicrobial agents against human pathogens [3]. These phytochemicals/phytonutrients are simply chemicals synthesized by plants. Different classes of phytochemicals are reported from the plant kingdom (alkaloids, terpenoids, flavonoids, and phenolic compounds being the most important ones). They have been used to treat various metabolic, immunological, and neurological disorders in humans throughout the globe as a part of traditional medicine [4]. Many naturally occurring phytochemicals possess pharmacological properties (antimicrobial, antioxidant, anticancer, anti-inflammatory, antimalarial, antiallergy, cytotoxic, antiviral, anti-hypertension, etc.) and are an increasingly attractive option for novel drugs. It is reasonably arguable to state herein that perhaps plant resources are possibly the best available repertoire for bioactive compounds that might result in the development of new pharmaceutical agents [5]. The article provides an overview of the plant-based antimicrobial, anti-inflammatory compounds and wound-healing mechanisms of medicinal plants. Additionally, this article also focuses on phytomedicine and bioprospecting for plant-based phytochemicals.

9.2 A short treatise on medicinal plants from the ethnobotany perspective

Precisely for primary needs and existence, people and cultures have depended on plants since the earliest origin and continued to use the evaluated plants with beneficial attributes [6]. From generation to generation, evolution of our enriched cultures is achieved by passing over more refined knowledge of plants and their respective usefulness [6]. Ethnobotany, a result of such continued knowledge, is the study of the use of indigenous (native) plants by the people of a particular culture and region around the world [8]. Apart from their use in ceremonial or spiritual rituals, ethnobotany reflects the application of plants for providing food, dye, medicine, shelter, oil, fiber, resin, soap, gum, latex, wax, tannin, and even the air we breathe [6] (Fig. 9.2). Human ancestors of all continents found thousands of indigenous plants attributed to various ailments like wound healing, curing of diseases, and easing troubled minds [6] (Fig. 9.1).

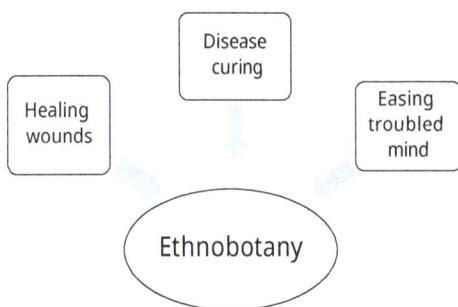

Fig. 9.1: Prehistorical application of ethnobotany.

The use of medicinal plants is evidenced to be around 60,000 years ago [7]. In the case of Egypt and China, it is as old as 3,500 and 2,700 BC [7]. During 1,550 BC, over 700 medicinal Egyptian formulas were scripted on the temples' walls that recorded the knowledge of illnesses and cures [7]. Hippocrates (460–380 BC), who is also known as the "Father of Medicine," developed the herbal system (300–400 species) of diagnosis and prognosis [7]. In Athens, the listed description of medicinal plants was started by philosopher "Aristotle" and his pupil "Theophrastus" [7]. Romanian Dioscorides, the writer of medicinal five-volumed "De Materia Medica" in AD 60, describes approximately 500 plants including drug preparation from about 1,000 plants [7]. Indian Ayurvedic knowledge is 2,500 years old and has described illness as an imbalance in the body which can be cured with dietary control and herbs to restore equilibrium [7].

The knowledge of Ayurvedic principles is also retained by traditional folk healers, which they have received from their ancestors for the appropriate usage of surrounding natural resources. The "Glossary of Drugs and Food Vocabulary," having 1,400 drug

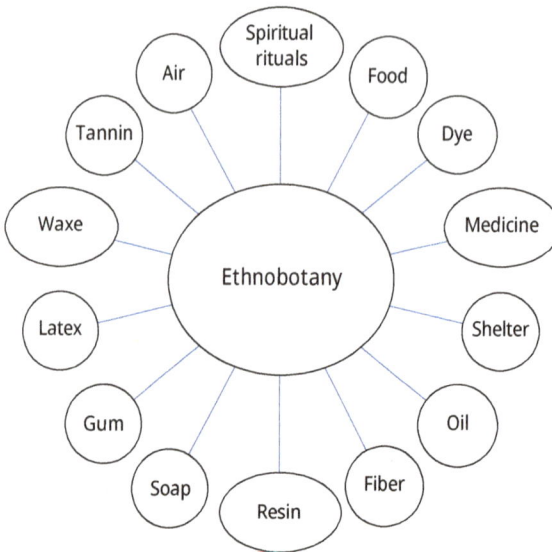

Fig. 9.2: Modern application of ethnobotany.

formulations, was written by Arabian Abdullah Ben Ahmad Al Bitar (1021–1080 A.D) [7]. Aelius Galenus, a Greek physician, surgeon, and philosopher, developed medicines from herbs collected from around the world. With time, traditional practitioners began focusing on medicine and healing skills [7]. By the 1530s, Paracelsus, a Swiss physician and alchemist, changed the attitude of Europe toward the health care system [7]. His work was immensely recognized from the notebook "Doctrine of Signatures," which marked the outward appearance of a plant with the problem it would cure [7]. In 1785, Dr William Withering published his work on Foxglove with a portion of its therapeutic uses and he described 200 cases including uses of foxglove (glycosides digoxin and digitoxin) to treat dropsy and heart failure [7]. Morphine, extracted from *Papaver somniferum*, was identified by Frederich Serturner in Germany and became one of the first plant-based drugs used in 1803 [9]. Later, similar techniques were used to produce emetine from ipecacuanha, aconitine from monkshood, quinine from Peruvian's bark, and atropine from deadly nightshade [9]. In 1852, the active ingredient of willow's bark, that is, slicin was synthesized for the first time and thereafter, in 1899, a pharma company called Bayer converted it to acetylsalicylic acid and launched it as "aspirin" into our modern society [9]. In the following 100 years, the plant-derived synthetic age was born. However, many synthetic versions were found to have less therapeutic effects or negative side effects compared to whole plant source [9]. Nearly 40% of the drugs used in the Western world for centuries are plant-derived. For instance, quinine, which is used to treat malaria, is extracted from *Cinchona calisaya*, and licorice root (*Glycyrrhiza glabra*) has been a part of cough drop's ingredient for 3,500 years [9]. In Elizabethan times, mints, which have had a medicinal interest since 100 A.D, were used to cure over

40 ailments [7]. However, the efficacy or toxicity of many of these herbals has never been seriously tested. Interestingly, since 1994, the use of any such herbal or traditional supplement is not regulated by FDA of US [7].

9.3 Physiologically active bio-compounds in plants

Natural physiologically active bio-compounds (PABs) with significant therapeutic effects, such as alkaloids, terpenoids, polyphenols, glycosides, and polysaccharides, are being synthesized and accumulated in medicinal plants [10]. The alkaloids are compounds exhibiting great structural diversity. Alkaloid compounds mainly originate from the amino acid precursors and characteristically contain nitrogen in heterocycle ring(s). Morphines and atropins are common examples of such type of compounds. The "protoalkaloids" (e.g., mescaline and adrenaline) usually have amino acids as precursors, but do not contain the nitrogen hetero rings [11]. Conversely, terpene-, purine-, or steroid-based alkaloids such as caffeine or theobromine neither originate from amino acid precursors nor contain heterocycle nitrogen ring(s) [12]. The ornithine, lysine, tyrosine, phenylalanine, and tryptophan are the most frequent amino acid initiators of the ring formations. The phenylethylamine and imidazole alkaloids originated from tyrosine and histidine, respectively. In contrast, isoquinoline, quinolone, tropane, indole, and phenanthrene are classified on the basis of skeletal analysis. The ricinine and trigonelline are grouped together due to the presence of pyridine rings. Similarly, ephedrine and tyramine exhibit phenylethylamine moiety, and therefore, they are classified as the same group [11, 12].

The terpenoids are derived from lipid-based compounds having the characteristic isoprene rings (the five-carbon ring structure with numerous structural variations). The most common types of terpeniods are steroids, carotenoids, and terpenes. Characteristically, the steroids and carotenoids both exhibit isoprene group with six carbon rings. But in case of steroid the fifth isoprene exhibits five carbons while the others are bonded with six carbon atoms. On the other hand, the carotenoids have isoprenes at both ends but the rings are interconnected with long chain fatty acids with double bonds [13]. The terpenes have the isoprenoid ring which can be classified based on the number of carbon ring moieties. Typical examples of terpene are as follows: monoterpenes (have two rings), sesquiterpenes (have three rings), diterpenes (have four rings), sesterpenes (have five rings), triterpenes (have six rings), tetraterpenes (have eight rings), and polyterpenes (have above hundred rings). Monoterpenes are further divided into cyclic (closed ring, e.g., limonene and pinenes) and acyclic (non-closed ring, e.g., geraniol) monoterpenes. Cyclic monoterpenoids are classified into mono-, bicyclic-, tricyclic-, and tetracyclic monoterpenes (e.g., carotenoids). Sesquiterpenes are also grouped as mono-, di-, tri-, and polycyclic and acyclic (e.g., sesquilavandulol) sesquiterpenes [13].

Polyphenols with significant structural variants are also abundant in medicinal plants. They are synthesized by plants through the shikimate or polyketide mechanism. They are the largest family of secondary metabolites and are mainly subgrouped into phenolic acids and flavonoids. Apart from these, stilbenes and lignans are two less common subgroup found as natural PABs.

Polyphenols are classified based on the number of phenol rings and the structural elements interlinking these rings between each other. Phenolic acids have relatively simple structures comprising two subclasses: hydroxybenzoic acids (gallic, vanillic, and salicylic acids) and cinnamic acids (ferulic and caffeic acids). The most versatile and largest group of phenolics is flavonoids. The most common example of flavonoid is diphenylpropanes (C6–C3–C6). It has two aromatic rings bonded via three carbon atoms which frequently form a heterocyclic ring. The flavonoids are further subdivided into 14 classes based on the degree of oxidation of the heterocycle: aurones, chalcones, dihydrochalcones, flavonols, dihydroflavonols, flavones, isoflavones, flavanones, flavonones, flavanols, flavandiols, proantocyanidins, and anthocyanidins [14]. Glycosides comprise one or more sugar moieties linked with nonsugar groups. The sugar moiety is termed glycone, while the nonsugar entity is referred to as genin or aglycone. In the case of O-glycosides, the glycosidic linkages are synthesized by the interaction of hemiacetal functional group of the glycone with the hydroxyl group of the aglycone. Similarly, glycosides are linked via C-, N-, or S-glycosidic bond to yield C-glycosides, glycosylamines, and thioglycosides, respectively. Alternatively, glycosides are classified on the basis of aglycone portion as flavonoid glycosides, triterpene glycosides, β-sistosterol, iridoid, anthraquinon glycosides, phenylpropanoid kaempferol glycosides, and saponine [15].

The major polysaccharides used in therapeutic practices are pectins, resins, and mucilages. Pectins comprise the constituents available in ointments along with boric and salicylic acids. They are typically used as emulsifiers and stabilizers. Resins and mucilaginous extracts from plant kingdom have significant applications in the pharmaceutical industry such as the production of mucilaginous solutions and emulsion fluids (for different drugs) [10]. In addition to the above, the major PABs are commonly characterized according to their pharmacological activities such as antimalarial, anticancer, antimicrobial, anti-inflammatory, cytotoxic, and neurotoxic [10–15].

9.4 Plant antimicrobials: mode of action and chemistry

Since ages, humans have been relying on plants and their derivatives to fulfill their day-to-day needs. The use of plants in treating many diseases was widespread in several human civilizations, and some of the systems of medicine, like Indian (Ayurveda) and Chinese systems, are still in practice [16, 17]. Plants produce a range of phytochemicals in their primary and secondary metabolism, of which numerous possess therapeutic values, especially antimicrobial properties. Many were isolated antimicrobial agents (chemical groups that included phenolics, terpenes, alkaloids, and polypeptides) and showed various modes of action against disease-causing microorganisms [18–20].

Almost 250,000–500,000 species of plants are there on Earth, and only 1–10% are eaten as food by human beings and other animal species [21]. The idea of utilizing plants as a source of the medicine is very important because plants produce a huge number of extremely specialized metabolites as a character of their defense mechanism [22].

9.4.1 Mode of action of plant antimicrobials

Antimicrobial action is mostly linked to two strategies: inhibiting chemically the formation and functions of essential elements of bacteria and evading traditional antibacterial resistance mechanisms. Plant antimicrobials target a wide range of antibacterial targets from the molecular to the organism and community levels and biofilms in some circumstances. These phytochemicals interact with macromolecules covalently and noncovalently during their antibacterial action at the molecular level, rendering them nonfunctional. Antimicrobial agents inhibit/kill the pathogen by disrupting the operation of cellular components such as capsule, cell wall, cell membrane, and mitochondria [23, 24]. The mechanism outlined below corresponds to antibacterial medicines that are currently on the market (Fig. 9.3).

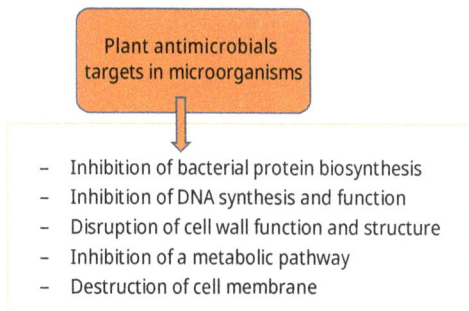

Plant antimicrobials targets in microorganisms

- Inhibition of bacterial protein biosynthesis
- Inhibition of DNA synthesis and function
- Disruption of cell wall function and structure
- Inhibition of a metabolic pathway
- Destruction of cell membrane

Fig. 9.3: Mode of action of plant antimicrobials on microbial cell [adapted from Khameneh et al. 2019 (open access)] [25].

9.4.2 Chemistry of plant antimicrobials

Phenolic compounds

Phenolic compounds are recognized as the most diverse classes among the bioactive secondary metabolites synthesized by medicinal plants. They are frequently used to conflict with pathogenic microbes. Flavones, flavanols, flavonoids, quinones, and tannins are phenolic substances. Against various microbial strains, these chemicals demonstrated a variety of modes of action [26].

Alkaloids

Alkaloids are heterocyclic nitrogen molecules with chemically varied structures that have analgesic, antispasmodic, and antibacterial properties. These chemicals are often used to play a specific function in the treatment of various infectious diseases [27]. However, quinine (alkaloids) is well-known for its antiprotozoal effect against the malarial parasite [28]. Most of the alkaloids show EP inhibitory action, for example, berberine and reserpine.

Organosulfur compounds

Many studies have shown that antibacterial and antifungal activities of sulfur-containing compounds were derived from plants. Sulfur-containing substances including allicin, ajoene, dialkenyl and dialkylsulphides, S-allyl cysteine and S-allyl-mercapto cysteine, and isothiocyanates show antibacterial activities against Gram-positive and Gram-negative bacteria [29, 30].

Terpenes

Terpenes, also known as isoprenoids, are the most expanded group of natural chemicals. They are present practically in all forms of life and perform distinct roles including participation in the main construction of cells (cholesterol and steroids in cell membranes) and contributing to cell functions. Several terpenes and their derivatives are confirmed to have essential defenses mechanism against herbivores and infections. Generally, Gram-positive bacteria showed high sensitivity to terpenes than Gram-negative bacteria. The lipophilic properties of terpenes contributed to their antibacterial action. Monoterpenes preferentially attack membrane structures by intensifying fluidity and permeability, remodeling protein structure, and creating an interruption in the respiratory chain [31].

9.5 Plant anti-inflammatory compounds: mode of action and chemistry

Plants have been playing a major role in the human health care regime for ages. They produce a variety of biologically active compounds or chemicals, which act as a defensive barrier for pathogens and environmental stress. These organic compounds are produced during secondary metabolism and have a variety of biological functions. Anti-inflammatory properties are one of them. Due to the excessive chemical and biological diversity of medicinal plants as well as the presence of several compounds with intriguing biological activity, they are currently the prime concern subject in modern research [32]. The main benefits of utilizing herbal medicines are their cost-effectiveness, accessibility, and fewer side effects. Inflammation is a dynamic process that is triggered by mechanical injuries, burns, microbial infections, and other unpleasant stimuli, which may lead to the onset and course of a variety of diseases. It is a natural defense mechanism that protects our bodies against tissue damage and pathogens. Although inflammation is a natural part of the body, it can have serious consequences for patients including pain, swelling, fever, and other symptoms. If a severe inflammatory reaction occurs, anti-inflammatory medicines must be used to control or suppress the inflammation. These medications typically have fewer or no side effects, necessitating the hunt for novel substitutes. Anti-inflammatory medications of various types are used to counteract these side effects. Anti-inflammatory medicines are divided into two categories: nonsteroidal anti-inflammatory drugs (NSAIDs) and corticosteroids. The action of the cyclooxygenase (COX) enzyme is inhibited by NSAIDs. Corticosteroids, on the other hand, inhibit the expression of the COX enzyme.

Most analgesic and anti-inflammatory medications available on the market work by inhibiting the COX pathways of arachidonic acid metabolism, which are responsible for prostaglandin formation. Since majority of analgesic and antiphlogistic agents displayed various adverse effects such as gastric lesions, NSAID-related peptic ulcers, unfavorable thrombotic effects for selective COX-2 inhibitors, and opiates stimulate tolerance and addiction, the current drugs exacerbate the conditions, thereby reducing these agents' effectiveness. Hence, natural products with anti-inflammatory activity, especially those derived from plants, are considered promising sources for new treatment medicines in this aspect.

Plant extracts' anti-inflammatory properties are mainly attributed to flavonoid, polyphenolic, proanthocyanidin, alkaloid, terpenoid, and steroid components. These secondary metabolites act on a variety of inflammatory pathway targets. Several flavonoids exhibit anti-inflammatory properties *in vitro* and *in vivo*. Flavones, flavanes, flavonols, flavanols (catechins), anthocyanidins, and isoflavones are some of the most common subfamilies of flavonoids. Terpenoids are another class of anti-inflammatory chemicals found in many medicinal plants. Triterpenes of the lupine type have anti-inflammatory properties because they prevent the production of prostaglandin E2 (PGE2) and nitric oxide. A tri-terpene, for example, lupenol, which was derived from the bark of *Pterodone marginatus* Vogel, belonging to the family Fabaceae, was found to suppress the proinflammatory cytokines like interleukin-2 (IL-2), interferon-γ (IFN-γ), and tumor necrosis factor-∞ (TNF-∞). Copalis acid and essential oils containing sesqui-terpenes also showed anti-inflammatory properties by inhibiting inflammatory media-tors IL-1 and TNF-∞ [33, 34]. Anti-inflammatory cytokines, which manage the actions of proinflammatory cytokines for homeostasis, are also significant players in the inflamma-tory response. The yellow-colored pigment viz., curcumin of *Curcuma longa* (turmeric) has pleiotropic mechanisms of action like anti-inflammatory, antioxidant, anticancer/ proapoptotic, and antibacterial properties. Curcumin inhibits critical proinflammatory signaling pathways. It also downregulates the secretion of prominent cytokines like TNF, IL-1, and IL-6 [35] and stops the expression of cell adhesion molecules like intercellular adhesion molecule-1, which are required for the interaction of leukocytes with endothe-lial cells [36]. The majority of these therapeutic plants is employed as crude extracts or isolated compounds. Hence, the research is critical to authenticate and establish the effi-cacy of these plants and to validate the compounds that are responsible for biological activity and understand the mode of actions underlie in each plant's pharmacological action.

9.6 Wound healing: a biochemical and physiological perspective

The cellular and molecular mechanisms involving tissue repair and their failure to cure are under investigation, and the currently available medications are very much limited. Globally, millions of people are affected annually by poor wound healing, especially those with chronic disease, surgery, posttrauma, or acute illness. Healthy tissue repair responses include inflammation, angiogenesis, matrix deposition, and cell recruitment. Improper functioning of any of these cellular cascades leads to severe clinical condi-tions such as vascular disease, diabetes, or aging, and they are all associated with heal-ing pathologies. Pursuing novel clinical strategies to improved body's natural repair mechanisms needs to be based entirely on comprehensive knowledge of the fundamen-tal biology related to repair and regeneration mechanisms [37]. In normal biological process, wound healing is attained following four perfect and systematically pro-grammed phases viz. homeostasis, inflammation, proliferation, and remodeling. A wound is successfully healed, when these four phases occur in the proper sequence co-ordination in a given time frame. However, there are many factors which restrict these four phases from occurring in proper and systemic sequences, thus causing improper or impaired wound healing. These factors are oxygenation, infection, age and sex hor-mones, stress, diabetes, obesity, medications, alcoholism, smoking, and nutrition. An upgraded knowledge of the range of effectiveness of these factors on repair mechanisms could make it possible to develop therapeutics which can enhance wound-healing activities [38]. Cutaneous wound repair consists of a complete series of events, which begin when the tissue entity is broken. Regardless of the existence of extreme diversities in the several classes of tissue damages (wounds) and their degree of injury, the wound healing or repair mechanism has formulaic set of responses involving migration, proliferation, differentiation, and apoptosis of cell types within the wound environment. These sets of responses strictly follow a precise temporal sequence unless obstructed by different pathological states such as diabetes, poor arterial perfusion, malnutrition, or sepsis. Following cutaneous injury, the body's first and foremost target is to stop hemorrhage from blood vessel injury. And it is done by the action of platelets and clotting cascades. Subsequently, the wounded area is predominated by inflammatory cells responding to platelet-released growth and chemotactic factors. Generally, in an ideal wound, it is seen that within the next 48 h of the tissue damage, monocytes and fixed tissue macrophages increase their population to maximized state. The later phase of inflammation overlaps with a period of proliferation characterized by the migration of fibroblasts and endothelial cells. Epithelialization is the primary repair mechanism in partial loss of skin thickness, whether from accidental abrasions, thermal or chemical injuries, or surgically planned skin graft harvest. The resurfacing process is of minor consequence in wounds with sutures and is short-lived [39].

9.6.1 Factors affecting wound healing

Zinc, an essential trace element (micronutrient), plays a vital role in human physiology. It acts as a cofactor for many metalloenzymes required to repair cell membrane, cell proliferation, growth and immune system function. The pathological effects of zinc deficiency in the body include skin lesions, growth retardation, weak immune function, and impaired wound healing [40]. The pH value within the wound is also a criterion that influences directly or indirectly all biochemical reactions taking place in this process of healing. Interestingly, it is a neglected parameter for the overall outcome [41]. Cellular recruitment and activation can be directly affected by alterations in the microenvironment, such as changes in mechanical forces, oxygen level, chemokines, extracellular matrix, and growth factors synthesis, which as a result, can lead to impaired wound healing [42].

9.7 A treatise on modern phytomedicine: a biotechnological perspective

Human civilizations and utilization of plant-based products, as natural medicines, are intricately connected. The traditional approaches to select phytomedicines, however, have many shortcomings, such as slow growth of medicinal plants, drugs from specific plant organs, low natural yield, high production cost, supply constraints, and overexploitation-driven extinction of medicinal plants. Modern biotechnological interventions (Fig. 9.4) can overcome these limitations and lead to massive production of drugs, facilitate new drug discovery, quickly identify main bioactive component in crude extracts or mixture of ingredients, responsible for medicinal properties, and decipher their biological mechanism of actions. These phytomedicines can range from antiparasitic, anti-cancerous, and anti-hypertensive to immunosuppressive, immunomodulatory, and anti-inflammatory drugs.

Plant cell/tissue culture is a biotechnological approach for mass cultivation as well as *de novo* synthesis of many bioactive substances as phytomedicines. The commonly used culture/micropropagation strategies for mass production of phytobioactive compounds are suspension culture, callus culture, and hairy root culture techniques. Many phytopharmaceutical products, for example, as anticancer compounds are subjects of research and commercial use. Some anticancer drugs developed from plants are vinblastine, vincristine, and vindeline from *Catharanthus roseus*; taxol/paclitaxel from *Taxus* sp. such as *Taxus baccata*; camptothecin from *Camptothecaacuminatae, Ophiorrhizamungos*, and *Merriliodendronmegacarpum*; podophyllotoxin from *Podophyllum peltatum* and *Podophyllum hexandrum*; and curcumin from *Curcuma longa* [43]. Other examples include ginenoside, diosgenin, and shikonin [44].

Pharmaceuticals based on plant natural products come from a complex series of metabolic pathways displayed by plants. Many important phytochemicals can be produced in heterologous hosts through metabolic engineering. Metabolic engineering is a biotechnological manufacturing alternative for the sufficient and timely supply of phytopharmaceuticals. Metabolic engineering focuses on the development of omics-based (genomics, phylogenomics, transcriptomics, and metabolomics) methodologies to aid in the identification and characterization of genes from complex phytochemical biosynthetic processes. The discovery of biosynthetic pathways of several bioactive compounds in plants is under investigation to enhance their production. Some notable examples include lignans – podophyllotoxin; alkaloids – vindoline and catharanthine, taxol, quinine, and camptothecin. Recently, nearly complete biosynthetic route of colchicines, an alkaloid of predominantly liliaceae family of plants, was discovered. Over the last few years, the combination of metabolomics and transcriptomics data analysis with functional genomics has emerged as a useful and effective tool for revealing biosynthetic pathways of phytopharmaceuticals. When the complete route is identified, the associated biosynthetic genes can be transferred into heterologous hosts, such as *Escherichia coli, Saccharomyces cerevisiae,* or plants like *Nicotiana ben-thamiana,* to generate engineered cell factories that produce phytochemicals of inter-est. One example is biosynthesis of tropane alkaloids of solanaceae family in yeast as cell factories. Some prominent examples of medicinal tropane alkaloids include atro-pine, hyoscyamine, and scopolamine [45]. Moreover, bioinformatic techniques pro-vide critical tools for identifying genes and pathways linked to bioactive components from medicinal plants. Data mining approaches enhance the coupling of data from

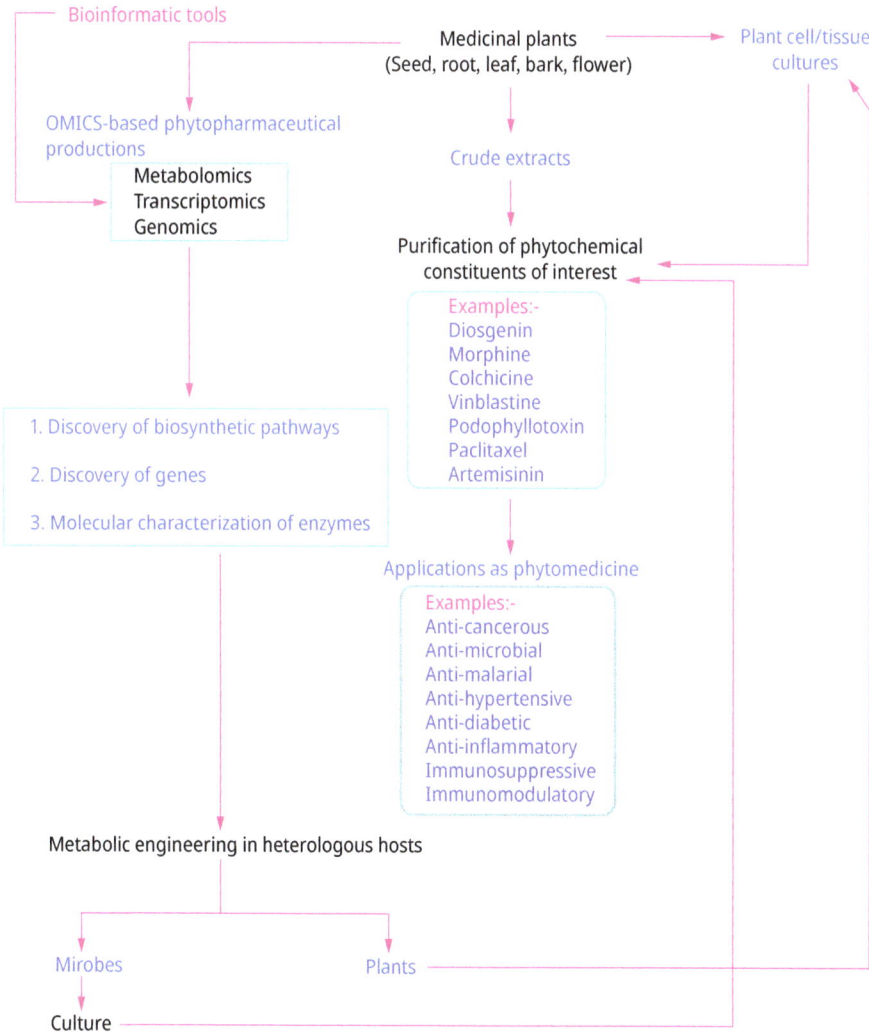

Fig. 9.4: Biotechnological approaches for phytomedicines.

transcriptomics and metabolomics, which opens up various possibilities for studying metabolic pathways of phytomedicines.

The production of phytopharmaceuticals in plant cell/tissue culture can also use heterologous overexpression pathways and gene-editing techniques in heterologous expression system. Only a few phytochemicals have been produced in heterologous expression system like yeast because enzymes involved in secondary metabolites production frequently require the presence of particular CYRs and posttranslational modifications. Because cytosolic CYRs and other posttranslational modification mechanisms are available in plant cell cultures, this constraint can be circumvented. Plant cell cultures, such as of *Nicotiana tabacum* and *Arabidopsis thaliana*, can be engineered and developed using multigene transformation approaches and CRISPR-Cas9 gene-editing system. Plant metabolic engineering to manufacture artemisinin, t-resveratrol, and taxanes are some examples that illustrate the capacity of biotechnological applications of cell cultures to produce phytopharmaceuticals. Understanding the routes for the formation of complex metabolites in plant cells allows for overproduction and purification of commercially important phytochemicals inside plants or heterologous hosts [44].

9.8 Bioprospecting for phytochemicals: current trends and newer frontiers

Plants serve as an essential natural resource for traditional and modern healthcare systems worldwide. Mankind has used herbal items to cure pain and various illnesses for over 60,000 years [46]. According to World Health Organization, nearly 80% of the population worldwide uses traditional herbal medicines for some aspect of primary health care [47]. In the current scenario, more than 50% of the natural drugs used in medicine are subsequent from plant sources [47].

Several studies have reported that various phytochemical compounds are accessible for use of plants as a medicine in the treatment of diseases in Indian and the worldwide [47]. Phytochemicals are a huge group of secondary metabolites formed by plant organisms involved in a complex network of ecological connections within an ecosystem. Over 35,000 plant species have the potential for healing due to their secondary plant material. These chemical components are functionally divided into primary and secondary metabolites [47]. The primary metabolites crucial for plant growth, growth, and reproduction are nucleic acids, proteins, lipids, and carbohydrates. Secondary metabolites formed in response to plant environmental involvement or as a defense mechanism against antagonistic agents can be represented by many compounds such as alkaloids, flavonoids, and terpenoids [47].

Numerous researchers have advocated that phytochemical compound extracts and secondary metabolites isolated from inflorescences and leaves showed significant pharmacological potential, such as hepatoprotective, antimicrobial, anti-inflammatory, antithrombotic, antidiabetic, antihypertensive, antioxidant, cytotoxic, antimalarial, anti-pyretic anticoagulating anti-cholinesterase, analgesic, anti-schistosomicidal, antifertility, anticancer, antidiarrheal, and pesticidal activity are some of the pharmacological properties this plant possesses [48].

The present study has focused on how the phytochemicals compound has produced fruitful outcomes for the pharmaceutical industry for drug discovery and development. Bioprospecting or biodiversity and traditional knowledge come together to explore and technological advances in the modern world have provided essential nutrition for this vast development of this new age field [49]. A bioprospecting research approach can be used to further investigate the natural products of phytochemicals. In addition, bioprospecting covers the study of genetic resources or biochemicals for new commercial leads and includes three main areas such as chemical prospecting and genetic prospecting and bionic prospecting.

9.8.1 Chemical prospecting

The chemical screening and novel bioactive compounds were identified from various plants for drug and pharmaceutical discoveries. Chemical prospecting of wild plant resources is increasingly being applied to pesticides (biopesticide and pesticide), drugs and drugs, cosmetics, proteins, enzymes, food additives, and other industrially valuable chemicals [49].

9.8.2 Gene prospecting

The genetically enhanced crops develop even more applicable for it aims at an everlasting improvement of agronomic qualities of interest. For many years, plant genetic improvement programs have been based on an empirical selection of target traits. However, great improvement has been ended in recent years. In terms of genetic improvement strategies, molecular research has been critical in determining which genes are relevant for each given agronomic feature such as abiotic stress tolerance [49]. However, a combination of biotechnology and bioinformatics tools is used to identify stress-related genes and develop GM crops by silencing and/or overexpressing specific genes.

9.8.3 Bionic prospecting

Bionic prospecting is a new field where modern designs, patterns, models, and technologies are developed based on natural biodiversity. New sensor technologies, architectures, bioengineering, and biomodeling are some of the interesting areas of bionic prospecting [49].

In the connection with prospecting of photochemical utilized in several beneficial rhizosphere microbial flora applications include increasing the development of valuable medicinal plants in both traditional and stressed situations. Plant and human health benefit from low-cost, biologically friendly behavior that has no negative influence on soil fertility. Even microbiological techniques boost plant growth in a variety of ways, both directly and indirectly, through various abiotic stress relief mechanisms. Thus, the plant growth-promoting rhizobacteria provides a solution to the problem in a cost-effective and environmentally friendly way. Even the dangerous fertilizer can improve the overall growth, yield, and quality of plant components of the medicinal plants. Whereas the phytochemicals are not only a source of health care but also an important product of world trade. In recent years, the trade-in of medicinal plants has increased significantly due to the low price, few side effects, and easy availability of medicinal herbs. Several studies have proved that photochemical products have been treating various human illnesses and cure of several human diseases [47, 50]. The growing demand for plants as raw materials in the pharmaceutical industry threatens plant biodiversity. Hence, it is highly important to utilize better screening and investigating methods from plants and other natural resources while developing and characterizing natural medicines [47, 50].

Based on the study, it plays that there are some significant differences in the way of bioprospecting agreements or agreements for access to genetic resources are employed across sectors. The agricultural and pharmaceutical sector significantly relies on partnerships with academic or research organizations, and most agreements involve a variety of monetary and nonmonetary incentives. However, in this chapter, we have suggested a fresh viewpoint on how access and benefit-sharing are handled in a variety of industries that all use genetic or biological resources as inputs for using the plant photochemical. We expect that by combining data from several scientific domains such as agricultural, medicine, botany, and pharmaceutical, we would be able to estimate the economic potential of bioprospecting globally. The collaboration between the pharmaceutical companies and medical resources will increase scientific awareness of bioprospecting while also providing various important commercial and mutually beneficial relationships.

9.9 Challenges and opportunities in medicinal plant research

Medicinal plant research has remarkably contributed in the evolution and origin of multiple herbal therapies. In the past few decades, the medicinal plants have been the subject of interest among the populace due to the boast of herbal medicines, lesser side effects, and lower cost in comparison to allopathic medicine. Furthermore, increasing human population has made the need of meeting the requirement of herbal medicine.

Despite the success of the formulation of several herbal medicines from medicinal plants in past two to three decades, future endeavors face many challenges. Considering the present condition of rising demand for plant-based drugs, challenges can be seen in fulfilling the requirements of conserving bioresources. And due to the rise in demand for plant-based drugs, heavy pressures on some selected highly valued medicinal plant populations are created, which unfortunately leads to the threat or extinction of the species [51]. Many of these medicinal plants showed slow growth rates, low population densities, and narrow geographic range, making them more prone to extinction [52]. Destructive way of harvesting underground parts of slow-growing, slow-reproductive rates, and climate-specific medicinal plants species is the major challenge of sustainability [53]. Insufficient data on highly threatened and endangered medicinal plant species is also another major challenging issue in medicinal plant research. Abundance of medicinal plants in any particular area including rare species is restricted by natural pathogens, herbivores, and seed predators.

Less than 10% information is available for the propagation of medicinal plants and agro-technology is made available for only 1% of the total known plants. This data shows that the development of agro-technology should be given priority in medicinal plant research [54]. Another challenging task is cultivation of medicinal plants as some of the species has lengthy cultivation cycle. People are not interested to cultivate the medicinal plants as their cost is lesser than many seasonal vegetables [55]. The entire process of new drug discovery (from compound extraction to drug discovery) needs an estimated time of 10 years costing nearly or beyond 800 million dollars. The crude plant extract is formulated mainly in tablet and capsule forms and, to some extent, as oral liquid. However, the dosage optimization is unsuccessful because of poor absorption, therapeutic efficacy, and poor adherence. There are some obstacles that are faced while designing the crude plant extract in capsule or tablet form, viz., powdering of crude herbs, particle size that affects process of blending, compression, and filling and high moisture content as well. Furthermore, crude plant extract has hygroscopic nature, poor solubility, and adhesive nature, which makes it difficult to formulate them in the solid dosage form. Since medicinal plant research consumes lots of time, better and quick techniques or methods for the collection of plant material, bioassay screening, compound extraction, and development must be adopted. The designing of high-throughput screening assays is a challenging task.

Presently, bioprospecting preferentially proceeds in an environment of unsettled and burgeoning frictions amidst bio-piracy and rights of sharing benefits between developing and developed countries. Protecting the legal status of indigenous knowledge and settlement in terms of compensation of the traditional herbal practitioners are the major issues for concern. In different countries such as India, United States, Europe, Canada, and other countries, different ways and systems are followed for awarding patents on medicinal plants [56]. But, in some countries, the plants and inventions made in connection to plants and their products are still ineligible for patent filing.

9.9.1 Opportunities

The development of high diversity medicinal plant has opened the door for exploring the medical efficacy, value addition, and use in curing various old and new diseases. The establishment of low cost and high-quality generic drugs in the global market can be used as an important tool for the marketing of medicinal plant-based products. Bioprospecting demands several indispensable requirements that needs to be coordinated, such as a team of scientific experts with expertise in an array of human endeavors, including international laws and legal understanding, social sciences, politics, and anthropology. Ayurveda and traditional knowledge of medicinal plants, rich genetic resources, high diversity, and associated ethnomedical knowledge are vital components for sustainable bioprospecting and value-addition processes [57]. To reduce the cost and time of the repeat collection of plant materials, it is of utmost necessity to build a library consisting of natural product extracts only. Also, there is a necessity for the availability of adequately encoded and preserved extracts in large numbers of biological screening with regards to high-throughput screening and obtaining hits within a short period.

Government policies and government, nongovernment institutional support have focused the medicinal plant research in various aspects such as production of quality planting material and strengthening research to improve it. The funding opportunities have been made available for those who are enthusiastic to work and build capacity in herbal medicine sector.

9.10 Conclusion

Anti-inflammatory, antibacterial, and wound-healing properties of medicinal plants have been surveyed and summarized in this chapter. Numerous "ancient" and "traditional" medicinal herbs have been proven to be effective in clinical studies. Humans have identified and used local plant resources for food and medicine since millennia. Validation studies have revealed an astonishing number of phytochemicals that may lay down the very foundation for drug discovery. Antioxidant and antibacterial properties are inherent in almost all therapeutic plants. The efficacy of medicinal herbs in expediting the healing process of wounds necessitates further studies; however their true potential in healing is enormous. The wound-healing process can be improved with the application of active substances. Inflammation can be reduced by the use of medicinal herbs and active substances. The wound-healing cascade in mammals also has many of the same biological targets and pathways. There are still more mechanisms that need to be discovered for a better comprehensive understanding.

References

[1]   A. G. Fakim. Medicinal plants: Traditions of yesterday and drugs of tomorrow. Mol Aspects Med. 2006, 27, 1–93.

[2]   J. L. S. Taylor, T. Rabe, L. J. McGaw, A. K. Jäger, J. V. Staden. Towards the Scientific Validation of Traditional Medicinal Plants. 2001, 34, 23–37.

[3]   R. Barbieri, E. Coppo, A. Marchese, M. Daglia, E. S. Sanchez, S. F. Nabavi, S. M. Nabavi. Phytochemicals for Human Disease: An Update on Plant-derived Compounds Antibacterial Activity 196. 2017, pp. 44–68.

[4]   A. Bansal, C. Priyadarsini. Medicinal properties of phytochemicals and their production. (Ed.). Natural Drugs from Plants, 2021. https://doi.org/10.5772/intechopen.98888

[5]   M. M. Phukan, R. S. Chutia, R. Kumar, D. Kalita, B. K. Konwar, R. Kakati. Assesment of antimicrobial activity of bio-oil from *Pongamia Glabra, Mesua Ferrea* and *Parachlorella* Spp deoiled cake. Int J Pharm Biol Sci. 2013, 4(4), 910–918.

[6]   USDA. Ethnobotany. https://www.fs.fed.us/wildflowers/ethnobotany/index.shtml [Retrieved on 12.05.2022]

[7]   USDA. Medicinal botany. https://www.fs.fed.us/wildflowers/ethnobotany/medicinal/index.shtml [Retrieved on 12.05.2022]

[8]   M. M. Iwu, J. Wootton, (Eds.), Ethnomedicine and Drug Discovery. Elsevier, Amsterdam, Netherland, 2002.

[9]   M. M. Iwu. Ethnobotanical approach to pharmaceutical drug discovery. In Iwu, M., Wootton, J., (Eds.). Ethnomedicine and Drug Discovery. Elsevier, Amsterdam, 2002, Vol. 1, pp. 309–320.

[10] M. Y. Lovkova, G. N. Buzuk, S. M. Sokolova, N. I. Kliment'eva. Chemical features of medicinal plants. Appl Biochem Microbiol. 2001, 37(3), 229–237.

[11] R. Eguchi, N. Ono, A. H. Morita, et al. Classification of alkaloids according to the starting substances of their biosynthetic pathways using graph convolution neural networks. BMC Bioinf. 2019, 20, 380.

[12] B. R. Lichman. The scaffold-forming steps of plant alkaloid biosynthesis. Nat Prod Rep. 2021, 38(1), 103–129.

[13] A. N. M. Alamgir. Secondary metabolites: Secondary metabolic products consisting of C and H; C, H, and O; N, S and P elements; and O/N heterocycles. In Alamgir, A. N. M., (Ed.), Therapeutic Use of Medicinal Plants and Their Extracts: Volume 2. Springer, 2018, pp. 165–309.

[14] A. Tresserra-Rimbau, R. M. Lamuela-Raventos, J. J. Moreno. Polyphenols, food and pharma. Current knowledge and directions for future research. Biochem. Pharmacol. 2018, 156, 186–195.

[15] H. Khan, A. Pervaiz, S. Intagliata, et al. The analgesic potential of glycosides derived from medicinal plants. DARU J Pharm Sci. 2020, 28, 387–401.

[16] P. R. diSarsina. The social demand for a medicine focused on the person: The contribution of CAM to healthcare and healthgenesis. Evid Based Complement Alternat Med. 2007, 4, 45–51.

[17] F. M. C. Sharples, R. van Haselen, P. Fisher. NHS patients' perspective on complementary medicine: A survey. Complement Ther Med. 2003, 11(4), 243–248.

[18] H. O. Edeoga, D. E. Okwu, B. O. Mbaebie. Phytochemical constituents of some Nigerian medicinal plants. Afr J Biotech. 2005, 4, 685–688.

[19] U. Anand, N. Jacobo-Herrera, A. Altemimi, N. Lakhssassi, A. Comprehensive. Review on medicinal plants as antimicrobial therapeutics: potential avenues of biocompatible drug discovery. Metabolites. 2019, 9, 258.

[20] G. M. Rossolini, F. Arena, P. Pecile, S. Pollini. Update on the antibiotic resistance crisis. Curr Opin Pharmacol. 2014, 18, 56–60.

[21] N. C. C. Silva, A. J. Fernandes. Biological properties of medicinal plants: A review of their antimicrobial activity. J Venom Anim Toxins Incl Trop Dis. 2010, 16(3), 402–413.

[22] M. M. Cowan. Plant products as antimicrobial agents. Clin Microbiol Rev. 1999, 12, 564–582.

[23] A. P. Magiorakos, A. Srinivasan, R. B. Carey, et al. Multidrug-resistant, extensively drug-resistant and pandrug-resistant bacteria: an international expert proposal for interim standard definitions for acquired resistance. Clin Microbiol Infect. 2012, 18, 268–281.

[24] A. A. Velayati, M. R. Masjedi, P. Farnia, P. Tabarsi, J. Ghanavi, A. H. ZiaZarifi, S. E. Hoffner. Emergence of new forms of totally drug-resistant tuberculosis bacilli: super extensively drug-resistant tuberculosis or totally drug-resistant strains in Iran. Chest. 2009, 136, 420–425.

[25] B. Khameneh, M. Iranshahy, V. Soheili, B. S. F. Bazzaz. Review on plant antimicrobials: A mechanistic viewpoint. Antimicrob Resist Infect Control. 2019, 8(1), 1–28.

[26] A. Kurek, A. M. Grudniak, A. Kraczkiewicz-Dowjat, K. I. Wolska. New Antibacterial Therapeutics and Strategies. 2011, Pol. J. Microbiol, Vol. 60, pp. 3–12.

[27] T. P. T. Cushnie, B. Cushnie, A. J. Lamb. Alkaloids: an overview of their antibacterial, antibiotic-enhancing and antivirulence activities. Int J Antimicrob Agents. 2014, 44, 377–386.

[28] M. M. Iwu, A. R. Duncan, C. O. Okunji. New antimicrobials of plant origin. In: Janick, J., (Ed.), Perspectives on New Crops and New Uses. ASHS Press, Alexandria, VA, USA, 1999, pp. 457–462.

[29] R. Barbieri, E. Coppo, A. Marchese, M. Daglia, E. Sobarzo-Sánchez, S. F. Nabavi, S. M. Nabavi. Phytochemicals for human disease: An update on plant-derived compounds antibacterial activity. Microbiol Res. 2017, 196, 44–68.

[30] D. Sobolewska, I. Podolak, J. Makowska-Wąs. Allium ursinum: Botanical, phytochemical and pharmacological overview. Phytochem Rev. 2015, 14(1), 81–97.

[31] R. Paduch, M. Kandefer-Szerszeń, M. Trytek, J. Fiedurek. Terpenes: Substances useful in human healthcare. Arch Immunol Ther Exp. 2007, 55(5), 315–327.

[32] R. Yang, B. C. Yuan, Y. S. Ma, S. Zhou, Y. Liu. The anti-inflammatory activity of licorice, a widely used Chinese herb. Pharm Biol. 2017, 55(1), 5–18.

[33] W. F. de Moraes, P. M. Galdino, M. V. M. Nascimento, et al. Triterpenes involved in the anti-inflammatory effect of ethanolic extract of *Pterodonemarginatus* Vogel stem bark. J Nat Med. 2012, 66(1), 202–207.

[34] K. B. Santiago, B. J. Conti, B. F. M. T. Andrade, et al. Immunomodulatory action of Copaiferaspp oleoresins on cytokine production by human monocytes. Biomed Pharmacother. 2015, 70, 12–18.

[35] Y. Kim, P. W. Bayona, M. Kim, et al. Macrophage lamin A/C regulates inflammation and the development of obesity-induced insulin resistance. Front Immunol. 2018, 9, 696.

[36] S. Purohit, A. Sharma, W. Zhi, et al. Proteins of TNF-α and IL6 pathways are elevated in serum of type-1 diabetes patients with microalbuminuria. Front Immunol. 2018, 9, 154.

[37] S. A. Eming, P. Martin, M. Tomic-Canic. Wound repair and regeneration: Mechanisms, signaling, and translation. Sci Transl Med. 2014, 6(265), 265sr6–265sr6.

[38] S. A. Guo, L. A. DiPietro. Factors affecting wound healing. J Dent Res. 2010, 89(3), 219–229.

[39] C. J. Schaffer, L. B. Nanney. Cell biology of wound healing. Int Rev Cytol. 1996, 169, 151–181.

[40] P. H. Lin, M. Sermersheim, H. Li, P. H. Lee, S. M. Steinberg, J. Ma. Zinc in wound healing modulation. Nutrients. 2017, 10(1), 16.

[41] L. A. Schneider, A. Korber, S. Grabbe, J. Dissemond. Influence of pH on wound-healing: A new perspective for wound-therapy? Arch Dermatol Res. 2007, 298(9), 413–420. doi:.

[42] M. Rodrigues, N. Kosaric, C. A. Bonham, G. C. Gurtner. Wound healing: A cellular perspective. Physiol Rev. 2019, 99(1), 665–706.

[43] P. Patel, V. Patel, A. Modi, S. Kumar, Y. M. Shukla. Phyto-factories of anti-cancer compounds: A tissue culture perspective. Beni-Suef Univ J Basic Appl Sci. 2022, 11(1), 1–21.

[44] T. Wu, S. M. Kerbler, A. R. Fernie, Y. Zhang. Plant cell cultures as heterologous bio-factories for secondary metabolite production. Plant Commun. 2021, 2(5), 100235.

[45] V. Courdavault, S. E. O'Connor, M. K. Jensen, N. Papon. Metabolic engineering for plant natural products biosynthesis: New procedures, concrete achievements and remaining limits. Nat Prod Rep. 2021, 38(12), 2145–2153.

[46] H. Yuan, Q. Ma, L. Ye, G. Piao. The traditional medicine and modern medicine from natural products. Molecules. 2016, 21(5), 559.

[47] A. Rizvi, B. Ahmed, M. S. Khan, H. S. El-Beltagi, S. Umar, J. Lee. Bioprospecting plant growth promoting rhizobacteria for enhancing the biological properties and phytochemical composition of medicinally important crops. Molecules. 2022, 27(4), 1407. https://doi.org/10.3390/molecules27041407.

[48] R. Dubey, S. Rajhans, A. U. Mankad. Phytochemicals of *Jatropha gossypiifolia* (Linn.): A review. Int J Innov Sci Technol. 2020, 5(3), 904–911.

[49] P. Pushpangadan, T. P. Ijinu, V. M. Dan, V. George. Trends in bioprospecting of biodiversity in new drug design. Pleione. 2015, 9(2), 267–282.

[50] S. G. Bhat. Medicinal plants and its pharmacological values. In El-Shemy, H., (Ed.), Natural Medicinal Plants, IntechOpen, 2021.

[51] P. K. Samal, A. Shah, S. C. Tiwari, D. K. Agrawal. Indigenous healthcare practices and their linkages with bioresource conservation and socio-economic development in Central Himalayan region of India. Indian J Tradit Knowl. 2004, 3, 12–26.

[52] A. B. Sharma. Global Medicinal Plants Demand May Touch $5 Trillion by 2050. Indian Express, 2004, p. 29.

[53] S. K. Ghimire, D. McKey, Y. Aumeeruddy-Thomas. Heterogeneity in ethnoecological knowledge and management of medicinal plants in the Himalayas of Nepal: Implication for conservation. Ecol Soc. 2005, 9, 36.

[54] I. A. Khan, A. Khanum. Role of Biotechnology in Medicinal and Aromatic Plants. 1998.

[55] A. K. Gupta, S. K. Vats, B. Lal. How cheap can a medicinal plant species be? Curr Sci. 1998, 74, 565–566.

[56] B. Koo, C. Nottenberg, P. G. Pardey. Plants and intellectual property: An international appraisal. Science. 2004, 306, 1295–1297.

[57] P. Pushpangadan, B. Kumar. Ethnobotany, CBD, WTO and the biodiversity act of India. Ethnobotany. 2005, 17, 2–12.

Walid Daoudi*, Abdelmalik El Aatiaoui, Selma Lamghafri,
Abdelouahad Oussaid and Adyl Oussaid

Chapter 10
Antioxidant activity of medicinal plants

Abstract: This chapter is part of the contribution to the valorization of extracts and essential oil of medicinal aromatic plants by the scientific justification of their medicinal reputation, the analysis of their antioxidant activity, and the identification of phytochemical substances that are believed to be responsible for this activity.

10.1 Introduction

During oxidative phosphorylation within the cell, energy is produced in the form of biochemical molecules called ATP. This physiological process, which uses organic matter in the presence of molecular oxygen, releases metabolic waste in the form of free radicals. The latter, in small quantities, are essential for the best functioning of human cells because they help the body to fight effectively against microbes and viruses, whereas in high concentrations and because of their chemical instability, they can interact with cell membranes, which causes the aging and death of these cells and consequently the appearance of damage to several of their constituents, particularly DNA.

In addition, the exposure of these cells to external aggressors of environmental and/or dietary origin increases the level of these radical species, which increases the risk of the appearance of various health disorders: Alzheimer's disease, cancer, cardiovascular or chronic inflammatory pathologies of the digestive system, etc.

In order to overcome this process of harmful chain reactions caused by these free radicals and to protect cells against aging, the human body develops its immune system and triggers its antioxidant defenses composed of enzymes, vitamins, trace elements, and proteins with an antioxidant potential that differs from one person to another depending on their diet and lifestyle.

Previous studies have shown that individuals with low levels of antioxidants are much more likely to develop serious diseases than those with normal levels, hence

---

*Corresponding author: Walid Daoudi, Laboratory of Molecular Chemistry, Materials and Environment (LCM2E), Department of chemistry, Multidisciplinary Faculty of Nador, University Mohamed I, 60700, Nador, Morocco, e-mail: walid.daoudi@ump.ac.ma
Abdelmalik El Aatiaoui, Abdelouahad Oussaid, Adyl Oussaid, Laboratory of Molecular Chemistry, Materials and Environment (LCM2E), Department of chemistry, Multidisciplinary Faculty of Nador, University Mohamed I, 60700, Nador, Morocco
Selma Lamghafri, Laboratory of Applied Sciences, National School of Applied Sciences Al-Hoceima, Abdelmalek Essaâdi University, Tetouan, Morocco

https://doi.org/10.1515/9783110791891-010

the importance of these compounds in our lives in preventing cellular damage and maintaining good health.

10.2 Antioxidants and applications

10.2.1 What is an antioxidant?

10.2.1.1 Definition

Antioxidants [1] are chemical substances that have the ability to break the chain of formation of free radicals that are generated by the body during natural processes or by certain environments by transforming them into new harmless species and thus promote the reduction or even the stop of aging and degeneration of cells caused by these radical derivatives and consequently protect our body.

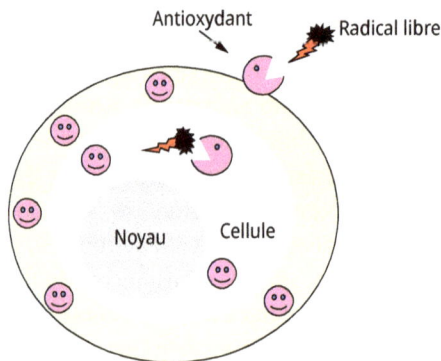

Fig. 10.1: Antioxidant defense.

10.2.1.2 Mode of action

It has been shown that the formation of radicals in the body has two origins: endogenous (linked to internal biological processes) and exogenous (due to external environmental and/or dietary factors).

The radical formed approaches the cell membrane, tearing off a single electron, which causes the loss of the normal shape of the cell, leading to the damage of its membrane and thus favors the entry of substances harmful to the cell, allowing its damage and dysfunction.

To ensure a good protection of the cell and consequently of the organism, the antioxidant reaction due to the presence of antioxidants is an adequate remedy in order to stop these radical reactions and avoid the appearance of deadly pathologies.

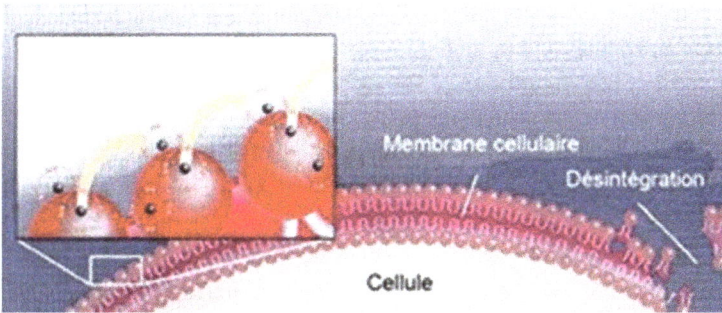

Fig. 10.2: Mode of action of free radicals on the plasma membrane of a cell.

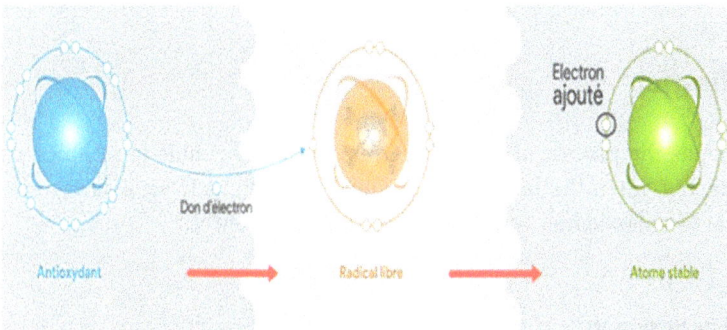

Fig. 10.3: Mode of action of antioxidants on free radicals.

The presence of an antioxidant molecule (reducer) close to an "oxidizing" free radical allows its neutralization by giving it an unpaired electron, which stops its aggressiveness toward healthy cells.

10.2.1.3 Efficiency

In order to highlight the full spectrum of antioxidant activity of an antioxidant and its compliance with industry expectations, a series of tests can be considered:
-   Peroxide index (measures the initial products of oxidation)
-   Rancimat testing (evaluates the sub-products of oxidation)
-   Oxipres testing (evaluates the sub-products of oxidation)

Economic market for antioxidants
Since the early 1990s, the number of scientists interested in the health effects of antioxidants has grown steadily. In 30 years, the large number (more than 4,000) of

scientific studies that have been published clearly demonstrate the importance of using antioxidants as medicines and in moderate concentrations on human health.

According to latest research and due to the increasing demand for antioxidants (especially those of natural origin which are being used more and more worldwide), the global market will soon reach millions with an increase of millions of dollars over the year 2020 and the Compound Annual Growth Rate (CAGR) increases to a figure of 6.0% during the forecast period (2019–2024). For example, the American Heart Association and the U.S. Agricultural Research Service encourage the population to consume more antioxidants in order to avoid a number of pathologies, particularly cardiovascular diseases. On the other hand, those of synthetic origin have recognized a reduced consumption and their use has been limited in recent years because of their probable side effects.

10.2.2 Classification

In order to overcome the process of oxidative stress, the body is able to produce its own antioxidants, but when the level of free radicals is high, it is necessary to increase the amount of antioxidants in the body from different sources:

10.2.2.1 Natural antioxidants

These antioxidants can be distinguished into two important categories:
– Endogenous antioxidants: correspond to those synthesized directly by the body:
    – Glutathione

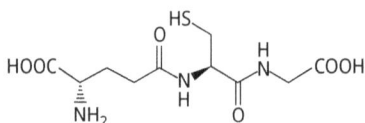

Fig. 10.4: General structure of glutathione.

    – N-acetylcysteine

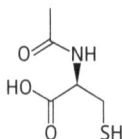

Fig. 10.5: General structure of N-acetylcysteine.

– coenzyme Q10

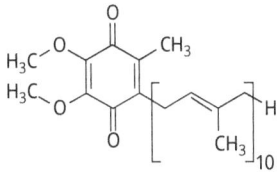

Fig. 10.6: General structure of coenzyme Q10.

– Enzymes
  Enzymes are bio-molecules synthesized by the body and can act as antioxidants.
– Copper/zinc and manganese
  These are essential trace elements that can intervene in many cellular mechanisms within the body to regenerate enzymes and antioxidants.
– Superoxide dismutase (SOD)
  It is an enzyme that has the property of eliminating free radicals from the human body and can transform the superoxide anion $O_2^{\cdot}$ into neutral molecular oxygen O_2 with the formation of water oxygen H_2O_2.
– Bilirubin

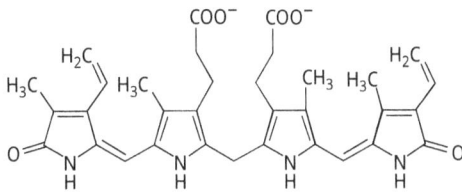

Fig. 10.7: General structure of bilirubin.

– Melatonin

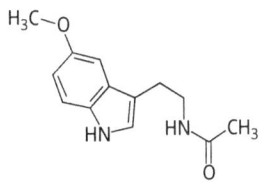

Fig. 10.8: General structure of melatonin.

- Nitric oxide.
  It is a chemical compound made up of one oxygen and one nitrogen atom and is produced in small quantities by many organisms, including the human body.

- Exogenous antioxidants: represent those which are brought from outside:
  - Vitamin C (also called ascorbic acid)
    Vitamin C or ascorbic acid containing antioxidant properties is an organic acid that water-soluble. It has the possibility of reacting in the intra as well as the extra-cellular.

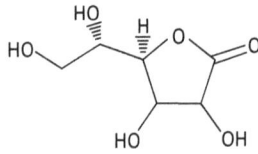

Fig. 10.9: Structure of vitamin C (L-ascorbic acid).

- Vitamin E (namely tocopherol).
  It is a vitamin capable of solving fat; it is known as an antioxidant that can be easily inserted in cell membranes: red blood cells, cells, muscle cells, neurons (it is the only antioxidant of the central nervous system), etc.

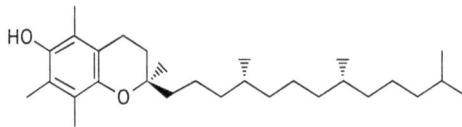

Fig. 10.10: Structure of vitamin E (α-Tocopherol).

- Polyphenols
  This family of water-soluble organic molecules is in the form of a complex assembly of smaller molecules, the phenols, with a benzene ring and hydroxyl functions. They are frequently used for their powerful antioxidant power while giving a hydrogen atom to the free radicals present and subsequently, their oxidized form will be stabilized by resonance or dimerization.

  These derivatives are subdivided into several subfamilies: simple phenols, phenolic acids, coumarins, flavonoids, and polymerized forms, such as tannins and lignin.

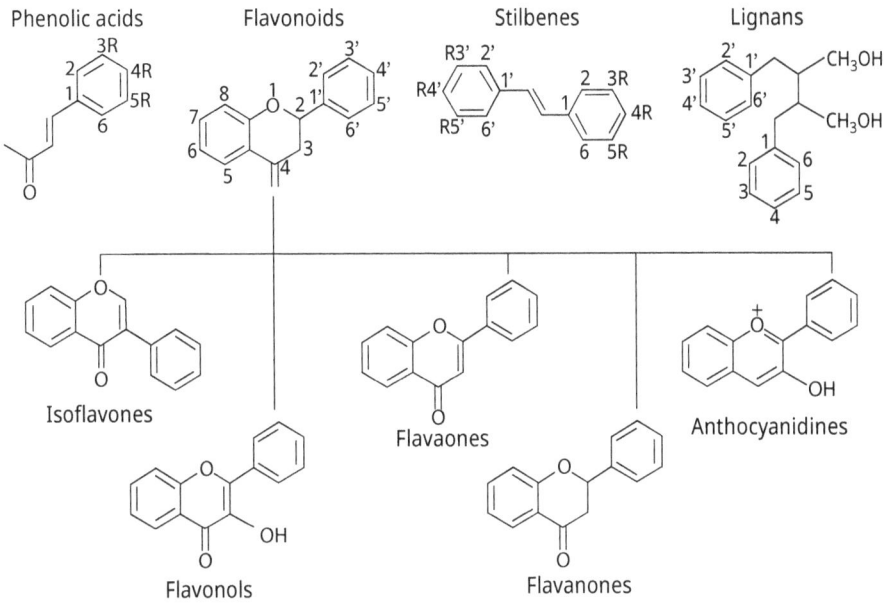

Fig. 10.11: Structures of different polyphenols.

- Carotenoids (lycopene, lutein, etc.).
  Currently, the carotenoid family is the basis of many scientific researches thanks to its important antioxidant property. It is composed of about 600 molecules, the most important and best known of which is β-carotene, which allows the neutralization of singlet oxygen while avoiding radical mechanisms within the cell.

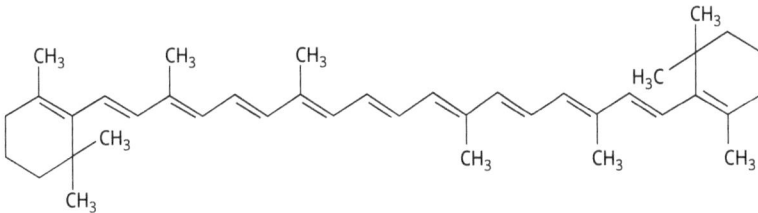

Fig. 10.12: Structure of β-carotene.

- Minerals and trace elements.
  Minerals (magnesium) and trace elements (zinc, selenium, and manganese) have antiradical properties and intervene in the case of oxidative stress to prevent the appearance and development of certain pathologies in the body.

10.2.2.2 Synthetic antioxidants

These are generally synthesized phenolic compounds containing a phenol unit substituted with alkyl or ester function of which the best known are:

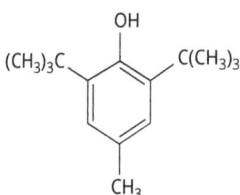

butylhydroxytoluene (BHT).

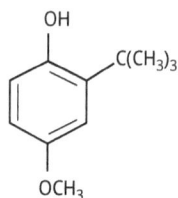

butylhydroxyanisole (BHA).

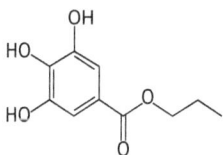

Gallate de propyle (PG).

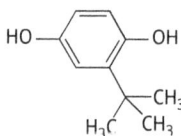

tert-butylhydroquinone (TBHQ).

Fig. 10.13: Examples of synthetic antioxidants.

10.2.3 Areas of application

10.2.3.1 Pharmaceutical industry (health)

Preparations in the pharmaceutical industry are considered stable if they do not change their properties for a specific period of time. In fact, the exposure of these drugs to various factors (such as molecular oxygen in air) can lead to oxidations and consequently to the loss of its required pharmacological activities. Therefore, the use of antioxidants (ascorbic acid, sodium metabisulfite, etc.) is essential in order to slow down or even stop these chain reactions and to prolong the duration of efficacy of these drug products.

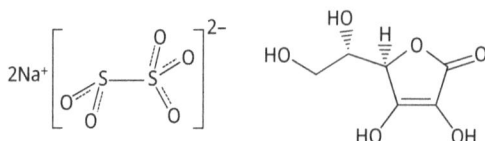

Fig. 10.14: Examples of antioxidants used in the pharmaceutical industry.

10.2.3.2 Agri-food industry (nutrition)

In the food industry, these chemical molecules can be used to slow down the oxidation of fats and lipids contained in foods, thus avoiding their degradation and prolonging their shelf life.

On the same subject, previous works [2] have demonstrated the important activity of aromatic and medicinal plants toward the preservation of foodstuffs and their protection against radical reactions thanks to the presence of a diversity of molecules with great antioxidant properties.

10.2.3.3 Cosmetics industry (beauty)

In the cosmetics industry, which is very sensitive, oxidation reactions can take place within these commercial preparations, causing the degradation of the fatty acids contained in these formulations and ending up with the deterioration of their odor and appearance, which promotes significant negative effects on the global economy of these care and beauty products. For this reason and in order to overcome these undesirable consequences, the use of antioxidants, both lipophilic and hydrophilic, remains an inevitable choice.

In this sense, and as an example of natural antioxidant commonly used in cosmetics, we can recall the Amiox ER, which is an antioxidant originating from Rosemary extract and has approved its effectiveness in stabilizing synthesized formulas, protecting the fatty phases from oxidation. It is considered a natural alternative to synthetic antioxidants and is particularly effective.

10.2.3.4 Nutraceutical industry (food supplements)

The preparation of food supplements in the nutraceutical industry is an essential way to ensure food formulas rich in antioxidants and to guarantee the consumer a sufficient daily intake of natural antioxidants and maintain good health. (example of IMMUNIA syrups concentrated in antioxidants with a number of 34 polyphenols).

10.3 Natural antioxidants from aromatic medicinal plants

10.3.1 Composition of aromatic medicinal plants

The term *aromatic medicinal plants* refers to a plant or part of a plant that possesses substances called active principles that can be used for therapeutic purposes such as antioxidant activity, without harmful effects at the recommended doses; this type of plants is commonly used in traditional medicine, furthermore, they can be utilized in cooking and in phytotherapy for the aromas they give off, and their essential oils that can be extracted. Moreover, the beneficial effects of plants on health are due to the fact that they contain substances called active principles, responsible for their therapeutic effects.

Currently, natural molecules extracted from medicinal plants constitute the major components of a large part of our pharmaceutical, cosmetic, and agro-food products. The action mechanisms of plants on the organism can be better understood by evaluating their chemical composition through different extraction, isolation, and identification techniques.

Generally, we can distinguish two types of chemical compounds: the volatile organic compounds, in particular the most widespread chemical family of essential oils, and the nonvolatile organic compounds which are mainly divided into two large chemical families, namely the phenolic compounds and the alkaloids.

10.3.1.1 Volatile organic compound

The specific aroma of plants and their flavor and fragrance are extracted from essential oils which are complex mixtures of volatile and semi-volatile organic compounds from a single botanical source.

In order to verify essential oils quality, and explain its properties and predict its potential toxicity, it is primordial to identify its chemical composition and exact components that is very complex and subject to many variables.

Gas chromatography coupled with mass spectrometry techniques are mainly used to analyze the composition of an essential oil. In terms of chemical consideration, the essential oil is a various mixture, very complex and consisting of several compounds. The difficulty of showing their activities is due to this complexity.

"Chemotypes" are the main chemical component of most essential oils. They allow to distinguish an essential oil extracted from the same type but with a different chemical composition.

There are two different chemical classes of essential vegetable oils constituents namely terpenes and phenylpropanoids. Terpene substances may be classified into two principal categories: on the one hand, terpenes containing a hydrocarbon structure, specially mono-, sesqui- and diterpenes. On the other hand their oxygenated derivatives, such as, oxides, alcohols, aldehydes, acids, ketones, phenols, esters, and lactones [3].

Furthermore, essential oils can be extracted from plant material by different methods, the main ones are based on volatility, solubility, expression, and steam distillation.

The choice of method depends on the extraction process used and the intended target, among other things, the nature of the plant material to be processed, the physical and chemical properties of the essence to be extracted, as well as the yield, applications and uses [4].

Among the extraction methods are: extraction by ultrasound or microwave irradiation or older: distillation, expression, incision, or enfleurage [5].

Indeed, the method that remains the most used and easiest is hydro distillation. Figure 10.15 shows the different ways of extraction of essential oils.

Fig. 10.15: Methods of obtaining essential oils.

10.3.2 Nonvolatile organic compound

10.3.2.1 Phenolic compound

Phenolic compounds or polyphenols are molecules belonging to the secondary metabolism of plants and play a protective role against environmental threats. These phytomicronutrients are also responsible for several physiological processes such as fruit ripening, seed germination, and cell growth [6]. They are also pigments and therefore responsible for the colors and hues of different parts of plants (fruits, flowers, and leaves). Depending on the section and the species of the plant, polyphenols accumulate qualitatively and quantitatively in plants. They are soluble in polar solvents. Moreover, they constitute more than 10,000 compounds, categorized into various range in complexity from simple molecules (phenolic acids) to highly polymerized substances (tannins), split into several classes namely phenolic acids, flavonoids, saponins, tannins, sterols, lignans, and stilbenes. According to the literature, among this list, flavonoids, phenolic acids, and tannins represent the principal ranges [7]. This diversity allows them to possess different significant biological characteristics such as anticancer, antioxidant, anti-inflammatory, antimicrobial, and anti-atherosclerosis effects as well as cardiovascular protection, angiogenesis inhibition, and improvement of endothelial function.

10.3.2.2 Alkaloids

Among the various group of chemical entities, alkaloids plants represent the largest groups of natural products estimated about 12,000 belonging to this category. The term *alkaloids* consist of fundamental compounds that include one or more nitrogen atoms, generally in conjunction as element of a ring structure [8]. It is sometimes difficult to distinguish the fine line between alkaloids and other nitrogenous natural metabolites [9]. flowering plants is the principal origin of Alkaloids, mostly found in Papilonaceae (lupins), Papaveraceae (poppies), Renonculaceae (aconits), and Solanaceae (tobacco) [10]. Additionally, they can also be found in lower plants, insects, marine organisms, microorganisms, and animals. Furthermore, several marketed medicines contain alkaloid proprieties [11]. Latest studies are carried out on alkaloids for the discovery of new therapeutic approaches. Indeed, they have a wide spectrum of pharmacological activity, incorporating antibacterial, antiviral, anti-inflammatory, and anticancer characteristics [12]. As a result, two plant alkaloids, sanguinarine and berberine, demonstrate cytotoxic efficacy toward hematopoietic cancerous cells and generated apoptosis in the investigated cell lines [13]. In addition, inflammatory effects in mice have been demonstrated to be modulated by curine, bisbenzylisoquinoline alkaloid; it is a result of several factors, including inhibition of macrophage activation and cytokine production as well as neutrophil recruitment, also the diminishing of nitric oxide levels [14]. Unfortunately, consuming some alkaloids such as arecoline, found in betel nuts, can lead to toxic effects. Overconsumption of this substance can lead to carcinogenesis [15]. Plants exploit this toxicity in order to protect themselves against aggression from other organisms; this instinctive action is a crucial ecosystem feature.

10.3.3 Families of natural antioxidants

In the body, a variety of intracellular and extracellular antioxidants, enzymatic (endogenous) and nonenzymatic (mainly food), are active. Antioxidant enzymes contain SOD, catalase (CAT), and glutathione peroxidase (GPX). While a variety of LR deactivators such as vitamin A (retinol), vitamin C (ascorbic acid), vitamin E (tocopherol), and micronutrients (iron, copper, zinc, selenium, manganese) act as enzyme cofactors are included in nonenzymatic antioxidants category. The effectiveness of the antioxidant system depends on nutritional patterns (vitamins and micronutrients) and the modification of endogenous antioxidant enzymes production by physical exercise, nutrition, and aging.

10.3.3.1 Enzymatic antioxidants

SOD serves as the primary line of defense against threat superoxide radicals and the initial defensive line against oxidative stress. However, as already mentioned, it causes the appearance of H_2O_2. In all cells, at rest, mitochondrial SOD reduces most of the $O_2^{\bullet-}$ and the remaining amount diffuses into the cytosol. About 65–85% of SOD activity occurs in the cytoplasm of muscle cells.

10.3.3.1.1 Catalase
All cells contain CAT, which is notably present in peroxisomes, cellular structures that produce H_2O_2 by detoxifying toxic substances in presence of oxygen [16]. In attendance of catalase, H_2O_2 is converted into water and oxygen; also it is used to detoxify certain toxic substances via a peroxidase reaction that requires a substrate such as phenol, alcohol, or formic acid.

10.3.3.1.2 Glutathione peroxidase
H_2O_2 is converted into water by mitochondria. And the glutathione is transformed into oxidized glutathione in the same reaction. The action of GPX and CAT on H_2O_2 are the same; however, GPX which is present in the cytosol cell, is more effective with a high concentration of ROS and CAT has a significant action with a lower concentration of H_2O_2 [17].

10.3.4 Nonenzymatic antioxidants

10.3.4.1 Vitamin E (tocopherol)

The tocopherol family of vitamins includes fat-soluble vitamin E that is composed of numerous isoforms. The most active and abundant type is α-tocopherol [18].

Due to its abundance in cells and mitochondrial membranes and its efficacy in acting directly on ROS, vitamin E is noted for being one of the most significant chain-breaking antioxidants. Many antioxidants, notably vitamin C, β-carotene, and lipoic acid, interact with this vitamin. These antioxidants can regenerate vitamin E from its oxidized form [19]. Vitamin E performs well in cell membranes because it avoids lipid peroxidation. In a lipid environment, the deactivation of ROS is facilitated by molecular structure of vitamin E.

10.3.4.2 Vitamin C (ascorbic acid)

Vitamin C is a water soluble vitamin and is likely the most crucial antioxidant in extracellular fluids and cytosol [20]. It is known to be more abundant in tissues where ROS production is greater. This phenomenon is characterized as an oxidative stress adaption. Vitamin C is capable of neutralizing ROS in fluids. Inside the cells, after reacting with ROS vitamin C strengthens the action of vitamin E and glutathione by renewing their active form. Additionally, it has the capacity to attract copper ions that have a potent oxidizing effect. Consequently, vitamin C supplementation has often been investigated.

10.3.4.3 β-carotene and vitamin A (retinol)

Vitamin A found in many lipid substances, is fat-soluble. When needed, the body can convert β-carotene, present in cell membranes, into vitamin A. However, its exact mechanism of action in vivo is obscure. β-Carotene is considered to deactivate ROS (particularly singlet oxygen and lipid radicals) and diminish lipid peroxidation. However, β-carotene, which is less important than vitamin E in the antioxidant system, in combination with vitamin A acts in the presence of C and E vitamin to protect cells from ROS.

10.3.5 Determination of the antioxidant power

10.3.5.1 Antiradical power

10.3.5.1.1 Free radicals

An element (atom, molecule, or ion) containing one or more unpaired electrons, on its outer layer, which makes it unstable, is described as a free radical. While free radicals are a particular form of chemical species (atoms or molecules) that have a single electron (or one paired). To ensure their stability, radicals react instantly with other components to capture the electron needed.

A chain reaction starts when a free radical attacks the nearest stable molecule by pulling out its electron, and this one becomes in its turn a free radical.

10.3.5.2 Evaluation of antioxidant reactivity by DPPH

A spectrophotometer was utilized to evaluate the DPPH free radical scavenging test; usually the reduction of the radical is detected by the changing of solution color from a violet color (DPPH-) to a yellow color (DPPH-H) measurable at 515 nm. A decrease in absorbance induced by antiradical substances determines the reduction capacity. The reaction can be summarized in the form of the equation:

$$DPPH + (AH)\ n \rightarrow DPPH - H + (A)n$$

10.3.5.3 Reducing power

The power of nonenzymatic antioxidants is assessed by the Ferric Reducing Antioxidant Power (FRAP) assay which is performed according to the method described by Sayah et al. (2017). Effectively, the presence of an antioxidant ferric iron (Fe^{3+}) present in the potassium ferricyanide complex $K_3Fe(CN)_6$ is indicated by yellow color which goes to blue-green when it is reduced to ferrous iron (Fe^{2+}). There is proportionality between the intensity of this color and the reducing power of the antioxidants in the medium.

10.5 Mechanism of action of antioxidants

The structural characteristics of phenolic compounds give them great antioxidant capacity. Numerous studies specify their antioxidant properties by different mechanisms involving electron transfer, proton transfer, or chelation of metal ions. They prevent damage caused by the presence of active oxygen entities ($O_2^-\bullet$, $HO\bullet$, $NO\bullet$, H_2O_2, $^1O_2$, $HOCl$, $RO\bullet$ et $ROO\bullet$).

The loss of proton of a polyphenol generates the formation of a radical strongly stabilized by mesomerism; which makes them good antioxidants. They intervene in particular at the level of lipidic peroxides LOO- by giving them a proton so that they stabilize in the form of hydroperoxides LOOH. The unpaired electron repartition on the aromatic ring and the absence of an oxygen attack site prevents the propagation of new radical reactions. They can also block the propagation phase by forming a stable complex with the lipid radicals formed according to the following reaction:

$$LO\bullet/LOO\bullet + A\bullet \rightarrow LOA/LOOA$$

The antioxidant activity of polyphenols allows delaying the beginning of oxidation because they are progressively consumed.

Transfert de proton (HAT)

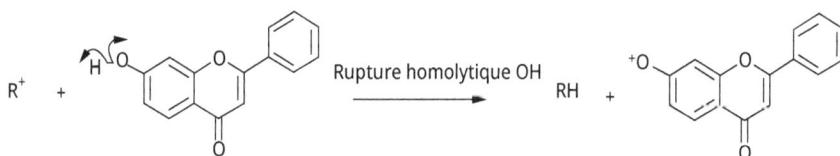

Transfert d'électron (SET)

Chélation des ions métaux de transition

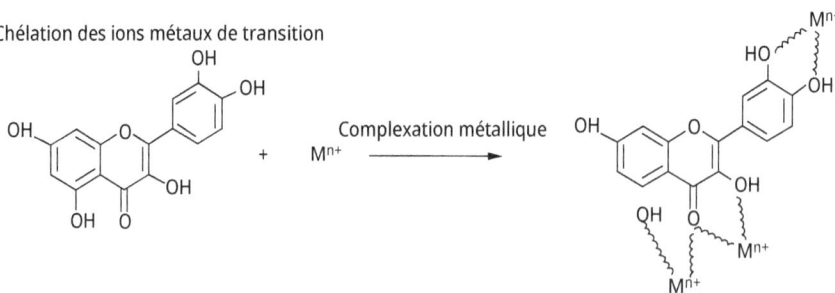

Fig. 10.16: Simplified presentation of the antioxidant mechanisms of polyphenols.

It has been concluded that the proximity of the alkyl and the number of hydroxyl groups present on the benzoic rings determines the reactivity of antioxidants. Flavonoids are known to be the most active due to their structure. Polysaccharides have also shown antioxidant activity through a metal ion chelation mechanism.

Table 10.1 summarizes aromatic medicinal plants that demonstrate promising antioxidant activity; also it indicates tests, solvents, and assays that have been used for the extraction.

Tab. 10.1: medicinal plants, their family, and different solvents and techniques used for them extraction.

Plants (family)	Parts used	Solvent	Assay	References
Acacia auriculiformis A. Cunn. ex Benth. (Mimosaceae)	Bark	HE, C, A, ET, ME	TPC, DPPH, HO, RP, TBARS	[21]
Achillea millefolium subsp. Millefolium Afan. (Asteraceae)	Essential oil	C, ME, Water	DPPH, SO, HO, TBARS	[22]
Aegle marmelos Correa (Rutaceae)	Fruit	ME	DPPH, RP, NO, SO	[23]
Allanblackia floribunda Oliv (Guttiferae)	Leaves, fruit	ME	DPPH, TPC, TF	[24]
Amaranthus lividus L. (Amaranthaceae)	Stem, leaves, flower	EA, ME Water	TEAC, DPPH, RP, Metal chelating, HO	[25]
Aporosa lindleyana Baill. (Euphorbiaceae)	Root	PE, C, EA, ME	DPPH, NO	[26]
Argyreia cymosa R. Sweet (Convolvulaceae)	Bark	PE, C, EA, ME	DPPH, HO, ABTS, NO, H2O2	[27]
Aristotelia chilensis Maqui (Elaeocarpaceae)	Fruit	EA, ME, Water	ORAC, FRAP, TPC	[28]
Azadirachta indica A. Juss var. siamensis Valeton (Meliaceae)	Leaves, raw fruit, ripe fruit, flower, stem bark	HE, ME, Water	DPPH, total antioxidant activity, TBARS	[29]
Azadirachta indica A. Juss var. siamensis Valeton (Meliaceae)	Leaves	ET	DPPH	[30]
Byrsonima crassifolia H. B. and K.(Malpighiaceae)	Leaves, bark, fruit	ME, Water	TPC, TF	[31]
Bergia suffruticosa (Delile) (Elatinaceae)	Whole plant	ME	TPC, DPPH, SO, RP	[32]
Burkea africana Hook (Leguminocaea)	Bark	PE, BT, EA, Water	DPPH, TBARS	[33]
Caesalpinia digyna Rottler (Caesalpiniaceae)	Root	PE, ME, Water	TPC, ABTS, DPPH, H2O2, NO, SO, HO, in *vivo*	[34]
Careya arborea Roxb (Barringtoniaceae)	Heartwood	PE, C, EA, ME, Water	DPPH, NO, In *vivo*	[35]

10.6 Conclusion

This chapter provides information about antioxidants, their application, classification, and the mechanism of action; it also highlights natural and potent antioxidants from aromatic medicinal plants and their activity. In addition, methods and techniques for assessing the reactivity have been discussed; the use of two different types of tests at least for antioxidant activity is recommended. This chapter is a concise material that aims to serve the antioxidant activity of medicinal plants experimenters.

References

[1]   B. Halliwell, J. M. Gutteridge. Free Radicals in Biology and Medicine. Oxford university press, USA, 2015.

[2]   I. Laib, M. Barkat. Composition chimique et activité antioxydante de l'huile essentielle des fleurs sèches de lavandula officinalis, 2011. http://dspace.univ-setif.dz:8888/jspui/handle/123456789/375 (accessed May 14, 2022).

[3]   F. Couic-Marinier, A. Lobstein. Composition chimique des huiles essentielles. Actualités Pharmaceutiques. 2013, 52, 22–25. https://doi.org/10.1016/j.actpha.2013.02.006.

[4]   M. Moghaddam, L. Mehdizadeh. Chapter 13 – chemistry of essential oils and factors influencing their constituents. In: Grumezescu, A. M., Holban, A. M., (Eds.), Soft Chemistry and Food Fermentation. Academic Press, 2017, pp. 379–419. https://doi.org/10.1016/B978-0-12-811412-4.00013-8.

[5]   M. D. Luque de Castro, M. M. Jiménez-Carmona, V. Fernández-Pérez. Towards more rational techniques for the isolation of valuable essential oils from plants. Trends Analyt Chem. 1999, 18, 708–716. https://doi.org/10.1016/S0165-9936(99)00177-6.

[6]   B. Nathalie, C. Jean-Paul. Méthode rapide d'évaluation du contenu en composés phénoliques des organes d'un arbre forestier – Institut National de Recherche en Agriculture, Alimentation et Environnement, (n.d.). https://hal.inrae.fr/hal-02669118 (accessed May 15, 2022).

[7]   A. King, G. Young. Characteristics and occurrence of phenolic phytochemicals. J Am Diet Assoc. 1999, 99, 213–218. https://doi.org/10.1016/S0002-8223(99)00051-6.

[8]   J. M. Barbosa-Filho, M. R. Piuvezam, M. D. Moura, M. S. Silva, K. V. B. Lima, E. V. L. Da-cunha, I. M. Fechine, O. S. Takemura. Anti-inflammatory activity of alkaloids: A twenty-century review. Rev Bras Farmacogn. 2006, 16, 109–139. https://doi.org/10.1590/S0102-695X2006000100020.

[9]   N. Bribi. Pharmacol Act Alkaloids: A Rev. 2018, 1, 1–6. https://doi.org/10.63019/ajb.v1i2.467.

[10]   H. Jing, J. Liu, H. Liu, H. Xin. Histochemical investigation and kinds of alkaloids in leaves of different developmental stages in *Thymus quinquecostatus*. Sci World J. 2014, 2014, 1–6. https://doi.org/10.1155/2014/839548.

[11]   B. Debnath, M. J. Uddin, P. Patari, M. Das, D. Maiti, K. Manna. Estimation of alkaloids and phenolics of five edible cucurbitaceous plants and their antibacterial activity. Int J Pharm Pharm Sci. 2015, 7, 223–227.

[12]   Z. Adamski, L. L. Blythe, L. Milella, S. A. Bufo. Biological activities of alkaloids: From toxicology to pharmacology. Toxins. 2020, 12, 210. https://doi.org/10.3390/toxins12040210.

[13]   A. Och, D. Zalewski, Ł. Komsta, P. Kołodziej, J. Kocki, A. Bogucka-Kocka. Cytotoxic and proapoptotic activity of sanguinarine, berberine, and extracts of chelidonium majus L. and Berberis thunbergii DC. toward hematopoietic cancer cell lines. Toxins. 2019, 11, 485. https://doi.org/10.3390/toxins11090485.

[14] J. Ribeiro-Filho, F. Carvalho Leite, A. Surrage Calheiros, A. de Brito Carneiro, J. Alves Azeredo, E. Fernandes de Assis, C. da Silva Dias, M. Regina Piuvezam, P. T. Bozza. Curine inhibits macrophage activation and neutrophil recruitment in a mouse model of lipopolysaccharide-induced inflammation. Toxins. 2019, 11, 705. https://doi.org/10.3390/toxins11120705.

[15] C.-H. Chang, M.-C. Chen, T.-H. Chiu, Y.-H. Li, W.-C. Yu, W.-L. Liao, M. Oner, C.-T. R. Yu, -C.-C. Wu, T.-Y. Yang, C.-L. J. Teng, K.-Y. Chiu, K.-C. Chen, H.-Y. Wang, C.-H. Yue, C.-H. Lai, J.-T. Hsieh, H. Lin. Arecoline promotes migration of A549 lung cancer cells through activating the EGFR/Src/FAK Pathway. Toxins. 2019, 11, 185. https://doi.org/10.3390/toxins11040185.

[16] F. Antunes, D. Han, E. Cadenas. Relative contributions of heart mitochondria glutathione peroxidase and catalase to H2O2 detoxification in in vivo conditions. Free Radical Biol Medi. 2002, 33, 1260–1267. https://doi.org/10.1016/S0891-5849(02)01016-X.

[17] J. Fuchs, S. Weber, M. Podda, N. Groth, T. Herrling, L. Packer, R. Kaufmann. HPLC analysis of vitamin E isoforms in human epidermis: Correlation with minimal erythema dose and free radical scavenging activity. Free Radical Biol Medi. 2003, 34. 330–336. https://doi.org/10.1016/S0891-5849(02)01293-5.

[18] J. S. Coombes, S. K. Powers, B. Rowell, K. L. Hamilton, S. L. Dodd, R. A. Shanely, C. K. Sen, L. Packer. Effects of vitamin E and α-lipoic acid on skeletal muscle contractile properties. J Appl Physiol. 2001, 90, 1424–1430. https://doi.org/10.1152/jappl.2001.90.4.1424.

[19] A. X. Bigard. Lésions musculaires induites par l'exercice et surentraînement. Sci Sports. 2001, 16. 204–215. https://doi.org/10.1016/S0765-1597(00)00037-X.

[20] F. M. Palmer, D. C. Nieman, D. A. Henson, S. R. McAnulty, L. McAnulty, N. S. Swick, A. C. Utter, D. M. Vinci, J. D. Morrow. Influence of vitamin C supplementation on oxidative and salivary IgA changes following an ultramarathon. Eur J Appl Phys. 2003, 89, 100–107. https://doi.org/10.1007/s00421-002-0756-4.

[21] R. Singh, S. Singh, S. Kumar, S. Arora. Evaluation of antioxidant potential of ethyl acetate extract/fractions of Acacia auriculiformis A. Cunn Food Chem Toxicol. 2007, 45, 1216–1223. https://doi.org/10.1016/j.fct.2007.01.002.

[22] F. Candan, M. Unlu, B. Tepe, D. Daferera, M. Polissiou, A. Sökmen, H. A. Akpulat. Antioxidant and antimicrobial activity of the essential oil and methanol extracts of Achillea millefolium subsp. millefolium Afan. (Asteraceae). J Ethnopharmacol. 2003, 87, 215–220. https://doi.org/10.1016/S0378-8741(03)00149-1.

[23] K. Dhalwal, Y. S. Deshpande, A. P. Purohit, S. S. Kadam. Evaluation of the antioxidant activity of sida cordifolia. Pharm Biol. 2005, 43, 754–761. https://doi.org/10.1080/13880200500406438.

[24] G. A. Ayoola, S. S. Ipav, M. O. Sofidiya, A. A. Adepoju-Bello, H. A. B. Coker, T. O. Odugbemi. Phytochemical screening and free radical scavenging activities of the fruits and leaves of allanblackia floribunda oliv (Guttiferae). Int J Health Res. 2008, 1, 87–93. https://doi.org/10.4314/ijhr.v1i2.47920.

[25] N. Ozsoy, T. Yilmaz, O. Kurt, A. Can, R. Yanardag. In Vitro Antioxidant Activity of Amaranthus Lividus L. Food Chemistry, Vol. 116, 2009, pp. 867–872. https://doi.org/10.1016/j.foodchem.2009.03.036.

[26] S. Badami, O. Prakash, S. H. Dongre, B. Suresh. In vitro antioxidant properties of Solanum pseudocapsicum leaf extracts. Indian J Pharmacol. 2005, 37, 251. https://doi.org/10.4103/0253-7613.16573.

[27] S. Badami, J. Vaijanathappa, S. Bhojraj. In vitro antioxidant activity of Argyreia cymosa bark extracts. Fitoterapia. 2008, 79, 287–289. https://doi.org/10.1016/j.fitote.2007.10.021.

[28] C. L. Céspedes, M. El-Hafidi, N. Pavon, J. Alarcon. Antioxidant and cardioprotective activities of phenolic extracts from fruits of Chilean blackberry Aristotelia chilensis (Elaeocarpaceae. Maqui Food Chem. 2008, 107, 820–829. https://doi.org/10.1016/j.foodchem.2007.08.092.

[29] P. Sithisarn, R. Supabphol, W. Gritsanapan. Comparison of free radical scavenging activity of siamese neem tree (Azadirachta indica A. Juss var. siamensis Valeton) leaf extracts prepared by different methods of extraction. Mpp. 2006, 15, 219–222. https://doi.org/10.1159/000092185.

[30] P. Sithisarn, R. Supabphol, W. Gritsanapan. Antioxidant activity of Siamese neem tree (VP1209). J Ethnopharmacol. 2005, 99, 109–112. https://doi.org/10.1016/j.jep.2005.02.008.

[31] J. N. S. Souza, E. M. Silva, A. Loir, J.-F. Rees, H. Rogez, Y. Larondelle. Antioxidant capacity of four polyphenol-rich Amazonian plant extracts: A correlation study using chemical and biological in vitro assays. Food Chem. 2008, 106, 331–339. https://doi.org/10.1016/j.foodchem.2007.05.011.

[32] S. Anandjiwala, H. Srinivasa, J. Kalola, M. Rajani. Free-radical scavenging activity of Bergia suffruticosa (Delile) Fenzl. J Nat Med. 2006, 61, 59–62. https://doi.org/10.1007/s11418-006-0017-7.

[33] E. Mathisen, D. Diallo, Ø. M. Andersen, K. E. Malterud. Antioxidants from the bark of Burkea africana, an African medicinal plant. Phytother Res. 2002, 16, 148–153. https://doi.org/10.1002/ptr.936.

[34] R. Srinivasan, M. J. N. Chandrasekar, M. J. Nanjan, B. Suresh. Antioxidant activity of Caesalpinia digyna root. J Ethnopharmacol. 2007, 113, 284–291. https://doi.org/10.1016/j.jep.2007.06.006.

[35] S. Badami, S. Moorkoth, S. R. Rai, E. Kannan, S. Bhojraj. Antioxidant activity of Caesalpinia sappan heartwood. Biol Pharm Bull. 2003, 26, 1534–1537. https://doi.org/10.1248/bpb.26.1534.

Anuragh Singh, Rushendran R., Siva Kumar B. and Ilango K.*

Chapter 11
Antidiabetic activity of selected Indian medicinal plants

Abstract: Diabetes is one of the metabolic disorders affecting all age groups due to lack of insulin production in the blood or hereditary. Diabetes affects more than 422 million people across the globe, with the population growth averaging 800 million by 2025. Multiple cases of diabetes in a family are a realistic possibility today, and the disease is quickly rising in prevalence as a result of its perceived ease of treatment. Till today scientists are unable to find a way for the treatment but some natural herbs are playing a major role knowingly or unknowingly such as *Adathoda vasica, Allium cepa, Ficus religiosa, Morus alba, Scoparia dulcis.* The different medicinal plant research reveals that vasicinol, vasicine, adhatonine, β-boswellic acid, linolenic acid, β-Sitosteryl-d-glucoside, Apigenin, Luteolin, Kampferol, Hentriacontane, Pentriacontane, phlobatannins. etc. are playing a crucial role in the treatment. This chapter deals with some important medical plants and their phytoconstituents involved in the antidiabetic activity.

11.1 Introduction

In the present scenario there has been a lot of interest in research on biofriendly, eco-friendly, and somewhat safer plant-based medications from the old system of medicine. More than 21,000 plants having important medical qualities have been identified by the WHO. India is said to be the world's botanical paradise and a treasure trove of therapeutic herbs. In India, there are over 2,500 medicinal species with a variety of pharmacological activities, of which more than 150 are commercially employed in the creation of newer medications [1]. Diabetes mellitus is an ailment preceded by hyperglycemia and alteration in lipid, carbohydrate, and protein metabolism [2].

It is the common chronic-metabolic disease, outlined by an upsurge in the levels of glucose sustained by relative or absolute insulin insufficiency. In the long term, the disease is connected with eye, renal, cardiovascular, and neurological complications such

Corresponding author: Ilango K., Department of Pharmaceutical Quality Assurance, SRM College of Pharmacy, SRM Institute of Science and Technology, Kattankulathur, 603203, Chengalpattu (Dt), Tamil Nadu, India, e-mail: ilangok1@srmist.edu.in
Anuragh Singh, Rushendran R., Department of Pharmacology, SRM College of Pharmacy, SRM Institute of Science and Technology, Kattankulathur, 603203, Chengalpattu (Dt), Tamil Nadu, India
Siva Kumar B., Department of Pharmaceutical Chemistry, SRM College of Pharmacy, SRM Institute of Science and Technology, Kattankulathur, 603203, Chengalpattu (Dt), Tamil Nadu, India

https://doi.org/10.1515/9783110791891-011

as weight loss, fatigue, polyuria, blurred vision, and delayed wound healing with elevation of urine glucose levels [3–5]. It is characterized by abnormal carbohydrate metabolism, an insufficient insulin level, and insulin resistance. Synthetic pharmaceuticals cause
unwanted side effects; treating diabetes using natural medicines, particularly medicinal
plants, has emerged as a viable alternative approach. Many secondary metabolites are
found in medicinal plants that play a crucial role in the treatment of diabetes [6].One of
the discrepancies of the immune system's regulation is the destruction of beta-cells in the
islets of Langerhans which leads to the development of insulin-dependent diabetes. Several environmental and genetic factors impact the immune system, triggering lymphocyte malicious activities, particularly lymphocyte attacks, and pancreatitis. Insulitis and
diabetes can be influenced by this inflammatory response [7, 8].

Diabetes affects more than 422 million people across the globe, with the population growth averaging 800 million by 2025. The condition produces severe morbidity, mortality, and long-term clinical manifestations, and it is a significant risk factor
for a variety of chronic diseases, appending cardiovascular disease [9, 10]. Neglect-
ing treatment can lead to problems with blood vessels, the heart, the kidneys and
the nervous system (including neuropathy). The management plan consists of physical activity, dietary changes and medication [11]. Metformin is the first-line drug for
mild to moderate diabetes mellitus, especially type 2, alone or in combination with a
sulfonylurea and/or α-glucosidase inhibitor; if it is inefficient, a third agent, such as di-
peptidyl-peptidase-4 (DPP4), thiazolidinedione, or glucagon-like peptide-1 (GLP-1) agonist, is desired [12–14]. The pathologic manifestations of diabetes, particularly type 2,
include both macrovascular and microvascular problems [15].

The use of insulin and hypoglycemic drugs is currently a most successful, effective,
and accepted treatment for diabetes, but these compounds have adverse side effects [16,
17]. Inhibiting α-glucosidase and α-amylase has been reported to be efficient in minimizing glucose uptake by decreasing monosaccharide [18]. α-glucosidase and α-amylase inhibitors likevoglibose, miglitol, and acarboseare are commercially marketed [19]. The
medicinal plants have a long history and are successfully employed to manage a different
ailment. There are different reasons why the implementation of medicinal plants is continuing to expand. Many plants across the world have been investigated for potential antidiabetic activity [20–24].The chapter gives information about a few important medicinal
plants having antidiabetic activity based on reliable clinical and laboratory evidence.

11.2 List of medicinal plants

Abelmoschus esculentus

The okra is also called as Lady's Finger, a flowering plant of the family malvaceae,
that originated in Ethiopia and has since spread to many other nations across the

world. Dharos, kacangbendi, okura,qiukui, okro, quiabos, quiabo,ochro, gumbo,okoro, quimgombo, bamieh, quingumbo, bamya, bendi, bamia, bhindi, and gombo are some other names for okra. Fruits have traditionally been used for cooling, stomachic, astringent, and aphrodisiac reasons, as well as in chronic condition of dysentery, gonorrhea, strangury, urinary discharges, and diarrhea. The decoction of tender pods is a demulcent, emollient, and diuretic used to treat spermatorrhea. The entire plant is demulcent and emollient; the seed may be utilized to stop the proliferation of cancer cells. Polysaccharides and quercetin, in particular, have the ability to inhibit the functions of -amylase and -glucosidase, which have antidiabetic and antioxidant properties [25].

Okra fiber contains 67.5% cellulose, 15.4% hemicellulose, 7.1% lignin, 3.4% pectic matter, 3.9% fatty and waxy matter, and 2.7% aqueous extract. Petals include 13 flavanoid glycosides, as well as glucosides of gossypetin and hibiscetin. Oxalic acid, protein, fat, minerals, glucose, calcium, and phosphorus are all found in fresh fruits. Flavonoids, galactose, rhamnose, and dalacturonic acid are all found in fruit mucilage. Ripe seeds contain 10–22% edible oil. The essential oil extracted from the pods and seeds contains terpenylacetate, citral and cyclohexanol, aliphatic alcohols, ptolualdehyde (in fruits); the nonvolatile neutral component comprises sitosterol and its 3 galactosides. The isolated biomolecules such as quercetin-3-osophoroside- I, 5,7,3',4'-tetrahydroxy flavonol-3- *O*- [d-rhamnopyranosil(1,2)] d-glucopyranoside and isoquercitrin are responsible for the *amylase* inhibitory activity. In addition, oligomeric proanthocyanidins present in unripe okra seeds were discovered to inhibit amylase and glucosidase activity. It was also shown that the monosaccharide (rhamnogalacturonan I backbone with type II arabinogalactan sidechains largely replaced at Rhap O4) exhibits antihyperglycemic activity [26].

Achyranthes aspera

Sometimes known as the Prickly Chaff flower, it is a tiny, branched, monoecious perennial subshrub found in the tropical and subtropical areas and belongs to the Amaranthaceae family. It is a popular folk and Ayurvedic herb whose leaves, flowers, and stems are used to cure a variety of maladies, including bronchitis, vomiting, piles, heart disease, stomach pains, itching, dyspepsia, ascites, diarrhea, sprains, hypertension, asthma, wound dressing, diabetes, etc. It has been established that the leaves' extract has a considerable antidiabetic effect, and pharmacological tests have revealed that the plant contains significant amounts of alkaloids, cardiac glycosides, tannins, flavonoids, steroids, terpenoids, saponin, and reducing sugar. It also possesses a variety of medical properties, including thyroid hormone stimulation, antifertility activity, antihyperlipidemic activity, antitumor promoter action, anti-inflammatory activity, immunostimulatory activity, and antibacterial activity.

An ethanol extract's antidiabetic efficacy in vivo against Streptozotocin-induced diabetic rats was investigated. Streptozotocin selectively kills beta-cells, which are vital in insulin production. A lack of insulin produces an increase in blood glucose.

Blood glucose levels were much reduced after treatment with aqueous leaf extract. As a result, *A. aspera* may work by either decreasing gut glucose uptake or enhancing blood glucose transport. Inhibiting -amylase and -glucosidase may enhance blood glucose management in type 2 diabetes mellitus patients following carbohydrate-rich meal intake. It could help prevent diabetic complications and act as a valuable adjuvant in the current situation [27].

Adhatoda vasica

It is also known as Malabar nut, has health benefits against different diseases, and has long been utilized in Ayurveda. Malaysia, India, the Himalayan area, Sri Lanka, and Burma are all known to have it. *Adhatoda vasica* contains major plant components like vasicinol, vasicine vasicol, adhatonine, and vasicinolone [28]. The leaves (methanolic extract) inhibit α-glucosidase and have also shown highest sucrase inhibitory activity with the specific phytochemical constituents such as vasicinol and vasicine from 40 Chinese medicinal herbs tested in rat intestinal α-glucosidase. The phytochemicals were screened in MS and NMR which shows IC_{50} values were 250 μM and 125 μM. The results suggest that this extract suppresses hyperglycemia [29].

Allium cepa

It might intensify hypoglycemia by inhibiting carbohydrate metabolizing enzyme α-glucosidase, resulting in a rise in adiponectin, that would delay absorption of carbohydrates and reduce insulin resistance. By stimulating activation of extracellular signal-regulated kinase, quercetin may enhance sugar-mediated insulin production while also providing significant protection to pancreatic β-cells from oxidative assaults. This action would therefore promote increased glucose tolerance by regulating glucose-induced insulin secretion, increasing the particular action of glucokinase, and increasing the number of pancreatic islets. The hypoglycemic actions are in order to clinically substantiate their recognized efficacies in diabetes management [30].

Allium sativum

It is a member of the Alliaceae family, a valuable spice and popular cure for a variety of diseases; physiological abnormalities. It originated in Central Asia and moved to China, the Near East, and the Mediterranean region before spreading west to Central and Southern Europe, Mexico, and Northern Africa (Egypt). It is used to treat a wide range of conditions such as high blood pressure, cancer, cholesterol, anthelmintics, hepatoprotective, anti-inflammatory, antifungal, antioxidant, bronchitis, tissue regeneration, tendinitis, joint pain, lumbago, bronchitis, backache, tuberculosis, persistent fever, malaria, rhinitis, obstinate skin disease including leucoderma, leprosy, indigestion, discoloration, enlargement of spleen, colic pain, piles, fracture of bone, fistula, gout, diabetes, urinary diseases, anemia, kidney stone, epilepsy, jaundice, cataract and night blindness. S-allyl-cysteine sulfoxide (allin), a precursor of several allyl sulfide elements of garlic oil, was demonstrated to have a hypoglycemic property similar to glibenclamide. In a streptozotocin-induced model, ethanolic extract of garlic reduces ALT, AST, creatinine, serum glucose, total cholesterol, triglycerides, uric acid, and urea levels, implying that this plant should be regarded a better option for future diabetes research [31].

Aloe vera

It is a perennial, pea green, xerophytic, succulent, stemless or very short-stemmed spreading by offsets that grows in tropical, semitropical, and arid climates worldwide. It has significant levels of anthraquinone chemicals which have been used for decades for treating cathartics, burns, wound healing, and purgatives. It contains 32 different types of anthraquinones and glycosidic derivatives, aloin or barbaloin being the most prevalent bioactive compound. Anthraquinones possesses antidiabetic activity by enhancing the glucose tolerance and insulin sensibility through upregulation of IRS-1 and PI3K. The sterols extracted from *Aloe vera* can activate peroxisome proliferator-activated receptors in a dose-dependent manner. These receptors act as transcription factors, controlling gene expression in lipid and carbohydrate metabolism. Saponins and triterpenes also play a crucial role in reducing blood sugar levels [32].

Andrographis paniculata

It is an herbaceous plant belongs to the Acanthaceae family called the "King of Bitters". It is dispersed in Southern Asia and utilized as a type of daily food medicine for ages. Leaves and roots are widely used as a folklore medicine in Asia and Europe. In modern pharmacology, it has been used for immune system stimulation as well as pharyngotonsillitis, myocardial ischemia, anti-inflammatory respiratory tract infections, anti-hyperglycemic, antimicrobial, oxygen radical scavenging, antimalarial, atherosclerosis, platelet aggregation, anti-HIV, and hepatoprotective effects. The aerial portions of it contains active phytochemicals such as diterpenoids, 2'-oxygenated flavonoids such as 14-deoxyandrographolide-19 – D-glucoside,14-deoxy-11,12-didehydroandrographolide, andrographolide, neoandrographolide, isoandrographolide 14-deoxyandrographolide, homoandrographolide. Andrographolide administration can lower plasma glucose via increasing the protein and mRNA levels of the subtype 4 form of the glucose transporter (GLUT4) in soleus muscle, which increases glucose consumption. Daily injection of 2 mg/kg andrographolide slows the evolution of diabetic nephropathy in mice via modulating the AKT/NF-B signaling pathway and decreasing renal oxidative stress, fibrogenesis, and inflammation. Notably, andrographolide inhibits -glucosidase and -galactosidase activity, which reported that the blood sugar levels, triglycerides, and LDL are reduced in a dose-dependent way [33].

Artemisia dracunculus

The *Asteraceae* plant *Artemisia dracunculus* has been utilized traditionally in Asian medicine, primarily in Iran, Pakistan, Azerbaijan, and India. In Asia, Europe, and the Americas, it is recognized as a spice species. Herb and leaf are the primary components obtained from this plant. The medicinal and/or spice characteristics of a plant are determined by the presence of essential oil with a very diverse composition, and also flavonoids, phenolic acids, coumarins, and alkamides (49). Specific bioactive compounds such as davidigenin, DMC-2, Sakuranetin, and 6-Demethoxy-capillarisin are measured by LC-MS having a capacity to treat diabetes and also having good bioavailability. Triglycerides and glucose levels are reduced in the blood stream; additionally it prevents the functioning of α-glucosidase and β-galactosidase [34].

Azadirachta indica

Azadirachta indica Linn is a tropical evergreen tree native to India and other southeast nations and locally known as Neem comes from the family Meliaceae. In India, neem is considered as "The village pharmacy" because of its healing versatility. Azadirachtin, Nimbidin, Nimbin, Nimbolide, Cyclic trisulfide, Gedunin, Mahmoodin, and other chemical constituents are utilized as anti-inflammatory, antipyretic, anti-gastric ulcer, antibacterial, spermicidal antifungal, antiarthritic, hypoglycemic, antimalarial, diuretic, immunomodulatory, and antitumor agents. Various plant elements have traditionally been used to alleviate eczema, epistaxis, eye disorders, intestinal parasitic expulsion, malnutrition, skin ulcers, biliousness, skin ailments such as ringworms, blistering, wounds, and itching. It's also utilized as an analgesic, alternative treatment, and cure for fevers and urinary tract infections. The aqueous extract of *Azadirachta indica* kernel has a high potential for producing silver nanoparticles. In an in vitro study, silver nanoparticles produced from aqueous kernel extract demonstrated reasonable antidiabetic effects [35].

Bauhinia purpurea

It is a purple orchid tree widely grown in India. The methanolic extract possesses antidiabetic activity, scientifically proved in the alloxan-inducing model (mice) by glucometer method compared to other extracts like petroleum ether, chloroform, ethyl acetate, acetone, and methanol. Flavonoids are playing a crucial role in the reduction of glucose levels [36].

Bauhinia strychnifolia

It is also known as Yanang Dang in Thai, is one of the possible plants for diabetes treatment. It contains a lot of phenolic and flavonoid compounds, which are regarded to be good for a number of biological functions, including antidiabetic properties. The stem bark contains kaempferol-3-glucoside, flavanone-4-O-L rhamnopyrosyl-D-glycopyranosides, betasitosterol(5, 7 dihydroxy and 5, 7 dimethoxy flavanone-4-O-L rhamnopyrosyl-D-glycopyranosides, and lupeol. Seeds include protein as well as fatty acids such as linoleic acid, oleic acid, stearic acid, and palmitic acid. Malvidin, cyanidin, kaempferol, and peonidin are all found in flowers and flavanol glycosides are found in the root. The ethanolic crude extract was prepared using a conventional approach and considered for further separation into several fractions, including BsH, BsD, BsE, and BsW. TPC and TFC were screened, as well as the BsE portion from ethyl acetate exhibited the maximum amount of phenolics and flavonoids, suggesting that it may have antidiabetic properties. Quercetin was most likely the main antidiabetic bioactive ingredient in the B. strychnifolia stem, and the flavonoid-rich fraction may be strong enough to be used as an alternative treatment for blood sugar control. BsE had the highest rate of glucose absorption increase and the strongest inhibitory effect against the enzymes -glucosidase and DPP-IV [36].

Boswellia serrata

Boswellia serrata is a moderate to large branching tree found in North Africa, India, and the Middle East. It comprises 3-O-acetyl-11-keto-boswellic acid, 3-O-acetyl β-boswellic acid, 11-keto – boswellic acid, and oils. The gum resin of Boswellia species and its pharmacologically active compounds, including 11-keto-ß-boswellic acids, have been shown to suppress the expression of proinflammatory cytokines in various immune-competent cells and may prevent insulitis and insulin resistance in type 1 and type 2 diabetes. It is hypothesized that molecularly, 11-keto-ß-boswellic acids act via interference with the IκB kinase/Nuclear Transcription Factor-κB (IKK/NF-κB) signaling pathway through inhibition of the phosphorylation activity of IKK. It has anti-hyperglycemia, peptic ulcer, antioxidant actions, and anti-inflammatory benefits in asthma, IBD, cancer, and osteoarthritis. In an STZ-induced diabetic rat model, oleo gum resin lowers blood sugar levels [37].

Brassica napus (canola meal)

Canola, generally known as rapeseed, is a major oilseed crop that has been grown since 1970 for edible oil and biofuel production. Canola oil is the third most often used cooking oil, high in polyunsaturated fatty acid and is considered a healthful component. It is a popular oilseed crop whose oil is used in both human nutrition and biodiesel manufacturing. Canola seeds have an oil content (40%), protein (20–25%), and protein (39–46%) in the meal. Furthermore, its fatty acid content is roughly, linoleic acid (20%), oleic acid (60%), and linolenic acid (10%). The polyphenol concentration of canola meal is highlighted, as its potential use in the recovery of antidiabetic chemicals. Water extracts and aqueous butanol revealed that it inhibits DPP-IV activity [38].

Bryonia dioica

It is a growing perennial, asparagus-like wild food plant that belongs to the Cucurbitaceae family. Algeria, Bosnia-Herzegovina, Iran, Iraq, Italy, Lebanon, Morocco, Portugal, Spain, and Tunisia are among the nations where it has grown and is utilized. Since the first century AD, this plant has been utilized for various purposes such as arthritis, bone pains, bruises, cancer, epilepsy, infections, intestinal worms, kidney diseases, lesions, rheumatism, toothache, and wounds. The plant extract at 30 mg/kg b.wt showed to have the most antidiabetic efficacy in acute studies by normalizing all investigated thresholds such as triglycerides, total cholesterol (serum), and serum urea levels in the neonatal streptozotocin inducing models [39]. Medicinal plants for antidiabetic activity with its phytoconstituents was listed in Tab. 11.1–11.3.

Tab. 11.1: Medicinal plants for antidiabetic activity with its phytoconstituents.

Plant name	Family	Phytoconstituents	Role
Adhatoda vasica	Acanthaceae	Vasicinol, vasicine, vasicinolone, vasicol, and adhatonine	Inhibits α-glucosidase
Allium cepa	Amaryllidaceae	Chlorogenic acid and chicoric acid	Increase glucose uptake in L6 muscular cells, upraise insulin secretion from the INS-1E insulin, α-glucosidase, rise in adiponectin, and reduce insulin resistance
Bauhinia purpurea	Fabaceae	Flavonoids	Reduction of blood glucose levels

Tab. 11.1 (continued)

Plant name	Family	Phytoconstituents	Role
Boswellia serrata	Burseraceae	β-boswellic acid, 3-O-acetyl-11-keto-β-boswellic acid, 3-O-acetyl-β-boswellic acid,11-keto-β-boswellic acid, and oils	Inhibits α-glucosidase
Brassica napus	Brassicaceae	Polyphenol	Inhibits DPP-IV activity
Bryonia dioica	Cucurbitaceae	Alkaloids	Inhibition of enzymes like aldose reductase, α-amylase, α-glucosidase, dipeptidyl peptidase-IV, and protein tyrosine phosphatase-1B
Ficus glomerata	Moraceae	Flavonoids and phenolic compounds	Inhibits α-glucosidase
Ficus religiosa	Moraceae	β-Sitosteryl-d-glucoside, tannins, saponins, polyphenolic compounds, flavonoids and sterols	Hypoglycemic activity
Gymnema sylvestre	Apocynaceae	Epicatechin, Apigenin, Luteolin, Kampferol, Hentriacontane, Pentriacontane, Phytin, Resin, Lupeol, β-amyrenerelated glycosides, alkaloid (conduritol), α and β- chlorophyll, Stigmasterol, d-quercitol, Nonacosane, Lignin	Inhibits DPP-IV activity and inhibits α-glucosidase
Lupinus albus	Fabaceae	γ-conglutin	Activate intracellular kinases and adaptor proteins, promote translocation of GLUT-4 receptors, and regulate muscle-specific gene transcription
Momordica charantia	Cucurbitaceae	Proteins, polysaccharides, flavonoids, triterpenes, saponins, ascorbic acid and steroids	Inhibits 11β-HSD1, decreasing local glucocorticoid aggregation, to ameliorate GSIS and β cells; stimulates adiponectin and the PPAR-α/PPAR-c pathway, upregulate ACO and leptin; downregulate resistin
Morus alba	Moraceae	Chlorogenic acid, isoquercitrin, rutin, and quercitrin	Amylase inhibitory actions and improve the glucose uptake

Tab. 11.1 (continued)

Plant name	Family	Phytoconstituents	Role
Peganum harmala	Zygophyllaceae	Alkaloids of β-carbolines	Inhibits α-glucosidase, improves glucose metabolism, insulin resistance, and β-cell dysfunction via GLUT4 regulation
Pterocarpus marsupium	Fabaceae	Phenols, polyphenols, flavones, flavonoids, tannin, terpenoids, alkaloids	Inhibits DPP-IV activity and inhibits α-glucosidase
Rauwolfia serpentina	Apocynaceae	Flavonoids	Inhibits DPP-IV activity and inhibits α-glucosidase
Saccharum officinarum	Poaceae	p-coumaric acid	inhibits DPP-IV activity and inhibits α-glucosidase
Scoparia dulcis	Scrophulariaceae	Diterpenes, triterpenes, apigenin, betulinic acid, luteolin, scopadulcic acid B, scoparic acid A, coxicol, glutinol, scoparic acid D and scutellarinand flavonoids	Inhibits α-glucosidase and PPAR-γ; regulate polyol pathway
Abelmoschus esculentus	Malvaceae	Quercetin, Polysacharides	α-amylase and α-glucosidase inhibition
Achyranthes aspera	Amaranthaceae	Flavonoids and phenolics	Significant reduction on serum Triglyceride
Allium sativum	Alliaceae	Caffeic acid 3-glucoside, polyphenol malonylgenistein, calenduloside E	DPP-4 (dipeptidyl peptidase-4) inhibitors, mproving pancreatic β-cell function
Aloe barbadensis miller	Asphodelaceae	Polysaccharides, phenolic compounds, arginine, proanthocyanidins, flavonoids,	Inhibition of α-amylase, α-glucosidase, and pancreatic lipase
Andrographis paniculate	Acanthaceae	Andrographolide	Inhibition of the enzyme alpha-glucosidase and alpha-amylase.
Artemisia dracunculus	Asteraceae	5-O-caffeoylquinic acid, Davidigenin; sakuranetin; 2′,4′-dihydroxy-4-methoxydihydrochalcone; 4,5-di-O-caffeoylquinic acid; and 6-demethoxycapillarisin	Inhibitory activity on PEPCK expression and aldose reductase activity; activating the PI3K pathway
Azadaricta indica	Meliaceae	Azadirachtin	Inhibits α-glucosidase, reduction in intestinal glucosidase
Bauhinia strychnifolia	Leguminosae	Kaempferol, luteolin, quercetin, gallic acid, naringenin, apigenin	Inhibits α-glucosidase, lipid synthesis, DPP-IV enzymes

Ficus glomerata

It is a member of the Moraceaceae family which is distributed throughout India. The phytoconstituents such as lupeol and quercetin were extracted and quantified using the HPTLC technique. The roots having a higher concentration of flavonoids and phenolic compounds, which have important medicinal properties such as antidiabetic, antioxidant, and anti-inflammatory properties. The antidiabetic activity of the diethyl ether fraction was greater than that of the other fractions of the alcoholic root extract [40].

Ficus religiosa

It belongs to the family Moraceae and commonly called as peepal in India. It is a large deciduous tree that can grow to be 35 meters tall. It has been historically used to cure a number of diseases in traditional medicine. Its bark, fruits, leaves, adventitious roots, latex, and seeds are all used as folk medicines, often in combination with other herbs. Stomatitis, gout, leucorrhea, inflammation, ulcers, and glandular swelling of the neck are all treated with this plant. *F. religiosa* has been linked to wound healing, antibacterial activity, and acetylcholinesterase inhibition. *F. religiosa* leaves has been used to treat diabetes in the traditional Ayurvedic system. In normal rabbits, β-sitosteryl-d-glucoside isolated from the bark of the plant demonstrated hypoglycemic activity. The presence of saponins, tannins, sterols, polyphenolic compounds and flavonoids was discovered during the phytochemical screening of the plant [41].

Gymnema sylvestre

It can be found from African nations to Saudi Arabia, Vietnam, Sri Lanka, and South China, including from Japan to the Philippines, Malaysia, Indonesia, Australia, and arid jungles across India. It is beneficial in the treatment of dyspepsia, constipation, hepatitis, hemorrhoids, renal and vesicle calculi, cardiopathy, asthma, bronchitis, amenorrhea, conjunctivitis, and leukoderma. Alkaloid (gymnamine), cinnamic acid, flavonoids, ascorbic acid, folic acid, and oleanines (gymnemic acid, gymnema saponins), dammarenes (gymnemasides), flavones, anthraquinones, phytin, pentatriacontane, tartaric acid, resin, hentriacontane, butyric acid, formic acid, β-amyrene, lupeol, related glycosides and anthraquinones, etc. In case of gymnema saponins, anti-sweet activity is also developed due to the presence of Acyl group, so it is an aglycone part of saponin. It also increases the amount of insulin in blood plasma. Antioxidants like ascorbic acid neutralize hydroxyl radicals, super oxide radicals, thereby proving its antioxidant nature and also showing blood sugar uptaking abilities. In contrast to that, antidiabetic compounds such as anthraquinone, flavones, flavonoids like epicatechin, Apigenin, Luteolin, Kampferol, Hentriacontane, Pentriacontane, Phytin, Resin,

Lupeol,alkaloid (conduritol), β-amyrene-related glycosides, α and β- chlorophyll, Stigmasterol, d-quercitol, Nonacosane, Lignin, etc. also have antioxidant activities [42].

Tab. 11.2: Photograph of selected medicinal plants with antidiabetic activity.

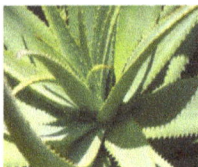

Abelmoschus esculentus Linn (Okra)	*Aloe vera*	*Bauhinia strychnifolia*
Achyranthes aspera Linn	*Andrographis paniculata*	*Boswellia serrata*
Adhatoda vasica	*Artemisia dracunculus*	*Brassica napus*
Allium cepa	*Azadirachta indica*	*Bryonia dioica*
Allium sativum (Garlic)	*Bauhinia purpurea*	*Ficus glomerata*

Tab. 11.2 (continued)

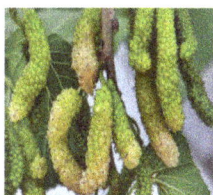

Ficus religiosa	*Peganum harmala*
Gymnema sylvestre	*Pterocarpus marsupium*
Lupinus albus	*Rauwolfia serpentina*
Momordica charantia	*Saccharum officinarum*
Morus alba	*Scoparia dulcis*

Lupinus albus

It belongs to the Fabaceae family. *Lupinus albus* is a natural or cultivated plant found in Europe, the Balkans, and Turkey, particularly in the Marmara and Aegean regions.

Lupine is a nutritive and diabetes-protective herb. The seeds are very plenty in dietary fiber and proteins, are almost independent of phytoestrogens and starch, and include many antioxidants, minerals, essential amino acids, and lipids. The seed proteins are also supposed to discharge some of the bioactive peptides playing prominent roles in scrambling cardiovascular diseases, diabetes, and obesity. LA. β-conglutin regulates the glucose which has been attributed to its insulin-mimetic capabilities. Gamma-conglutin was demonstrated to stimulate intracellular kinases and adaptor proteins involved in signaling of insulin, enhance GLUT-4 receptor trafficking to the cell membrane, and control gene transcription (muscle) similarly to the insulin. Despite the fact that lupin protein extract hydrolysate can directly activate insulin production in the pancreaticcells, it acts as an insulinotropic agent via a Gq protein signal transduction (Gq protein/phospholipase C/protein kinase C – Gq/PLC/PKC pathway) [43].

Momordica charantia

It is yearly climbing plant falls under the Cucurbitaceae family that is also known as bitter gourd, karela, or bitter melon. It is inherent to East India and is now extensively grown and consumed in subtropical, tropical, and temperate climates. It is having the capacity to increase glucose uptake and decrease insulin resistance by inhibiting Suppressors of cytokine signaling 3 (SOCS-3) and Jun N-terminal kinase (JNK) expressions, directly or indirectly activating the IRS-1/PI3K/Akt pathway, that leads to the biogenesis and translocation of GLUT4, and suppressing the Glycogen synthase kinase 3 (GSK-3), it alsoinhibited by interacting with biomolecules of this plant. Positive AMPK activity, which is influenced by increased adiponectin, thyroid hormones, and specific Phyto-agents that plays a pivotal role in GLUT4 activation. Furthermore, it stimulates the PPAR-/PPAR-c pathway and adiponectin which can increase leptin and Acetyl-CO, while decreasing resistin that suppresses insulin resistance. The inhibitory action of β-amylase, pancreatic lipase, and β-glucosidase, delays digestion and absorption of food in the small intestine. It also activates PI3K, which is linked with lowering glucose absorption specifically in the jejunum epithelium, and by suppressing GLUT2 related to decrease glucose reabsorption in the kidney.

It can also activate AMPK pathway by directly or indirectly to inhibit glycogenolysis, as well as PEPCK and G-6-Phaseare essential enzymes in gluconeogenesis. Simultaneously, it has the ability to regulate 11-HSD1, resulting in a decrease in glucocorticoids and a decrease in PEPCK, which inhibits gluconeogenesis. Through TAS2R or PLC-2, MC drugs enhance L-cell allosterism and proliferation, and depolarize L cells to increase GLP-1 secretion, which can improve the functioning of β-cell division and secretion of insulin. Furthermore, it blocks 11-HSD1, lowering local glucocorticoid agglutination, hence improving GSIS and cells. It also upregulates insulin increase serum TAOC, Pdx1 genes, and pancreatic GSH, and decrease pancreatic MDA to reduce β-cell damage and improve pancreatic function [44].

Tab. 11.3: Structures of specific compound possess antidiabetic activity.

Plant name	Phytoconstituents	Structure
Abelmoschus esculentus	Quercetin	
Achyranthes aspera	Apigenin	
Adhatoda vasica	Vascicine	
Allium cepa	Chlorogenic acid	
Allium sativum	Caffeic acid 3-glucoside	
Aloe barbadensis miller	Arginine	
Andrographis paniculata	Andrographolide	
Artemisia dracunculus	Davidigenin	
Azadaricta indica	Azadirachtin	

Tab. 11.3 (continued)

Plant name	Phytoconstituents	Structure
Bauhinia purpurea	Luteolin	
Bauhinia strychnifolia	Kaempferol	
Boswellia serrata	β-boswellic acid	
Gymnema sylvestre	Epicatechin	
Momordica charantia	Ascorbic acid	
Morus alba	Rutin	
Rauwolfia serpentina	Luteolin	

Tab. 11.3 (continued)

Plant name	Phytoconstituents	Structure
Saccharum officinarum	p-coumaric acid	
Scoparia dulcis	Betulinic acid	

Morus alba

It is commonly known as white mulberry and silkworm mulberry are all names for *Morus alba*. It is a small to medium-sized tree that grows up to of 10–20 m (33–66 ft). Therefore, some specimens have been found to survive for a longer period that is more than 250 years. It is native to India and Central China has been widely farmed and naturalized around the world including the Mexico, the United States, Kyrgyzstan, Australia, Argentina, Iran, Turkey, India, The starch breakdown by -amylase in *Morus alba* was investigated in vitro. In a concentration-dependent way, *Morus alba* demonstrated significant -amylase inhibitory actions. A modified Ellmann's technique was used to investigate the in vitro acetylcholine esterase inhibitory activity of a 50 percent methanolic extract of *Morus alba*. For 35 days, the animals were given different doses of mulberry leaf extract. Blood glucose, the relative body weight of the pancreas, the width of islets, and the number of cells in each group were also examined. According to the histology and biochemical findings, the plant extract may help to lower glucose levels by promoting regenerationβ-cell [45].

Peganum harmala

Since ancient times, *Peganum harmala* has been used traditionally and commonly utilized for therapeutic purposes. In Central Asia, North Africa, and the Middle East, this plant is extensively distributed and utilized as a medicinal herb. Several alkaloids found in the root and seeds are pharmacologically active and responsible for the effects. Chemical compounds such as steroids, alkaloids, and flavonoids are isolated from the seeds, flowers, leaves, roots, and stems as a result of phytochemical investigations. The alkaloids of β-carbolines derivatives are abundant in this plant. It has been demonstrated to have hypoglycemic and cytoprotective properties. It has been used to treat diabetes in several regions of the world for centuriesmight be due to antioxidant, enzyme inhibition, receptor agonist or antagonist action, or other mechanisms that have yet to be discovered. One possible explanation for the hypoglycemic activity of the hydroalcoholic extract of seed is increased insulin production from diabetic rats' remaining pancreaticcells. In streptozotocin-induced diabetic male rats reported antidiabetic and hypolipidemic properties. More research is needed to determine the extract's components and the mechanism(s) through which it exerts its benefits [46].

Pterocarpus marsupium

It's also known as Malabar Kino, Indian Kino, VijayaSar, or Venkai, and it's a medium to big deciduous tree that may reach 31 meters in height. It is indigenous to India. Phenols, polyphenols, flavones, flavonoids, tannin, terpenoids, alkaloids, and cardiac glycosides are among the phytochemical ingredients. Aldose reductase is a major enzyme in the polyol pathway that transforms aldose to polyol. It is scientifically proven to decrease aldose reductase activity in diabetic male albino rats produced by streptozotocin. Diabetes mellitus and cataract have a favorable link, according to researchers, and diabetic cataract is the leading cause of blindness. Only the surgical removal of the lens from the eye is approved as a therapy for diabetic cataracts. Oral antidiabetic medications and insulin are common treatments for diabetes mellitus, according to reports. Diabetes mellitus's secondary consequences are still uncontrolled. However, more clinical trials are needed to determine whether *Pterocarpus marsupium* extracts have a therapeutic impact [47].

Rauwolfia serpentina

The Indian snakeroot, also known as devil pepper or serpentine wood, is a milkweed flower that belongs to the *Apocynaceae* family. It is indigenous to the Indian subcontinent as well as East Asia (from India to Indonesia). It's a perennial undershrub that grows up to 1,000 meters in India's sub-Himalayan areas (3,300 ft). It includes dozens of indole alkaloids, including ajmaline, ajmalicine, reserpine, and serpentine, among others.Alkaloids, sugars,glycosides, flavonoids, phlobatannins, cardiac glycosides,saponins, resins, tannins, triterpenoids, and steroids are phytochemical components. Plants high in flavonoids have been shown to have powerful antidiabetic, hypolipidemic, hypotensive, anti-inflammatory, and antioxidative properties, as well as long-term hypoglycemic, weight-loss, and hypolipidemic benefits in alloxan-induced diabetic wistar mice (male). In the alloxan-induced diabetic mice, it has been improved glycemic, anti-atherogenic, and cardioprotective parameters, indicating that it is an efficient antidiabetic drug [48].

Saccharum officinarum

It is a huge, strong, and growing grass species of the *Saccharum* genus. Sucrose, a simple sugar that aggregates in the stalk internodes, is abundant in its stalks (thick). It originated in New Guinea and is currently grown for sugar, ethanol, and other goods in tropical and subtropical nations across the world. It's one of the most prolific and intensively farmed sugarcane varieties. Other sugarcane species, such as *Saccharum sinense* and *Saccharum barberi*, can interbreed with it. Complex hybrids make up the majority of commercial cultivars. Type 2 diabetes might be prevented or delayed by improving insulin-stimulated glucose absorption and reducing insulin resistance. AMPK is a crucial energy sensor involved in insulin metabolism, and its activation through phosphorylation can aid glucose absorption. However, in insulin-resistant cells, phosphorylation activity was decreased. p-coumaric acid improved 2-NBDG absorption in L6 skeletal muscle cells via activating AMPK, as previously observed. Future research should focus on sugarcane phenolics as well as other phytochemicals (triterpenoids, sterols, and lignins, among others) in sugarcane. Overall, the findings show that sugarcane bagasse can aid in the development of novel agents and the enhancement of agro-industrial revenue [49]. Antidiabetic medicinal plants was represented in the Fig. 11.1.

Scoparia dulcis

It is a flowering plant that belongs to the plantain family. Liquoriceweed and goatweed, are some of the common names for this plant. It is native to the Neotropics; however, it may be found in tropical and subtropical areas all over the world. All of the species of *S. dulcis* have been found to have antidiabetic or hypoglycemic effects. The antidiabetic function of this plant established by research based on several types of extracts of the plant discussed in this portion of the review. The levels of lipids such as triglycerides, phospholipids, cholesterol, and fatty acids were lowered in streptozotocin-induced animal model and also the levels of very low-density lipoprotein and low-density lipoprotein cholesterol, and the activity of 3-hydroxy-3-methylglutaryl (HMG) CoA reductase. Using the DPPH (1,1-diphenyl-2-picrylhydrazyl) test and the phosphor molybdenum assay, this study also found that this extract had modest antioxidant activity when compared to ascorbic acid. In addition to whole plant extracts and aerial portions of the plant.

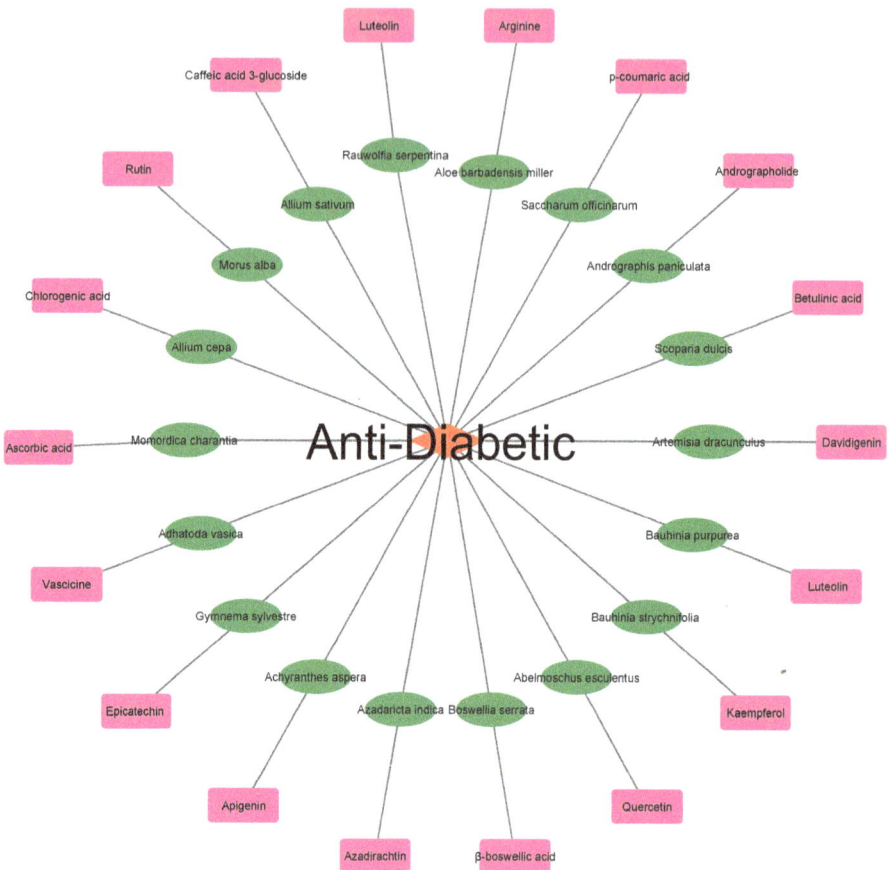

Fig. 11.1: Diagrammatic presentation of antidiabetic medicinal plants.

Inhibitory activities of glucosidase

The use of enzyme inhibitors causes carbohydrate malabsorption, which helps to keep blood sugar and insulin levels in check [42]. This trait, in particular, has been demonstrated to be an advantageous in the management of type 2 diabetes. SAD was also evaluated in vitro on STZ-treated rat insulinoma cell lines (RINm5F cells) and isolated islets. SAD elicited a twofold increase the secretion of insulin from isolated-islets at a dosage of 20 mg/mL, suggesting its insulin secretagogue action. SAD's cytoprotective properties were also established in this investigation [50].

11.3 Conclusion

The present chapter provides a picture of medicinal plants that has been studied for the treatment. Diabetes is a serious life-threatening disease and must be constantly monitored and effectively subdued with proper medication and by adapting to a healthy lifestyle. Uncontrolled diabetes leads too many chronic complications such as heart failure, blindness, and renal failure. However, its complications can be reduced through proper awareness and timely treatment. The specific phytoconstituents such as stigmasterol, palmitic acid, β-sitosterol, β-carbolines, vasicine, epicatechin, apigenin, luteolin, kaempferol, quercetin, and other flavonoids, alkaloids, polyphenols, and terpenoids are playing a crucial role in the management of diabetes by inhibiting α-glucosidase, α-amylase, lipid synthesis, DPP-IV enzymes, 11β-HSD1. It also improves glucose metabolism, insulin resistance, and β-cell dysfunction via GLUT4 regulation. In addition to the glycemic control, some of the herbs depicted effectiveness in the control of cardiovascular complications by reducing TG, cholesterol levels, and BMI. Herbal medicine are always preferred treatment options by patients or as adjunctive to conventional treatment for diabetes. More researches are needed in order to separate the active components of plants and molecular interactions of their compounds for analysis of their curative properties.

References

[1]   M. Modak, P. Dixit, J. Londhe, S. Ghaskadbi, T.p. J Clin Biochem Nut. 2007, 40(3), 163–173.
[2]   G. Goodman, J. Hardman, L. Limbird, G. A. Goodman. Pharmacol Basis Ther. 2006, 60, 1686–1710.
[3]   D. Islam, A. Huque, S. Sheuly, L. Chandra Mohanta, S. Kumar Das, A. Sultana, E. Parvin Lipy, U. K. Prodhan. J HerbmedPharmacol. 2018, 7(3), 148–154.
[4]   M. Rahimi-Madiseh, P. Karimian, M. Kafeshani, M. Rafieian-Kopaei. Iranian J Basic Med Sci. 2017, 20(5), 552–556.
[5]   N. Mohamed Maideen, R. Balasubramaniam. J Herbmed Pharmacol. 2018, 7(3), 200–210.
[6]   W. Adhi Putra, N. Fakhrudin, A. Nurrochmad, S. Wahyuono. J Appl Pharmaceu Sci. 2022, 12(01), 41–54.
[7]   S. Kazemi, H. Shirzad, M. Rafieian-Kopaei. CurrPharmaceu Des. 2018, 24(14), 1551–1562.
[8]   M. Rahimi-Madiseh, E. Heidarian, S. Kheiri, M. Rafieian-Kopaei. Biomed Pharmacotherapy. 2017, 86, 363–367.
[9]   J. P. Boyle, A. A. Honeycutt, K. V. Narayan, T. J. Hoerger, L. S. Geiss, H. Chen, J. Theodore, M. S. Thompson. Diabetes Care. 2001, 24(11), 1936–1940.
[10] M. Rachpirom, L. R. Barrows, S. Thengyai, C. Ovatlarnporn, C. Sontimuang, P. Thiantongin, P. Puttarak. Pharmacog Res. 2022, 14(1), 89–99.
[11] F. Azizi, H. Hatami, M. Janghorbani. Epidemiology and Control of Common Diseases in Iran. Eshtiagh Publications, Tehran, 2000, pp. 602–616.
[12] A. Artasensi, G. Pedretti, L. Vistoli Fumagalli. Molecules. 2020, 25(8), 1–20.
[13] P. Aschner, B. Nielsen, B. Henning, B. Peter, C. Andrew, C. Ruth, F. Stephen, G. Marion, G. Roger, J. JoseIcon, P. Home, M. McGill, S. Manley, S. Marshall, J. C. Mbanya, A. Neil, A. Ramachandran, K. Ramaiya, G. Roglic, N. Schaper, L. Siminerio, A. Sinclair, F. Snoek, P. Van Crombrugge, G. Vespasiani, V. Viswanathan. Diabetes Res Clin Prac. 2014, 104(1), 1–52.

[14] N. P. Skliros, C. Vlachopoulos, D. Tousoulis. HellenikeKardiologikeEpitheorese. 2016, 57(5), 304–310.

[15] G. Orasanu, J. Plutzky. J American College Cardiol. 2009, 53(5), 35–42.

[16] J. E. Graham, D. G. Stoebner-May, G. V. Ostir, S. Al Snih, M. K. Peek, K. Markides, K. J. Ottenbacher. Health Quality Life Outcomes. 2007, 5(1), 1–7.

[17] B. Moradi, S. Abbaszadeh, S. Shahsavari, M. Alizadeh, F. Beyranvand. Biomed Res Ther. 2018, 5(8), 2538–2351.

[18] M. Telagari, K. Hullatti. Indian J Pharmacol. 2015, 47(4), 425–429.

[19] M. S. Hedrington, S. N. Davis. Exp Opin Pharmacotherapy. 2019, 20(18), 2229–2235.

[20] F. Jamshidi-Kia, Z. Lorigooini, H. Amini-Khoei. J HerbmedPharmacol. 2018, 7(1), 1–7.

[21] M. Asadi-Samani, N. Bagheri, M. Rafieian-Kopaei, H. Shirzad. Phytother Res. 2017, 31(8), 1128–1139.

[22] S. Karami, M. Roayaei, H. Hamzavi, M. Bahmani, H. Hassanzad-Azar, M. Leila, M. RafeieianKopaei. Int J Pharma Investig. 2017, 7(3), 137–141.

[23] Z. Rabiei, M. Gholami, M. Rafieian-Kopaei. Bangladesh J Pharmacol. 2016, 11(3), 711–715.

[24] L. Jalaly, G. Sharifi, M. Faramarzi, A. Nematollahi, M. Rafieian-Kopaei, M. Amiri, M. Fariborz. DARU J Pharma Sci. 2015, 23(1), 1–7.

[25] B. T. Ahmed, S. A. Kumar. Res J Biotech. 2016, 11(3), 34–41.

[26] M. T. Islam. Phytother Res. 2019, 33(1), 72–80.

[27] A. Kumar, K. Gnananath, S. Gande, E. Goud, P. Rajesh, S. Nagarjuna. J Pharmacy Res. 2011, 4(7), 3124–3125.

[28] S. Naz, S. Alam, W. Ahmed, S. M. Khan, A. Qayyum, M. Sabir, N. Alia, A. Iqbal, B. Yamin, S. Nisa, A. Salah Khalifa, A. F. Gharib, A. E. Askary. Saudi J Bio Sci. 2022, 29(2), 941–954.

[29] H. Gao, Y. N. Huang, B. Gao, B. Li, C. Inagaki, J. Kawabata. Food Chem. 2008, 108(3), 965–972.

[30] S. Sabiu, M. Madende, A. A. Ajao, R. A. Aladodo, I. O. Nurain, J. B. Ahmad. Bioactive Food Dietary Intervent Diabetes. 2019, 137–154.

[31] A. Eidi, M. Eidi, M. Esmaeili. Phytomed. 2006, 13(9-10), 624–629.

[32] F. Haghani, M. R. Arabnezhad, S. Mohammadi, A. Ghaffarian-Bahraman. Revista Brasileira de Farmacog. 2022, 1–14.

[33] A. E. Nugroho, M. Andrie, N. K. Warditiani, E. Siswanto, S. Pramono, E. Lukitaningsih. Indian J Pharmacol. 2012, 44(3), 377–381.

[34] D. M. Ribnicky, P. Kuhn, A. Poulev, S. Logendra, A. Zuberi, W. T. Cefalu, I. Raskin. Int J Pharmaceu. 2009, 370(1-2), 87–92.

[35] N. T. L. Chi, M. Narayanan, A. Chinnathambi, C. Govindasamy, B. Subramani, K. Brindhadevi, T. Pimpimon, S. Pikulkaew. Environmental Res. 2022, 112684.

[36] M. S. Meshram, P. Itankar, A. Patil. J PharmacogPhytochem. 2013, 2(1), 171–175.

[37] R. Praparatana, P. Maliyam, R. Barrows, P. Puttarak. Molecules. 2022, 27(8), 2393.

[38] S. Hussain, A. U. Rehman, H. K. Obied, D. J. Luckett, C. L. Blanchard. Separations. 2022, 9(2), 38. 1–16.

[39] E. Chekroun, A. Bechiri, R. Azzi, H. Adida, N. Benariba, R. Djaziri. Phytotherapy. 2017, 15(2), 57–66.

[40] M. Upadhye, U. Deokate, R. Pujari, M. Phanse. Indian J Pharmaceu Edu Res. 2022, 56(2), 470–478.

[41] R. Pandit, A. Phadke, A. Jagtap. J Ethnopharmacol. 2010, 128(2), 462–466.

[42] S. Laha, S. Paul. Pharmacog J. 2019, 11(2), 201–206.

[43] B. A. Yildirim, E. Ali, K. Saban, Y. Serkan. World J Adv Res Rev. 2020, 7(2), 7–16.

[44] Z. Liu, J. Gong, W. Haung, F. Lu. Evid Based Complement Alter Med. 2021, 1–14.

[45] B. Devi, N. Sharma, D. Kumar, K. Jeet. Int J Pharm Pharm Sci. 2013, 5(2), 14–18.

[46] G. Komeili, M. Hashemi, M. Bameri-Niafar. Cholesterol. 2016, 2016, 1–6.

[47] Y. Xu, Y. Zhao, Y. Sui, X. Lei. Biotech. 2018, 8(4), 1–7.

[48] M. B. Azmi, S. A. Qureshi. Adv Pharmacol Sci. 2012, 2012, 1–11.

[49] R. Zheng, S. Su, H. Zhou, H. Yan, J. Ye, Z. Zhao, L. You, X. Fu. Indust Crops Prod. 2017, 101, 104–114.

[50] G. Pamunuwa, D. Karunaratne, V. Y. Waisundara. Evid Based Complement Alternat Med. 2016, 2016, 1–11.

Shailendra Yadav*, Dheeraj Singh Chauhan and Vandana Srivastava

Chapter 12
Anti-coronavirus and antiviral activity of medicinal plants

Abstract: Viral infections are a major global public health concern. The Covid-19 pandemic is among the viral eruptions that have posed a daunting challenge to the worldwide healthcare systems. This pandemic has led to considerable casualties, economic impact, and considerably affected the lifestyles of human beings in general. This has brought the attention and focus of world scientists and researchers to explore the vaccination and cure. Here it is important to note that currently the mostly available and tested medical options are from the allopathic medications. However, the allopathic medication receives criticism due to the serious side effects it poses along with the beneficial aspects. Therefore, a comprehensive review of the medicinal plants that can potentially treat viral infections is mandatory. Accordingly, this chapter focuses on the ethno botanical knowledge of medicinal plants conventionally applied for the treatment of various viral diseases including Covid-19. There are numerous ways of treatment of viral diseases, and treatment by using the medicinal plant is an ancient method and can be improved by developing drug delivery systems in these perspectives. Medicinal plants contain natural products with microbial activities, but mostly these products exhibit a high potential via synergistic effect, but in these compounds, one or two compounds show preferred antiviral activity. During the Covid-19 pandemic, many tribal communities in different countries used medicinal plants as a possible treatment. This emphasizes the importance of medicinal plants for not only Covid-19 but also for other viral infections.

12.1 Introduction

Chemical medicines are traditionally considered as the most effective strategy to counter the viral infections [1, 2]. However, the application of allopathic treatment pathways has received considerable criticism due to the side effects of the modern medicine. To

*Corresponding author: Shailendra Yadav, Green Chemistry Lab., Department of Chemistry, Basic Science, AKS University, Satna 485001, Madhya Pradesh, India,
e-mail: syshailendra5@gmail.com
Dheeraj Singh Chauhan Modern National Chemicals, Second Industrial City, Dammam 31421, Saudi Arabia
Vandana Srivastava Department of Chemistry, Indian Institute of Technology (Banaras Hindu University), Varanasi 221005, Uttar Pradesh, India

https://doi.org/10.1515/9783110791891-012

address to these side effects, physicians generally advise multiple drug dosage, which makes the overall effect more complex. Further, the long-term dosage of allopathic medication leads to development of immunity against these drugs that requires stronger and increased doses of these medicines, which is not considered healthy [4–6]. Viral diseases and Covid-19 pandemic have caused a serious influence on the countries across the globe. This has prompted the focus of scientists and researchers to explore and achieve the possible solutions for both prevention as well as curing this pandemic [3].

The natural extracts derived from different parts of plants such as root, stem, leaves, fruits, flowers, fruit skin/ peels, and seeds possess several phytochemicals containing the capabilities to treat numerous diseases. This has a wide range from antimalarial, antifungal, antimicrobial, antiviral, anti-HIV, anticancer, etc. [7, 8]. The conventional allopathic medication has achieved praise due to the quick remedy against the targeted diseases. Contrariwise, the herbal medication has received criticism for relatively slower activity. However, a considerable difference that lies with the herbal medication is their environmentally benign nature, and a less likelihood to pose negative effects on humans. Positive physiological effect of herbal medicine can be improved by developing a drug delivery system for these.

Several research articles and books are available in the literature in the past couple of years covering the possible strategies to cure the Covid-19 pandemic [9–14]. However, most of the available literature focuses on the treatment of pandemic employing the chemical medicines. There is a scarcity of literature on the utility of herbal medication in the cure of Covid-19. This chapter provides a brief outline of the significance of antiviral activities of plant extracts with emphasizing the action of naturally derived medication for treating the Covid-19 pandemic. This chapter summarizes the availability of the naturally derived medicines with structure of important active chemical species. A literature survey is provided on the application of herbal medicines to counter viral diseases. Herein we have emphasized the potentiality of natural medicines in antiviral diseases and as a possible solution to the Covid-19 pandemic.

12.2 Viral diseases and currently available treatment

In living organisms many diseases are caused by various diseases causing agents, viruses are one of the nonliving disease causing agents, but all viruses are not disease causing. Structurally, viruses contain RNA or DNA genetic material encapsulated in a protein cover. Whenever viruses enter the cells of living organisms, they multiply themselves by using components of cell and this causes damage or destroys infected cells with specific symptoms and it is commonly called viral disease. All the viral disease are not transmittable; i.e., they are not all transmitted from human to human but some of them are transmittable like common cold, flue, HIV, herpes, and SARS-COV-2. Other viral diseases spread through different ways like insect bite, eating infected food or meat of animals. Different types of viral diseases are discussed here with symptom, effective common treatments and prevention (Tab. 12.1).

Tab. 12.1: Different viral diseases, transmission, prevention, and medicinal plants for their treatment.

Types Viral disease	Transmission	Treatment	Prevention	Disease and medicinal plants for treatment
Respiratory viral diseases	Spreading by inhaling air contained virus after coughing and sneezing of infected people and also by touching virus congaing surfaces	Various antiviral drugs like Tamiflue, Remdesivir, Lopinavir Oseltamivir, Ribavirin, Favipiravir, Betulinic Acid	By adopting good personal hygine and Vaccination	**Flu and Common Cold:** *Thymus vulgaris* [15], *Allium sativum L* [16], *Ocimum sanctum* [16], *Zingibernofficinale* [16], *Bombaxceiba L* [16] **Influenza:** *Aloe vera (L.) Burm.f* [16], *Calotropisgigantea (L.) Dryand* [17], *Mangifera indica L* [18] **SARS, MERS, COVID-19:** *Justiciaadhatoda L,* [19], *Allium sativum L* [16]. *Aloe vera (L.) Burm.f* [20], *Bombaxceiba L* [21], *Cyperusrotundus L.* ([22], *Azadirachta indica* [23], *Piper nigrum L* [24], *Aegle marmelos (L.)* [25], *Citrus sinensis (L.)* [26], *Zingiberofficinale Roscoe* [27], *Curcuma longa L* [28]., *Andrographi spaniculata, Ocimumtenuiflorum L., Ocimumbasilicum L., Cinnamomum verum Presl* [30]
Gastrointestinal viral diseases	Contaminated food and water with feces can spread virus and also by sharing household goods or personal objects with infected people.	Not any treatments are existing. Mostly are cured within 24–48 h. During infection period, drinking plenty of fluids is necessary for maintain electrolytic concentration.	By adopting good personal hygine and Vaccination	**Norovirus infection:** *Origanumvulgare, Thymus mastichina, Thymus vulgaris* [31] **Rotavirus infection:** *Rubiacordifolia L.* [32] **Adenovirus infection:** *Radix Lithospermi* [33], *Camellia sinensis Kuntze* [34] **Astrovirus infection:** *Achyrocline bogotensis* [35]

Tab. 12.1 (continued)

Types Viral disease	Transmission	Treatment	Prevention	Disease and medicinal plants for treatment
Exanthematous viral disease	In most of the case spread through cough or sneeze of infected person. Chicken pox and smallpox are transmitted by contact liquid with broken skin lesions of infected person.	Treatment based on symptoms management. Fever-reducing medication adopted. Antiviral drugs, given for chickenpox or shingles.	Rubella, chickenpox, Measles, shingles, and smallpox can be prevented via vaccination. Risk of chikungunya can be prevented by protection from mosquito bites.	**Measles:** *Ophiopogonjaponicus, Coriandrum sativum seed, elsholtziacristata or eclipta prostrate* [36] **Chickenpox:** *Azadirachta indica, rubiacordifera* [37] **Shingles:** *Cynodon dactylon, Oroxylum indicum, Drymaria cordata, Centella asiatica, Sesamum indicum* [38] **Roseola:** *Nepetacataria, Menthapiperita, Achilleamillefolium* [39] **Smallpox:** *Sarracenia purpurea, Azadirachta indica* [40] *Sarracenia purpurea* [41] **Chikungunya virus infection:** *Ipomoea aquatic, Persicariaodorata Rhapis excels* [42]
Hepatic viral diseases	Spread by blood contact and via needles or razors. Hepatitis B canalsobe spread through sexual contact.	Treatment for hepatitis B, C, and D are based on symptoms management and use of antiviral drugs.	Vaccines available for hepatitis A, B and C.	**Hepatitis A:** *Boerhaviadiffusa* [43], *Menthalongifolia* [44] **Hepatitis B:** *Curcuma longa Linn., Ganodermalucidum, Swertia chirayita* [45] **Hepatitis C:** *Acacia nilotica, mbeliaschimperi, Piper cubeba* [45] **Hepatitis D:** *Taraxacum Officinalis, Lepidium sativum, Azadirachta indica* [45] **Hepatitis E:** *Adiantumcapillusveneris, Equisetum debileRoxb, Morusnigra L.* [45, 46]

Tab. 12.1 (continued)

Types Viral disease	Transmission	Treatment	Prevention	Disease and medicinal plants for treatment
Cutaneous viral diseases	Transmitted through close physical contact	No cure for herpes available. Antiviral medications, such as acyclovir, can help shorten/ prevent outbreaks.	Good hygiene habits	**Warts, including genital warts:** *Allium sativum, Camellia sinensis* [47] **Oral herpes:** *Artemisia kermanensis, Saturejahotensis L and Rosmarinus officinalis* [48] **Genital herpes:** *Nauclealatifolia , Hibiscus sabdariffa, Avicenna marina* [49] **Molluscumcontagiosum:** *Azadirachta indica oil, Eucalyptusglobulus, Thymus vulgaris* [50]
Hemorrhagic viral diseases	Dengue and yellow fever transmitted via infected insect bite. Ebola spreads through contact with the blood or other body fluids, Lassa fever spreads via inhaling or consuming the dried feces or urine of a rodent with the virus	No specific treatment for hemorrhagic viral diseases.	Avoid being bitten by insects, proper protection required Protect himself against rodent	**Ebola:** *Eucalyptus, Allium sativum, Camellia sinensis, Phyllanthus amarus* [51] **Lassa fever:** *Moringaoleifera Elaeisguineensis Jacq. and Acacia nilotica* [52] **Dengue fever:** *Carica papaya and Euphorbia hirta, Azadirachtaindica* [53] **Yellow fever:** *Cymbopogoncitratus, Citrus aurantifolia, Enantiachlorantha, Carica papaya, Morindalucida, and Lawsoniainermis* [54] **Marburg hemorrhagic fever, Crimean-Congo hemorrhagic fever:** There is no specific medicinal plant for treatment for Marburg virus disease

Tab. 12.1 (continued)

Types Viral disease	Transmission	Treatment	Prevention	Disease and medicinal plants for treatment
Neurologic viral diseases	Transmission via bite of infected animal or bug, e.g. mosquito or tick. Poliovirus and other enteroviruses are quite contagious and spread through close contact with infected people.	No specific treatment for mild viral meningitis or encephalitis	Vaccine available for poliovirus and mumps virus (that causes meningitis and encephalitis). For protection good hygiene and avoiding contact with infected people are recommended.	**Polio:** *S. jollyanum, D. monbutensis, D. bateri* [55] **viral meningitis:** *no specific treatment for viral meningitis* **Rabies:** *Aegle marmelos, Ricinus communis, Mangifera indica* [56]

12.3 Significance of medicinal plants

All the plants are gift of god and fulfill many life requirements of animals and human. Some of plants have specific compounds containing medicinal values and are called medicinal plants. Many diseases are cured by using medicinal plants via different medical practices like Ayurveda and Unani; for example *allium sativum* [15] and *zingiberofficinale* [16] are effective against flu. Although most of the allopathic medicines are synthetic compounds, in most cases, basic material is obtained from plants; thus, medicinal plants are the bases of all medical practices. Indigenous people in many countries are still using direct medicinal plant or remedies made of plants for treatment of common diseases as well as they are using *justicia adhatoda L* [19], *azadirachta indica* [23], *piper nigrum L* [24], *citrus sinensis (L.)* [26], *curcuma longa L* [28], *withaniasomnifera (L.) Dunal* [29], *tinospor acordifolia (Willd.) Miers, andrographispaniculata, ocimumtenuiflorum L., ocimumbasilicum L., cinnamomum verum Presl* [30] during COVID-19 pandemic and being cured from disease.

12.4 Survey of literature on antiviral and anti-coronavirus activity of medicinal plants

Natural products like phenolic compounds, flavonoids, glycosides, steroids, saponins, essential oils present in some medicinal plants are effective against different viral diseases of human. These compounds either show medicinal activity alone or in various combinations via synergistic mechanisms against virus causing diseases. Some natural products like epigallocatechin [53] prevent attachment, entry, and membrane fusion of virus in the host cell while some others like ajoene inhibit replication of virus. Some of compounds present in plants are effective for treatment of symptoms of viral disease; for example thymol and menthol exhibit good inflammatory action. In a single medicinal plant, many active compounds are present and they act against a particular viral disease according to their nature. Table 12.2 exhibits active compounds in some medicinal plants having potential for treatment of different viral diseases.

Tab. 12.2: Structures of major antiviral agents and the target viral disease of some medicinal plants.

Thymol	**(+)-catechin**	**Ajoene**
Thymus vulgaris [15, 31]	*Sarraceniapurpurea* [30]	*Allium sativum L.* [16, 47, 51]
Flu and Common Cold Norovirus infection	**Smallpox**	**SARS,MERS,COVID-19**

Myricetin [41]	**Eugenol**	**Quercetin**
Sarraceniapurpurea	*Ocimum sanctum* [16], *Cinnamon mumverumPresl* [30]	*Ipomoea aquatic* [42]
Active against various virus dieses	**Flu and Common Cold, Covid-19**	**Chikungunya virus infection**

6-gingerol	**Anthraquinone(derivatives)**	**Kaempferol**
Zingiberofficinale [16, 27]	*Persicariaodorata* [42]	*Bombaxceiba L.* [16, 20]
Flu and Common Cold Covid-19	**Chikungunya virus infection**	**Flu and Common Cold Covid-19**

Tab. 12.2 (continued)

Vitexin	**Aloe-emodin**	**Caffeoyltartaric acid**
Rhapis excels [42]	*Aloe vera (L.) Burm. f.* [16, 21]	*Boerhaviadiffusa* [43]
Chikungunya virus infection	Flu and Common Cold Covid-19	Hepatitis
Uscharin	**Apigenin-7-o-glucoside**	**Mangiferin**
Calotropisgigantean (L.) Dryand [17]	*Menthalongifolia* [44]	*Mangiferaindica L.* [18, 56]
Enfluenza	Hepatitis	Enfluenza, Rabies
Ergosterol	**Anisotine**	**Swertiamarin**
Ganodermalucidum [45]	*Justiciaadhatoda L* [19]	*Swertiachirayita* [45]
Hepatitis	Covid-19	Hepatitis

Tab. 12.2 (continued)

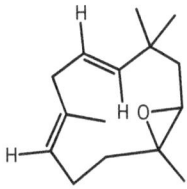	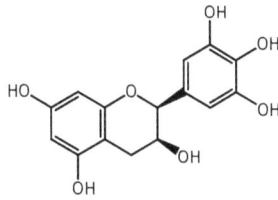	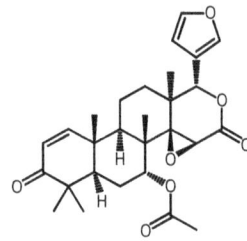
Humulene epoxide *Cyperusrotundus L* [22]	**Gallocatechin** *Acacia nilotica,*	**Gedunin** *Azadirachtaindica* [23, 37, 40, 45, 50, 53]
SARS	**Lasa fever**	**Covid-19, smallpox, hipatites etc**
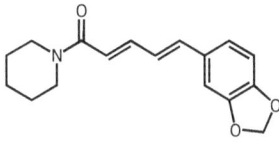	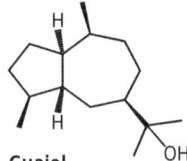	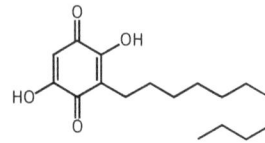
Piperine *Piper cubeba* [45]	**Guaiol** *Piper nigrum L* [24]	**Embelic acid** *Embeliaschimperi* [45]
Hepatitis	**Covid19, SARS**	**Hepatitis**
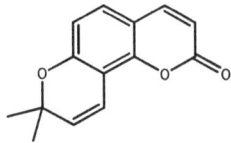	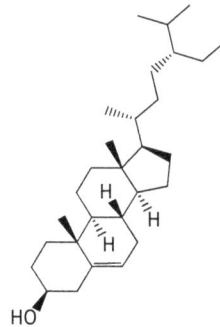	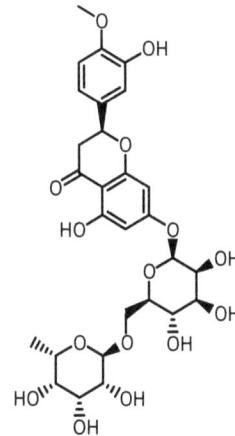
Seselin *Ale marmelos* (L.) [25]	**β-sitosterol** *Taraxacumofficinalis* [45]	**Hesperidin** Citrus sinensis (L.) [26]
SARS, COVID1-19	**Hepatitis**	**SARS, Covid-19**

Tab. 12.2 (continued)

γ-Tocopherol	Curcumin	Quercetin-3-O-glucoside
Lepidiumsativum [45]	*Curcuma longa L.* [28, 45]	*Adiantumcapillusveneris,* [46]
Hepatitis	**Covid-19, SARS, Hepatitis**	**Hepatitis**

		Tinosporin
Withaferin A	**7-methoxy coumarin**	*Tinosporacordifolia* (Willd.) Miers
Withaniasomnifera (L.) *Dunal* [29]	*Equisetum debileRoxb* [46]	[30]
Covid-19, SARS	**Hepatitis**	**Covid-19, HIV**

	Andrographolide	
Morusin	*Andrographispaniculata (Burm. f.)*	**Epigallocatechin-3-O-gallate**
Morusnigra L [46]	*Wall. ex Nees* [30]	*Camellia sinensis* [47, 51]
Hepatitis	**Covid-19, SARS**	**Ebola. Genital wart**

Tab. 12.2 (continued)

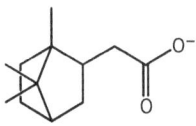

Bornylacetate
Ocimumtenuiflorum L.

Covid-19, SARS

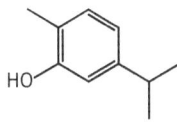

Carvacrol
Artemisia kermanensis [48]

Oral herpes

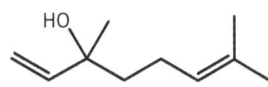

Linalool
Ocimumbasilicum L.

Covid-19

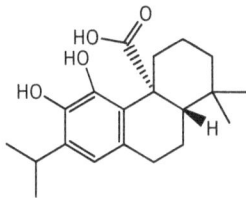

carnosic acid
Rosmarinusofficinalis [48]

HIV, herpes

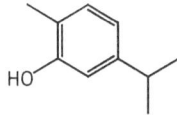

Carvacrol
Origanumvulgare, Thymus mastichina [31]

Norovirus infection

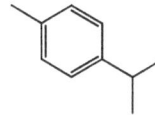

p-cymene
Saturejahotensis L [48]

herpes

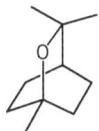

1,8-cineole
Thymus mastichina [31]

Norovirus infection

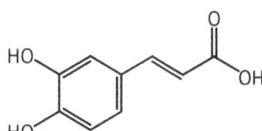

Caffeic acid
Nauclealatifolia [49]

Genital herpes

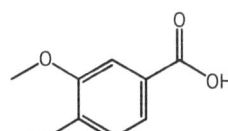

vanillic acid
Rubiacordifolia L [32].

Rotavirus infection

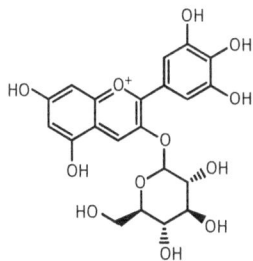

Delphinidin-3-glucoside
Hibiscus sabdariffa [49]

Genital herpes

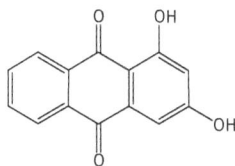

xanthopurpurin
Rubiacordifolia L. [32]

Rotavirus infection

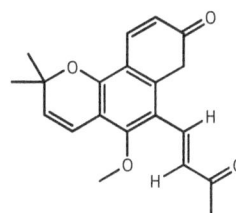

Avicennone
Avicenna marina [49]

Genital herpes

Tab. 12.2 (continued)

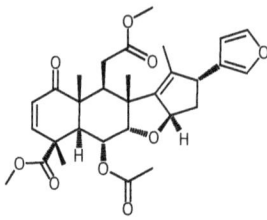

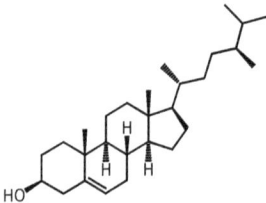

Nimbin *Azadirachtaindica*	**azadirachtin** [50, 51] *Azadirachtaindica*	**Shikonin** *Radix lithospermi* [33], *Camellia sinensisKuntze* [34]
Dengu	**Molluscumcontagiosum**	**Adenovirus infection**
ursolic acid *Phyllanthusamarus* [51]	**Catechins** *Camellia sinensisKuntze* [34]	**Gallocatechin** *Acacia nilotica* [52]
Ebola	**Adeno virus**	**Lassa fever**
Campesterol *Achyroclinebogotensis* [35]	**Alpha Tocopherol** *Elaeisguineensis* Jacq [52]	**Methylophiopogonanone A** *Ophiopogonjaponicas* [36]
Astrovirus	**Lassa fever**	**Measles**

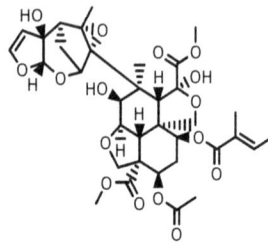

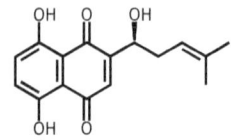

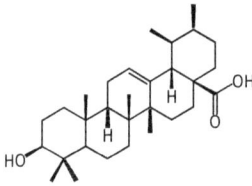

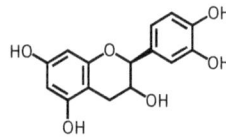

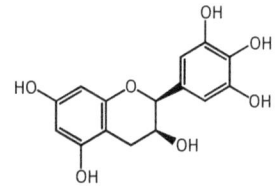

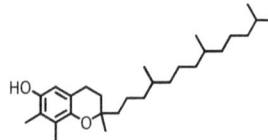

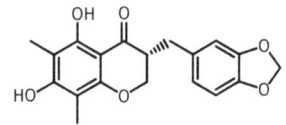

Tab. 12.2 (continued)

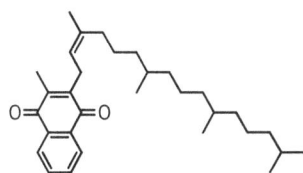	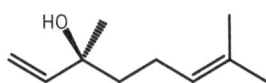	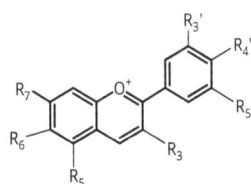
Phytonadione	**(S)-(+)-linalool**	**Anthocyanin(Basic Structure)**
Moringaoleifera [52]	*Coriandrumsativum* [36]	*Euphorbia hirta* [53]
Lassa fever	**Measles**	**Dengue**
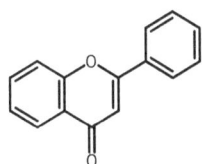	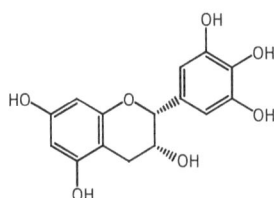	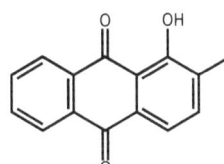
		1-hydroxy 2-methyl
Flavone	**Epigallocatechin**	**anthraquinone**
Elsholtziacristata [36]	*Carica papaya* [53, 54]	*Manjisthacordifera* [37]
Measles	**Dengue**	**Chekenpox**
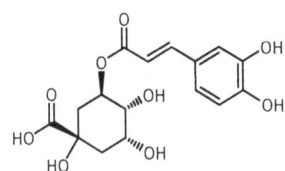	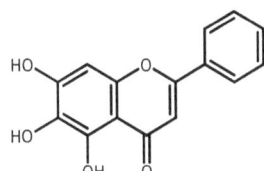	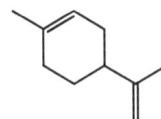
	Baicalein	
Chlorogenic acid	*Oroxylumindicum,*	**Limonene**
Cymbopogoncitratus	*Sesamumindicum* [38]	*Citrus aurantifolia* [54]
Yellow fever	**Shingles**	**Yellow fever**
	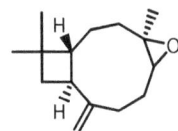	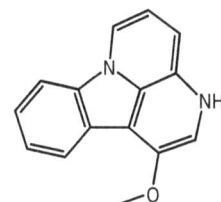
9, 12-octadecadienoyl chloride,	**Caryophyllene oxide**	**methoxycanthin**
(Z,Z) Cynodondactylon [38]	*Enantiachlorantha* [54]	*Drymariacordata* [38]
Shingles	**Yellow fever**	**Shingles**

Tab. 12.2 (continued)

Phenoxazine *Morindalucida* [54]	**Naphthoquinone (derivatives)** *Lawsoniainermis* [54]	**Rutin** *Centellaasiatica* [38]
Yellow fever	**Yellow fever**	**Shingles**
Quercetin *S. jollyanum* [55]	**Sesamin** *Sesamumindicum* [28]	**Anthraquinone** *D. monbutensis* [55]
Polio	**Shingles**	**Polio**
Diterpenes *Nepetacataria* [39]	**Marmelide** *Aegle marmelos* [56]	**Menthol** *Menthapiperita* [39]
Raseola	**Rebies**	**Raseola**
Lupeol *Ricinuscommunis* [56]	**Cynaroside I** *Achilleamillefolium* [39]	
Rebies	**Raseola**	

12.4.1 Antiviral activity

Antiviral activity of medicinal plants is due to antiviral agents present in the medicinal plants. These antiviral agents' activities are based on their specific structure, concentration, and their nontoxic nature for humans. Herbal antiviral agents express different mechanisms; for example *Andrographis paniculata* (Burm.f.) Nees exhibit antiviral effect through immunomodulation, *Withania somnifera* (L.) Dunal by immune stimulatory properties, *Allium sativum L.* inhibit HIV virus replication, and in some cases, medicinal plants are helpful in treatment of symptoms of viral diseases. Mostly exact mechanisms of activities of medicinal plants and remedies made of medicinal plants are not known, and in such cases, the synergistic effect of natural products is responsible for activities.

12.4.2 Anti-coronavirus activity

Corona viruses belong to a family of viruses, which cause diseases like common cold, SARS, MERS, and running pandemic Covid-19. Covid-19 is caused by SARS-COV-2 virus which is an enveloped RNA virus Fig. 12.1 and transmitted through inhaling virus containing air and touching surfaces on which the virus is already present. This virus spike has better capacity of binding with ACE-2 receptors and thus primarily causes infection in lungs as lungs contain ACE-2 receptors (Fig. 12.2). Virus enters lung cell, highjacks cell function; undergoes replication, and due to this toxic molecules release occurs (Figs. 12.3, 12.4). These diseases have caused 6,224,814 deaths all over the word until now and except vaccines no specific medicine has been developed still for

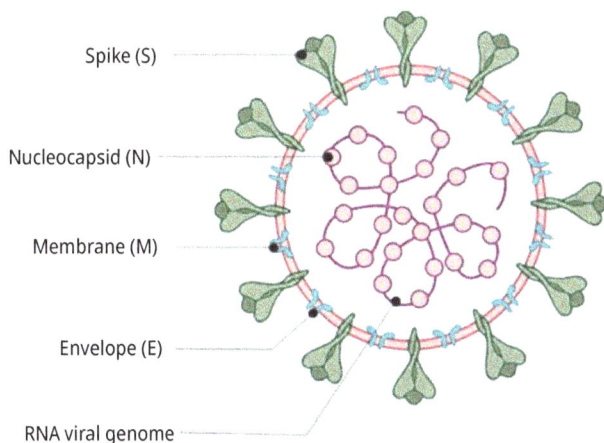

Fig. 12.1: Structure of SARS-COV-2. https://www.biophysics.org/blog/coronavirus-structure-vaccine-and-therapy-development.

treatment of this disease. During this pandemic if we look toward developing countries, there was no availability of good medical facilities in tribal areas. People of these areas used mostly medicinal plants or remedies made of plants for treatment of Covid-19 while people of developed countries like the USA, the UK, and Italy used antiviral medicines and other medical support for treatment until the vaccination and mortality rate was greater in those people who used antiviral medicine due to side effects of medicines. Thus common medicinal plants which are being used in treatment of coronavirus causing diseases are safe and effective are given in Tab. 12.1 and active ingredient of each plants are given in Tab. 12.2. Thus, medicinal plants are a safer solution for coronavirus diseases.

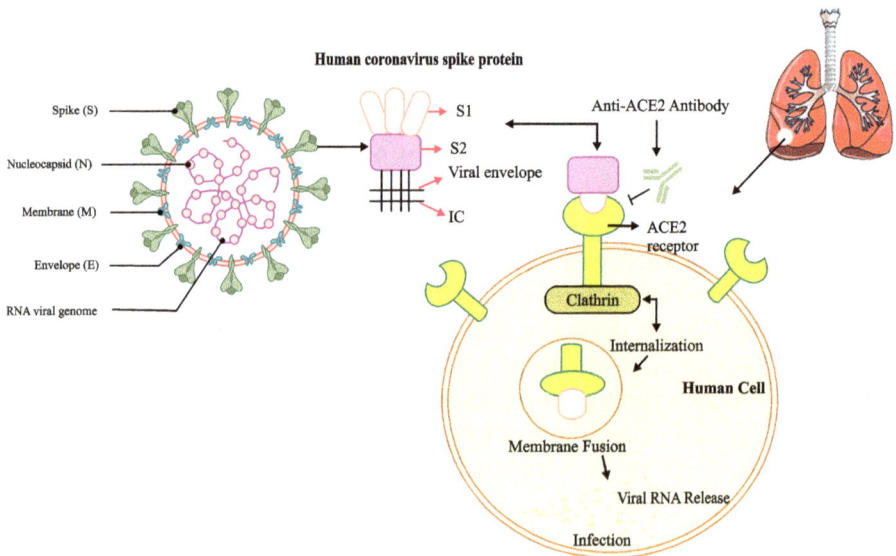

Fig. 12.2: SARS-COV-2 virus binding with ACE2 receptor in lungs [65].

12.5 Conclusions and prospects

Since 1892, many viral diseases have come to exist in plant and animal kingdom, and scientists and researchers are trying to search better treatments and prevention and it has resulted in many solutions in the form of allopathic medicines and vaccines. In spite of appreciable development of medical science virus causing fatal diseases like AIDS and Ebola are still in need of effective solution. Medicinal plants are resources of newer effective drugs as well as these are directly utilized for treatment of many viral diseases. Whenever new disease came in to existence researchers always moved toward synthesized medicine as well as medicinal plants and medicinal plants were used as safer

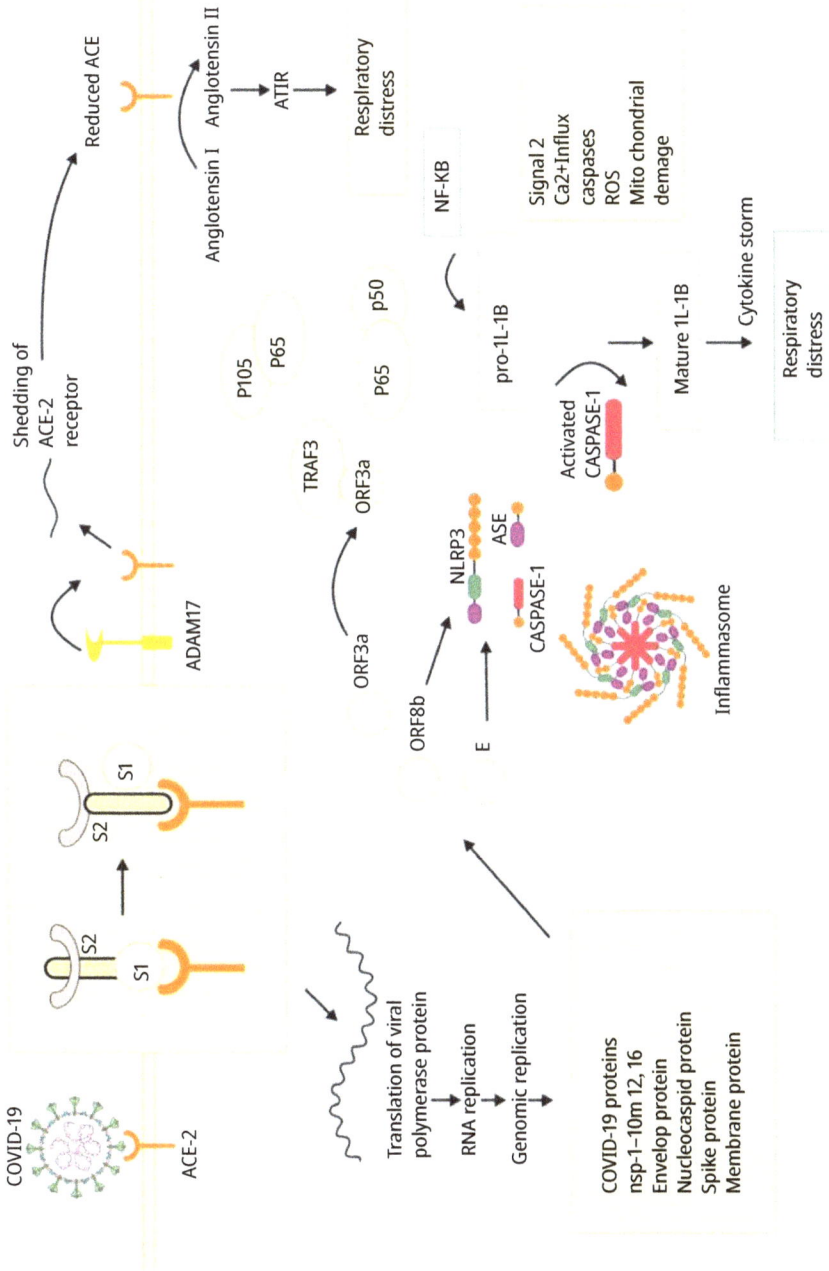

Fig. 12.3: SARS-COV-2 viral entry, replication, and damage caused by virus [65].

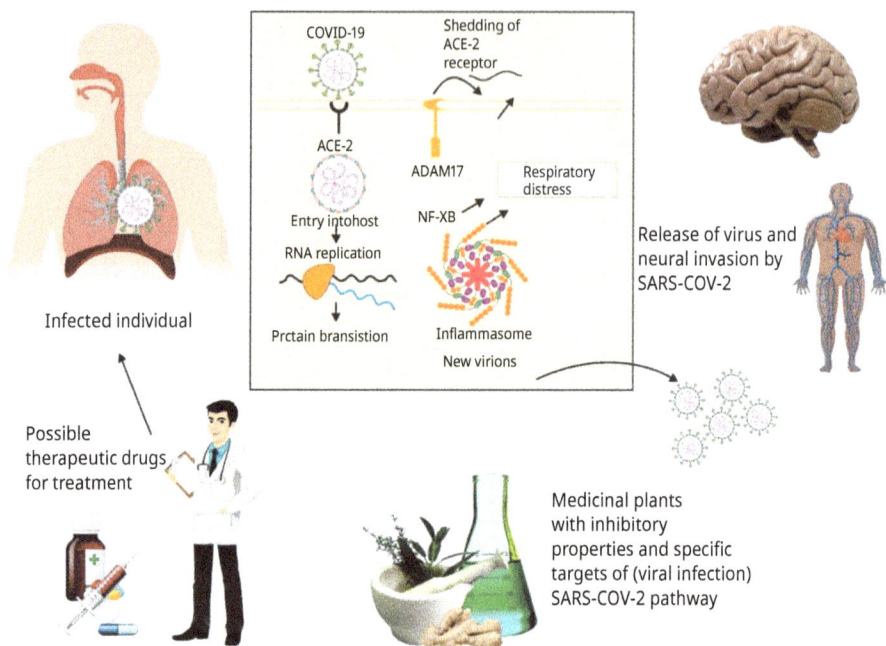

Fig. 12.4: Infection of SARS-COV-2 and possible treatments [65].

remedies. During the Covid-19 pandemic, researchers all over the world evaluated many plant species for treatment of Covid-19 and got potential ability in these species. Some of the plant species seeds, barks, leaves, root, flowers, fruits which are being used by human society as food also have potential antiviral activities of many virus causing diseases but due to lack of drug delivery system and required fixed effective doses they are not effectively working as that of their potential to treat. However, these plant species are being used to treat many diseases with effective results.

References

[1]   K. Yuki, M. Fujiogi, S. Koutsogiannaki. Clinical Immunology. 2020, 215, 108427. https://doi.org/10.1016/j.clim.2020.108427.
[2]   S. Salehi, A. Abedi, S. Balakrishnan, A. Gholamrezanezhad. Am J Roentgenol. 2020, 215(1), 87–93. https://doi.org/10.2214/AJR.20.23034.
[3]   E. De Clercq, G. Li. Clin Microbiol Rev. 2016, 29, 695–747. https://doi.org/10.1128/CMR.00102-15.
[4]   D. Bennadi. J Basic Clin Pharm, 2014, 5(1), 19–23. https://doi.org/10.4103/0976-0105.128253.
[5]   F. Attena. J Alternative Compl Med. 2016, 22, 343–348. https://doi.org/10.1089/acm.2015.0381.
[6]   S. R. Gawde, Y. C. Shetty, D. B. Pawar. Perspect Clin Res. 2013, 4, 175–180. https://doi.org/10.4103/2229-3485.115380.

[7]   C. A. Newall, L. A. Anderson, J. D. Phillipson. Herbal Medicines, A Guide for Health-care Professionals. The Pharmaceutical Press, London, 1996, 210211.

[8]   S. Verma, S. J. Singh. Current and future status of herbal medicines. Vet World. 2008, 1(11), 347–350.

[9]   C. Bailly, G. Vergoten. Pharmacol Ther. 2020, 214, 107618. https://doi.org/10.1016/j.pharmthera.2020.107618.

[10]   J. H. Beigel, K. M. Tomashek, L. E. Dodd, et al. New Engl J Med. 2020, 383(19), 1813–1836. https://doi.org/10.1056/NEJMoa2007764.

[11]   Q. Cai, M. Yang, D. Liu, J. Chen, D. Shu, J. Xia, X. Liao, Y. Gu, Q. Cai, Y. Yang. Engineering. 2020, 6(10), 1192–1198. https://doi.org/10.1016/j.eng.2020.03.007.

[12]   J. Chen, L. Xia, L. Liu, Q. Xu, Y. Ling, et al. Open Forum Infect Dis. 2020, 7(7), ofaa241. https://doi.org/10.1093/ofid/ofaa241.

[13]   Z. Sahraei, M. Shabani, S. Shokouhi, Saffaei. Int J Antimicrob Agents, 2020, 55(4), 105945. https://doi.org/10.1016/j.ijantimicag.2020.105945.

[14]   F. Touret, X. de Lamballerie. Antiviral Res. 2020, 177, 104762. https://doi.org/10.1016/j.antiviral.2020.104762.

[15]   M. H. Shahrajabian, W. Sun, Q. Cheng. Nat Prod Commun. 2020, 15(8), 1–10. https://doi.org/10.1177/1934578X20951431.

[16]   S. C. Bachar, K. Mazumder, R. Bachar, A. Aktar, M. Al Mahtab. Front Pharmacol. 2021, 12, 732891. https://doi.org/10.3389/fphar.2021.732891.

[17]   S. Parhira, Z. F. Yang, et al. PLOS One. 2014, 9, e104544. https://doi.org/10.1371/journal.pone.010454.

[18]   A. A. S. Al Rawi, H. S. H. Al Dulaimi. J Pure Appl Microbiol. 2019, 13, 455–458. https://doi.org/10.22207/jpam.13.1.50.

[19]   R. Chavan, A. Chowdhary. Int J Pharm Sci Rev Res. 2014, 25, 231–236. https://doi.org/10.1002/14651858.

[20]   J. G. Choi, H. Lee, Y. S. Kim, Y. H. Hwang, et al. Am J Chin Med. 2019, 47, 1307–1324. https://doi.org/10.1142/S0192415X19500678.

[21]   S. Schwarz, D. Sauter, K. Wang, R. Zhang, et al. Planta Med. 2014, 80, 177–182. https://doi.org/10.1055/s-0033-1360277.

[22]   R. M. Samra, A. F. Soliman, A. A. Zaki, et al. J Essent Oil Bearing Plants. 2020, 23, 648–659. https://doi.org/10.1080/0972060x.2020.1823892.

[23]   M. A. Alzohairy. Altern Med. 2016, 7382506. https://doi.org/10.1155/2016/7382506.

[24]   P. Pandey, D. Singhal, F. Khan, M. Arif. Biointerface Res Appl Chem, 2021, 11(4), 11122–11134. https://doi.org/10.33263/BRIAC114.1112211134.

[25]   P. Kaushik, V. K. Yadava, G. Singh, R. K. Jhaa. Indi Jour of Traditil Knowl. 2020, 19(Suppl), 19. S 153–S 157.

[26]   R. Battistini, I. Rossini, C. Ercolini Et Al Food Environ Virol. 2019, 11, 90–95. https://doi.org/10.1007/s12560-019-09367-3.

[27]   M. Haridas, V. Sasidhar, P., et al. Futur J Pharm Sci. 2021, 7(1), 13. https://doi.org/10.1186/s43094-020-00171-6.

[28]   R. K. Thimmulappa, K. J. T. Radhakrishnan, A. Bhojraj, S., et al. Heliyon. 2021, 7, e06350. https://doi.org/10.1016/j.heliyon.2021.e06350.

[29]   A. Saggam, K. Limgaokar, et al. Front Pharmacol. 2021, 12, 623795. https://doi.org/10.3389/fphar.2021.623795.

[30]   R. S. Singh, A. Singh, H. Kaur, et al. Phytother Res. 2021, 35(8), 4456–4484. https://doi.org/10.1002/ptr.7150.

[31]   J. Sarowska, D. Wojnicz, A. Jama-Kmiecik, M. Frej-Mądrzak, I. Choroszy-Król. Molecules. 2021, 26(15), 4669. https://doi.org/10.3390/molecules26154669.

[32]   Y. Sun, X. Gong, J. Y. Tan, L. Kang, D. Li. Front Pharmacol. 2016, 7, Article 308. https://doi.org/10.3389/fphar.2016.00308.

[33]   H. Gao, L. Liu, Z. -y Qu, et al. Biol Pharm Bull. 2011, 34(2), 197–202. https://doi.org/10.1248/bpb.34.197.

[34]   A. Karimi, M.-T. Morad, et al. J Complement Integr Med. 2016, 13(4), 357–363. https://doi.org/10.1515/jcim-2016-0050.

[35]   M. A. Téllez, A. N. Téllez, F. Vélez, J. C. Ulloa. Altern Med, 2015, 15(1), 428. https://doi.org/10.1186/s12906-015-0949-0.

[36]   Hoi. Pharmascope Inter Jour of Res Pharm Sci. 2020, 11(3), 3872–3877. https://doi.org/10.26452/ijrps.v11i3.2570.

[37]   https://www.planetayurveda.com/library/chickenpox/.

[38]   H. Khatiwada, A. Neupane, L. B. Thapa. J Ayurvedic Herb Med. 2021, 7, 66–70. https://doi.org/10.31254/jahm.2021.7203.

[39]   https://www.anticoagulationeurope.org/conditions/roseola/.

[40]   W. Arndt, C. Mitnik, S. White, et al. PloSOne. 2012, 7(3), e32610. https://doi.org/10.1371/journal.pone.0032610.

[41]   S. Garcia. Front Plant Sci. 2020, 11, Article 571042. https://doi.org/10.3389/fpls.2020.571042.

[42]   S. Kumar, C. Garg, et al. Indian J Pharmacol. 2021, 53(5), 403–411. https://doi.org/10.4103/ijp.IJP_81_19.

[43]   S. Sahoo, A. Das, G. Mahalik. Ind J Natural Sci. 2021, 12(69), 37559–37566.

[44]   K. H. Ali, A. A. El-Badry, J. Med. Biomed Sci. 2010, 2, 67–73.

[45]   Z. N. Samani, M. R. Kopaei. IJPSR Sr No: 2 Page No, 3589–3596.

[46]   A. M. Abbasi, M. A. Khan, M. Ahmad, M. Zafar, H. Khan, N. Muhammad, S. Sultana. Afr J Biotechnol. 2009, 8(8), 1643–1650.

[47]   Z. Villines January 21, 2022, https://www.medicalnewstoday.com/articles/321036.

[48]   S. Gavanji, S. S. Sayedipour, B. Larkia, A. Bakhtaric, et al. J Acute Med. 2015, 5(3), 62–68. https://doi.org/10.1016/j.jacme.2015.07.001.

[49]   A. Garber, L. Barnard, C. Pickrell, J. Evid.-Based. Integr Med. 2021, 26, 1–57. https://doi.org/10.1177/2515690X20978394.

[50]   S. B. Kalasannavar, M. P. Sawalgimath. Anc Sci Life. 2013, 33(1), 49–51. https://doi.org/10.4103/0257-7941.134606.

[51]   S. D. Jain, D. K. Birla, D. Mishra. PharmaTutor. 2018, 6(9), 20–23. https://doi.org/10.29161/PT.v6.i9.2018.20.

[52]   I. B. Abubakar, S. S. Kankara, I. Malami, et al. Eur J Integr Med. 2022, 49, 102094. https://doi.org/10.1016/j.eujim.2021.102094.

[53]   M. S. M. Saleh, Y. Kamisah. Biomolecules. 2020, 11(1), 42. https://doi.org/10.3390/biom11010042.

[54]   A. Abiodun, A. O. Adeyemi, J. O. Moody, et al. J Herbs Spices Med Plants. 2010, 16, 203–218. https://doi.org/10.1080/10496475.2010.511075.

[55]   J. O. Moody, V. A. Roberts. Viruses. https://www.ajol.info/index.php/njnpm/index.

[56]   S. Mukherjee, S. Roy, A. Chowdhary. Int J Curr Res. 2017, 9(10), 59564–59568. https://www.journalcra.com/article/vitro-anti-rabies-virus-activity-indian-medicinal-plants.

[57]   S. Ben-Shabat, L. Yarmolinsky, D. Porat, A. Drug Deliv Transl Res. 2020, 10, 354–367. https://doi.org/10.1007/s13346-019-00691-6.

[58]   W. Zhou, A. Yin, J. Shan, S. Wang, B. Cai, L. Di. Molecules, 2017, 22(4), 654. https://doi.org/10.3390/molecules22040654.

[59]   B. K. Kim, A.-R. Cho, D. J. Park. Food Chem. 2016, 206, 85–91. https://doi.org/10.1016/j.foodchem.2016.03.052.

[60]   H. Zhang, X. Yang, L. Zhao, Y. Jiao, J. Liu, G. Zhai. Drug Deliv. 2016, 23, 1933–1939. https://doi.org/10.3109/10717544.2015.1008705.

[61]   R. Yang, X. Huang, J. Dou, G. Zhai, L. Int J Nanomed. 2013, 8, 2917. https://doi.org/10.2147/IJN.S47510.

[62]   C. Xu, Y. Tang, W. Hu, R. Tian, Y. Jia, P. Deng, L. Zhang. Carbohydr Polym. 2014, 113, 9–15. https://doi.org/10.1016/j.carbpol.2014.06.059.

[63] Y. Jiang, F. Wang, H. Xu, H. Liu, Q. Meng, W. Liu. Int J Pharm. 2014, 475475–475484. https://doi.org/10.1016/j.ijpharm.2014.09.016.

[64] D. S. Chauhan, S. Yadav, M. A. Quraishi. Curr Res Green Sustain Chem. 2021, 4, 100114. https://doi.org/10.1016/j.crgsc.2021.100114.

[65] B. Vellingiri, K. Jayaramayya, M. Iyer, A. Narayanasamy, V. Govindasamy, B. Giridharan, S. Ganesan, A. Venugopal, D. Venkatesan, H. Ganesan, K. Rajagopalan, P. K. S. M. Rahman, S.-G. Cho, N. S. Kumar, M. D. Subramaniam. Sci Total Environ. 2020, 725, 138277. https://doi.org/10.1016/j.scitotenv.2020.138277.

Ankush Kerketta and Balram Sahu*

Chapter 13
Nanomaterials synthesis from medicinal plant extract

Abstract: Synthesis of nanomaterials by plant extracts is the utmost convenient, easy, cost-effective, and ecologically beneficial way which is devoid of the contribution of hazardous chemicals. Hence, a number of environmentally benign methods for the rapid synthesis of nanomaterials have been reported in recent years by employing aqueous extracts of plants. This chapter summarizes the synthesis of medicinal plant-based nanomaterials, plant components involved in its synthesis, and factors affecting nanomaterials' synthesis process and application. Many phytochemicals found in plants, including flavonoids, alkaloids, tannins, saponins, and other metabolites, play a vital role in the production of NMs and have important implications for the development of a variety of purposes. In order to create nanoparticles, various secondary metabolites included in the extracts work as stabilizing and/or reducing agents. Medicinal plant extracts offer a potential method for the production of nanomaterials through safer pathways. This chapter mainly focuses on the recent reports of nanomaterial synthesized by medicinal plant extract. Here, significant and recent developments for diverse biological uses of these plant-based nanomaterials are reviewed together with discussions. Recent developments in nanotechnology, with a focus on medicinal plant extract specifically, offer insight into their use as edible nanoparticles generated from plants. This chapter also briefly covers the application and future prospects of nanomaterials.

13.1 Introduction

In today's era of advanced technology, an interesting class of materials known as nanomaterials (NMs) is now in high demand for a variety of applications. Any particles that are at least 100 nm in size are known as nanoparticles or nanomaterials. Richard Feynman (a Nobel Prize winner and an American physicist) first time proposed the term *nanotechnology* in 1959 and seeded the idea of contemporary technology, and because of this, he is known as the father of modern nanotechnology. Particularly in the realm of biotechnology, nanotechnology is a growing area of

*Corresponding author: Balram Sahu, Department of Botany, Govt. Rani Durgawati College, Wadrafnagar, 497225, Chhattisgarh, India, e-mail: balramsahu@hotmail.com
Ankush Kerketta, Department of Botany, Govt. Kalidas College, Pratappur, 497223, Chhattisgarh, India

https://doi.org/10.1515/9783110791891-013

interdisciplinary research [1, 2]. Nanotechnology, which offers specialized nanomaterials with a lot of potential for producing things with noticeably higher performances, is an excellent example of an emerging technology. In NMs, molecules undergo fundamental changes in all of their characteristics, namely, biological, physical, and chemical. Due to their entirely new or improved properties based on size, distribution, and shape, the uses of NMs are expanding quickly on numerous fronts. It is rapidly advancing in a wide range of fields, namely, health care, cosmetics, biomedical, food and feed, drug-gene delivery, and environment. It is also being used in the chemical, electronic, space, energy, catalysis, nonlinear optical, and photo-electrochemical industries [3, 4].

Nanotechnology research is expected today not because of only application but also by synthesis [5–7]. It offers to expand research in various areas like electronics, energy, medicine, and life sciences. There are several ways to synthesize nanomaterials/nanoparticles, including chemical, physical, and biological methods [5, 8, 9]. The chemical technique of synthesis may produce a large number of nanoparticles [term nanoparticles (NPs) and nanomaterials (NMs) are used as synonyms in this chapter] in a short period of time with definite shape and size. However, there are many advantages of the chemical method but it is costly, complicated and inefficient and also produces hazardous wastes that are harmful for human health as well as environment [10–12]. The physical method of NP's synthesis is high energy consuming method. Therefore, more environmentally friendly and economically feasible techniques are required for NPs synthesis. The use of biological agents in the synthesis of NMs is gaining popularity because they provide green chemistry processes that are harmless and ecologically friendly [13]. In recent years, it has been shown that many biological systems, including plants and algae, diatoms, bacteria, yeast, fungi, and human cells, appeared as green alternatives to the synthesis of NPs [14–19].

Plant extracts from various medicinal plants have secondary metabolites which have unique medicinal values [20–22]. Plant secondary metabolites contain varieties of biomolecules such as alkaloids, saponins, terpenoids, phenolics, tannin, amino acids, proteins, enzymes, and vitamins, and these molecules have the potential to function as reducing and stabilizing agents that convert metal ions into NMs in a single-step green synthesis process [23]. There have also been several reports on the synthesis of NMs from different types of medicinal plants/plant extracts like AgNP by using *Ocimum Basilicum* and *Teucrium Polium* [24], *Cucumis prophetarum* [25], *Moringa oleifera* [26]. Rashid et al. [27] successfully synthesized silver nanoparticles (AgNPs) from various medicinal plants such as *Bergenia stracheyi*, *Bergenia ciliata*, *Rumex hastatus* and *Rumex dantatus* plant extracts. Gold nanoparticles (AuNPs) were synthesized by using *Capsicum annuum* [28], *Ligustrum vulgare* [29], and many more medicinal plant extracts. AuNP was synthesized by using various medicinal plants like *Camellia sinensis*, *Calophyllum inophyllum*, *Guazuma ulmifolia*, *Catharanthus roseus*, *Cassia auriculate*, *Aegles marmelos*, *Andrographis paniculate*, *Asparagus racemosus*, *Cinnamomum aromaticum*, *Annona muricate*, *Justicia gendarussa*, and *Piper nigrum*. Also, lanthanum, zinc oxide, platinum, nickel oxide, copper oxide, and palladium NPs were synthesized from different types of medicinal plant/plant extracts [30, 31].

The aim of this chapter is to provide information about the basic concepts, advances, and trends relating to the synthesis of NMs from medicinal plant extracts on a single platform by analyzing relevant data and discussing procedures of synthesis, characteristics, potential applications, and possible opportunities relating to the broad and fascinating area of NMs. Although it is difficult to include all the literature on NMs/NPs, this overview discusses significant works from the last four years and the present. Researchers may rapidly get a fundamental understanding from this chapter, which gathers the developments and characteristics of numerous kinds of medicinal plant extract-based NMs in a single chapter.

13.2 Synthesis of NMs from the medicinal plant

Since the beginning of life on earth, biological and inorganic elements have been in continual communication with one another. This constant interplay allowed life on this planet to coexist with a well-organized mineral deposit. In recent years, scientists' interest in the interactions between inorganic molecules and biological organisms has grown. Numerous microbes, enzymes, and plant extracts have been shown in studies to be capable of producing inorganic NMs through extracellular or intracellular pathways. In many aspects, NMs made using biological means are significantly better to those made using chemical techniques. Although the latter techniques may generate vast numbers of nanoparticles with certain sizes and shapes in a short amount of time, they are difficult, out-of-date, expensive, and ineffective, and they result in dangerous toxic wastes that are bad for both the environment and human health. The biological method is both cost-effective and ecologically beneficial since it does not need the use of costly chemicals. With the biological approach, the use of expensive chemicals is eliminated as this method doesn't employ harmful chemicals and does not produce toxic byproducts thus it is an environmentally safe synthetic process and that's why this method is in more and more demand in the present day [11]. NMs synthesized by using plants or plant extracts provide a better platform for NM synthesis as they are produced by natural capping agents which are devoid of hazardous chemicals. Table 13.1 briefly shows the list of some recent reports regarding the synthesis of NPs/NMs from medicinal plants.

Chakravarty et al. [32] used *Syzygium cumini* fruit extract for the synthesis of spherical silver nanoparticles size of ~47 nm. To generate silver nanoparticles, 10 ml of extract was combined with 50 ml of silver nitrate [AgNO$_3$] solution and stored in a dark environment to avoid photooxidation. After some time, a visible change in the solution's color showed the silver nanoparticles' development. 33, utilized the traditional medicinal plant *Azadirachta indica* to synthesize AgNPs with sizes ranging from 19.27 to 22.15 nm

Tab. 13.1: Different types of medicinal plants used for the synthesis of NMs/NPs.

Medicinal plants	Nanoparticles	Size (nm)	Shape	References
Syzygium cumini	Ag	47	Spherical	Chakravarty et al. [32]
Azadirachta indica	Ag	19.27–22.15	Spherical	Chi et al. [33]
Hyptis suaveolens	Cu	5.8–7	Spherical	Shubhashree et al. [34]
Ocimum sanctum	Ag	10–30	Spherical	Baruah et al. [35]
Adhatoda vasica	ZnO	15–20	Orthogonal/nanorod	Pachaiappan et al [36]
Elaeagnus angustifolia	ZnO	36	Spherical	Iqbal et al. [145]
Aegle marmelos	Ag	47	Irregular	Devi et al. [37]
Asparagus racemosus Sophora interrupta	Ag	8–12	Spherical	Satyanarayana et al. [38]
Ocimum tenuiflorum	Fe	–	–	Kaur [39]
Calliandra haematocephala	ZnO	–	Flower-shaped	Vinayagam et al. [40]
Aloe socotrina	ZnO	15–50	Spherical	Fahimmunisha et al. [41]
Anacardium occidentale	AO-Cu	30	Spherical and scattered	Chandra and Khan [42]
Rosmarinus officinalis	palladium	15–90	Spherical	Rabiee et al. [43]
Litsea cubeba	Au	10–18	Spherical	Doan et al. [44]
Pimenta dioica	Au	13	Spherical	Kharey et al. [45]
Tangerine peels	TiO	50 & 150	–	Rueda et al. [46]
Ocimum tenuiforum	Se	15–20	Spherical	Liang et al. [47]
Nigella sativa	Pt	3.47	Spherical	Aygun et al. [48]

in spherical form. In a 250 ml Erlenmeyer flask, 20 mL of *A. indica* aqueous kernel extract has been combined with 80 mL of 1 mM $AgNO_3$. The reaction solution was then incubated at room temperature for 24 h in the absence of light. After incubation, a color shift was detected, which was regarded as preliminary confirmation of AgNPs production. 34, reported copper nanoparticles (CuNPs] manufacturing using an aqueous leaf extract of *Hyptis suaveolens* (L.) in order to allay environmental worries. Wild tulsi (*Ocimum tenuiflorum*) aqueous extract (10 mL) was employed as the reactive solution to react with copper (II) sulfate pentahydrate (5 g) aqueous solution (200 mL) with constant magnetic stirring at 400 rpm and warming at 80 °C for two hours. The obtained NPs were spherical in shape and their size lies between 5.8 and 7.0 nm. Baruah et al. [35] demonstrated the green production of AgNPs using tulsi leaf extract. The yellowish plant extract (150 mL) was then employed as a reactive solution in a room temperature reaction with an aqueous $AgNO_3$ solution (5 mL, 2 mM). The researchers kept the ultimate pH of this reacting combination at 7.4 by adding diluted aqueous NaOH solution externally. After 24 h of reaction, the authors noticed a brown color appearance in the reacting mixture. The nanoparticles are spherical in shape, and the diameter was 10–30 nm. Sharma et al. [49] used *Ocimum tenuiflorum* leaves extract to create copper oxide nanoparticles (CuONPs]. The studies found that the produced copper particles appeared spheroidal in form and ranged in size from 6 to 18 nm. The ready tulsi leaf extract was then used as a reacting solution with an aqueous solution of copper (II) acetate monohydrate for 4 h at 70 °C with constant stirring to produce CuONPs. Using an aqueous solution of diluted KOH, the authors kept the pH of the reaction mixture at 11.5 throughout the NPs. The reaction mixture was changed into a brown-black precipitate after 4 h of reaction, according to the authors. As a result, the precipitate was cooled to room temperature, rinsed with deionized water and ethanol, and afterwards left to dry at 55–60 °C in a hot air oven. 36, investigated the synthesis of zinc oxide nanoparticles (ZnONPs] using zinc acetate dehydrate, zinc sulfate, and zinc nitrate as precursors and extract of *Justicia adhatoda* leave as a reducing agent. In this synthesis method, 11.5 g of zinc sulfate was put in 40 mL of water and well stirred with a magnetic stirrer. Employing a sterile injection, 10 mL of *Adhatoda vasica* leaf extract had been administered drop by drop to this combination. The resultant mixture was continually swirled for at least 10 min. Dropwise additions of sodium hydroxide solution (2 M) were made to the aforementioned solution. Finally, the mixture was then centrifuged at 3,000 rpm for 15 min and rinsed with deionized water several times before drying at 150 °C in a hot air oven. It was important to obtain a fully dried zinc oxide nanopowder. To achieve the desired tiny ZnONPs, the powder has been ground into a fine powder using a mortar and pestle. Likewise, to make zinc oxide nanoparticles, 11.8 g of zinc nitrate and 8.8 g of zinc acetate dihydrate had been employed. The shape of NPs was orthogonal/nanorod and size ranges from 15–20 nm. Iqbal et al. [146] used *Elaeagnus angustifolia* leaf extracts to generate effortless and commercially feasible ZnONPs. To make ZnO NPs, 1 g of zinc-nitrate hexahydrate [$Zn(NO_3)2 \cdot 6H_2O$] was mixed with 100 mL of *E. angustifolia* leaf extract. Furthermore, the mixture has been heated continually (60 °C/2 h). The solution became yellowish black after 40 min, suggesting the formation of ZnO NPs. ZnONPs exhibited a spherical shape and size was approximately 36 nm. Tailor et al (2020) have published the green production of AgNPs using *Ocimum sanctum* leaf extract. The author developed AgNPs by reacting $AgNO_3$ solution (90 mL, 1 mM) with 10 mL of tulsi leaves aqueous extract for 15 min at 80 °C. The produced AgNPs were rod-like and spherical in form, with a mean particle size of 15.76 nm (6.13–-32.04 nm). Devi et al. [37] investigated *Aegle marmelos* fruit extract for the synthesis of AgNPs. For the synthesis of AgNPs, a 1 mM water solution of $AgNO_3$ was produced. The synthesis of AgNPs from $AgNO_3$ solution is accomplished by co-precipitation with *A. marmelos* fruit extract, which functions as a reducing and capping agent. The size of NPs was 47 nm and morphology was irregular. Satyanarayana et al. [38] studied leaf extracts of *Asparagus racemosus* and *Sophora interrupta* for the synthesis of AgNPs. In the real studies, 5 mL of each leaf extract was mixed with 95 mL of 1 mM AgNO$_3$ solution for the reduction of Ag$^+$ ions at 45 °C while shaking at 500 rpm continuously for 24 h in an orbital shaking incubator. The NPs are spherical in shape and their size was between 8 and 12 nm. 39, developed iron nanoparticles (FeNPs) by reacting an aqueous phase of FeSO$_4$ with a liquid leaf extract of *Ocimum tenuiflorum* at 25 °C and 8.0 pH. Significant parameters of nanoparticle synthesis, like reaction temperature (25–45 °C), pH of the reaction mixture (2.0–8.0), and amount of Fe precursor solution (0.1–1.0 M), were improved in this study utilizing central composite design and response surface methods. Vinayagam et al. [40] proposed an environmentally friendly technique by using extract of *Calliandra haematocephala* leaves for green synthesis of zinc oxide nanostructures. Employing scanning electron microscope (SEM) examination, flower-shaped nanoparticles have been discovered. Nanoparticles have been created by combining 0.05 M zinc acetate dihydrate and 1 M sodium hydroxide in a 2.5:1:1 (v/v) ratio and carefully blending. The resulting admixture had been heated at 80 °C for 60 min. The initial creation of a chalky solution quickly led to the production of an off-white precipitate. The technique resulted in the creation of zinc oxide nanoparticles, as predicted. Fahimmunisha et al. [41] developed research synthesized ZnONPs from *Aloe socotrina* extract. *A. socotrina*'s aqueous leaf extract [10 mL] had been mixed with 1 mM of aqueous zinc acetate after the pH had been raised to 12. The resulting mixture had a light white tone. The precipitate had been recovered and washed with distilled water following centrifugation. The pellet was then dried for 3 h in a hot air oven at 80 °C, and the dry powder was then calcined for 3 h at 350 °C and stored for further research. The nanoparticles in spherical in shape and size range from 15 to 50 nm. Chandra and Khan [42] proposed a revolutionary environmentally friendly copper nanoparticle [CuNPs] synthesis method that uses waste material from the cashew plant, namely, the testa of *Anacardium occidentale*. To prepare 0.1 M CuCl$_2$H$_2$O solution, 0.8524 g CuCl$_2$H$_2$O is dissolved in 50 mL of Merck Millipore water. *A. occidentale* testa extract and 0.1 M CuCl$_2$H$_2$O solution were combined in a 3:1 ratio. The aforesaid solution was then heated in Tarson's Digital Spinot for 15 min at 60–70 °C with steady stirring at 500 rpm for full precursor reduction until the orange-colored solution turned dark brown which indicates the synthesis of nanoparticles. The nanoparticles were slightly spherical and scattered with a size range of 30 nm. Rabiee et al. [43] aimed for a green and environmentally friendly production of palladium nanoparticles (PdNPs) employing *Rosmarinus officinalis* leaf extracts. To generate PdNPs using *Rosmarinus officinalis* leaf broth, 10 mL of the leaf broth had been mixed with 20 mL of palladium acetate solution (0.1 M). The mixture had been swirled at room temperature for 24 h, which is required for the NP synthesis step. The NPs have spherical shapes and sizes ranging from 15–90 nm. Doan et al. [44] offered a straightforward single-step green fabrication of colloidal AuNPs employing *Litsea cubeba* fruit extract as a reducing and stabilizing agent. The biogenic AuNPs were synthesized using chloroauric acid solution and aqueous *Litsea cubeba* fruit extract. To avoid any undesired photochemical reaction for freshly generated AuNPs, 1 mL of the extract was combined with 20 mL of Au^{3+} ion solution while vigorously swirling in the dark. The alteration in coloration of the solution after the reaction was completed served as a visible indicator of the synthesis process's success. The nanoparticles were spherical in shape and size ranging from 10 to 18 nm. Kharey et al. [45] work on gold nanoparticles using aromatic medicinal herb *Pimenta dioica* leaf extract by simple and fully green at a low cost. The AuNPs synthesized appeared spherical, with a mean particle size of 13 nm. Rueda et al. [46] demonstrated the bio-mediated synthesis of TiO_2 nanocrystals by the low-cost method using tangerine [*Citrus reticulata*] peels extract. In this method, titanium isopropoxide and tangerine peel extract were mixed and maintained pH 5. The sample was agitated at 700 rpm for 3 h; the aqueous phase had been separated by centrifugation, and the precipitated samples had been dried at 100 °C for 8 h before being calcined at 600 °C for 3 h to yield TiO_2 nanocrystals powder samples. Liang et al. [47] demonstrated the production of selenium nanoparticles [SeNPs] utilizing *Ocimum tenuiforum* leaf extract. For the SeNPs production, the precursor and extract quantities were kept between 10 and 50 mM and 1 to 5%, correspondingly. When a % *O. tenuiforum* leaf extract had been combined with a 10 mM liquid of sodium selenite, optimal findings for the synthesis were obtained by keeping a 9:1 ratio at room temperature. The pH was 9 and the synthesis was carried out with constant agitation at 130 rpm for 75 h. SeNPs had been synthesized that were 15–20 nm in diameter and spherical in morphology. Aygun et al. [48] synthesized biogenic platinum nanoparticles (PtNPs] by utilizing black cumin seed extract (*Nigella sativa* L.) as a reduction agent. To make the nanoparticles, a 15 mL *Nigella sativa* L. extract has been mixed with 85 mL of deionized water, and afterwards, $PtCl_4$ was introduced to the mixture. For two days, the resultant mixture had been agitated at 200 rpm and 75 °C. The transformation of the colorless solution to black implies that $PtCl_4$ was reduced to nanosized and transformed to PtNPs. The resulted NPs were 3.47 nm in size and spherical in morphology.

13.3 Phytochemicals responsible for the synthesis of NMs

Different properties of NMs like size, shape, and amount are strongly influenced by the biomolecules present in the plant extract. Medicinal plants are the richest source of plant secondary metabolites like terpenoids, flavonoids, and alkaloids. The plant's biomolecules serve as reductant and capping agents during the synthesis process of NMs [50]. The biomolecules such as flavans, flavanol, and flavanonol present in the *Shorea robusta* leaf extract are responsible for the reduction of silver ions into silver nanoparticles within 20 min [51]. The largest production of AgNPs was achieved after 20 min, according to Habeeb Rahuman et al. [11] who also noted that the alkaloid, flavonoids, and terpenoids found in the leaf extract of *Carissa carandas* are responsible for the decrease and stabilization of AgNPs. Table 13.2 briefly shows the list of some important biomolecules present in medicinal plants that play role in synthesis of NMs/NPs.

Aktepe and Baran [52] prepared AgNPs from *Cumcuma maxima* fruit extract. A knife has been used to slash the *Cucurbita maxima* fruit in half. Other than the edible, the fibrous structure of the inside (waste part] had been excluded and dried at room temperature. The dried fibrous structure had been afterwards weighed and 300 g of

Tab. 13.2: Phytochemicals present in medicinal plants.

Medicinal plants	Phytochemicals	References
Cucurbita maxima	Alcohol, amines, and carbonyl groups of phytochemicals	Aktepe and Baran [52]
Oil palm kernels	Phenols, flavonoids, carboxylic acids, and proteins	Irfan et al. [53]
Chromolaena odorata Manihot esculenta	Glycerin and beta-sitosterol	Essien et al. [54]
Simarouba glauca	Phenols	Sivaselvi et al. [144]
Capsicum chinense	Flavonoids and phenolic	Lomelí-Rosales et al. [55]
Hibiscus tiliaceus	Phenols and flavonoids	Viswanath Konduri et al. [56]
Hypericum perforatum	phenolic acids and flavonoids	Pradeep et al. [57]
Cattleya intermedia	Mangiferin	Aboyewa et al. [58]
Ocimum tenuiflorum	Flavonoids	Baruah et al. [35]
Tropaeolum majus	Tannins, alkaloids, terpenoids, flavonoids, and cardiac glycosides	Bawazeer et al. [59]
Sideritis argyrea, Sideritis brevidens, Sideritis lycia	Chlorogenic acid	Ceylan et al. [60]

Tab. 13.2 (continued)

Medicinal plants	Phytochemicals	References
Syzygium aromaticum	Polyphenolic compounds, flavonoids, hydroxybenzoic acid, hydroxycinnamic acid, hydroxyphenyl propane, and various terpenoids	Rodríguez-Jiménez et al. [61]
Capparis deciduas	Phenols, ascorbic acid, flavonoids, polyphenolic, citric acid, alkaloids,	Iqbal et al. [145]
Cucumis prophetarum	Tannins, alkaloids, triterpenoids, phenol and saponins	Hemlata et al. [25]
Chromolaena odorata	Glycerin, β-sitosterol	Essien et al. [62]
Allium Sativum	Triterpenoids, terpenoids, steroids, tannin, glycoside, Saponin and alkaloids	Velsankar et al. [63]
Myristica fragrans	Isoeugeol, caratol, methoxy eugenol, hydroxy benzofuranone, lauric acid, myristic acid, palmitic acid, oleic acid, stearic acid, stigmasterol and licarin a	Sasidharan et al. [64]
Aloe barbadensis	Flavanol, quercetin, Kaempeferol, myricetin and fisetin	Jeevanandam et al. [65]

deionized water (DW) had been mixed and boiled. The procured plant extract had been then stored in a refrigerator at + 4 °C after cooling at room temperature and filtering for use in scientific investigations. Alcohol, amines, and carbonyl groups are detected phytochemicals that are responsible for bioreduction. 53, demonstrated the plant-mediated fabrication of AuNPs using aqueous extract of oil palm kernels, which was discovered to be a nontoxic, simple, and environmentally friendly procedure. Powdered oil palm kernel [OPK] had been stored in an airtight vial for future use. To make the OPK extract, 10 g of finely ground OPK had been mixed with 100 mL of DW and heated at 90 °C for 20 min. The resulting mixture had been filtered utilizing Whatman No. 1 filter paper via gravitational filtration. OPK aqueous extract had been maintained in a glass vial and refrigerated. The availability of phytochemicals in the extract of OPK, such as phenols, flavonoids, carboxylic acids, and proteins, works well to predict the biosynthetic reaction, resulting in stable AuNPs. Essien et al. [54] found and support the biological formulation of ZnONPs utilizing aqueous leaf extracts of *Chromolaena odorata* and *Manihot esculenta*. Each powdered sample (24 g) had been mixed with 250 mL distilled water individually, sealed, and heated in a water bath for 15 min. The mixtures had been then cooled to room temperature and filtered to produce aqueous extracts, whereby it has been then retained in the refrigerator at 5–10 °C for future use. Both extracts contained considerable quantities of two phytochemicals, glycerin and beta-sitosterol, which may have played competitive dominant functions in the creation of the ZnONPs. Sivaselvi et al.

(2022) [144] used phytochemicals present on *Simarouba glauca* for the synthesis of AgNPs. Approximately 1 kg of *Simarouba glauca* leaves had been washed, dried, and ground thoroughly for further processing. It had been then sieved and transported for further processing. For the maceration extraction, 5 × 200 g of ground powder had been weighed and treated with chosen solvents such as ethanol (C_2H_5OH), methanol (CH_3OH), ethyl acetate ($CH_3-COOCH_3$), chloroform ($CHCl_3$), and water (H_2O). The solvents containing the leaf were stored for four days before being filtered. This study indicated that phytochemicals with high phenolic group concentrations perform an important role in nanoparticle yield. Lomelí-Rosales et al. [55] studied the biological synthesis of gold (AuNPs] and silver (AgNPs) nanoparticles using aqueous extracts of *Capsicum chinense* root, stem, and leaf. For extract preparation, 18.04 g, 3.33 g, and 24.3 g of raw sliced stems, roots, and leaves, respectively, have been infused with 30 mL of double-distilled water (ddH_2O) over a heat plate at 80 °C for 20 min. Following that, aqueous extracts had been gravity filtered through a Whatman general-purpose filter paper. After centrifuging the filtered extract at 4,000 rpm for 15 min at room temperature, the solution had been filtered through a 0.45 mm nylon Acro disc filter. Filtered stem and root extracts had been diluted to an ultimate volume of 50 mL with ddH_2O, while filtered leaf extracts had been diluted to a final volume of 250 mL with ddH_2O. In the study, it is hypothesized that the flavonoids and phenolic chemicals found in the *Gnidia glauca* extracts contribute to the creation and stabilization of the AuNPs. Viswanath Konduri et al. [56] targeted this study to make ZnONPs from an aqueous leaf extract of *Hibiscus tiliaceus*. *H. tiliaceus* aqueous leaf extract has been made by boiling 6 g of leaf powder in 100 mL of deionized water for 15 min at 80 °C. The extract had been then filtered through Whatman no.1 filter paper and preserved at 4 °C in the fridge for future studies. The existence of phenols and flavonoids in the phytochemical analysis of *Hibiscus tiliaceus* disclosed that they are important contributors to ZnONPs reduction and also capping agents. Pradeep et al. [57] used *Hypericum perforatum* extracts for the synthesis of AgNPs. In short, 2 grams of powdered biomass had been incorporated into an Erlenmeyer flask incorporating 20 mL of 80% methanol and thoroughly mixed in an incubator shaker at 150 rpm for 1 h at 25 °C. The combination had been then sonicated for 10 min in an ultrasound water bath and the homogenate had been centrifuged at 15,000 g for 30 min. The supernatant had been dried in a rotary evaporator under vacuum at 35 °C, and the subsequent residue had been dissolved in 100 mL of ultrapure water to achieve a turbid % *H. perforatum* aqueous methanolic extract. Among the phytochemicals found in the extract, phenolic acids and flavonoids were found to participate in reduction of Ag^+ ions, so although xanthones and phloroglucinols behave as capping agents and naphthodianthrones have been associated in both steps. Aboyewa et al. [58] demonstrate for perhaps the first time the formulation of AuNPs from a *Cattleya intermedia* extract. In double distilled water, standard solutions of dried plant materials have been ready. The plant extract was made by heating 100 g of plant leaves in 1,000 ml of deionized water and leaving it on a magnetic stirrer without heat for 24 h. Following that, the potion was filtered across Whatman No 1 and 0.45 m filters. The supernatant that resulted was freeze-dried.

The dried extracts have been stored in sterile sealed containers at 4 °C for future use. The plant extracts containing high amounts of mangiferin were utilized to synthesize AuNPs. Baruah et al. [35] made tulsi leave extract for the synthesis of nanoparticles. In 100 ml deionized water, 5 g of dried tulsi leaves powder had been introduced. For almost 60 min, the formulation had been stirred on a hot magnetic stirrer set to 60 °C. The resulting solution had been kept at room temperature, filtered through Whatman filter paper 1, and a pale-yellow extract had been achieved. The findings from this study show that tulsi extract as well as its constituent flavonoids have the capability to reinforce as a reducing and capping agent in the fabrication of AgNPs. Bawazeer et al. [59] prepared AgNPs using *Tropaeolum majus* extract. Tropaeolum majus dried powder was immersed in methanol for three days and tested for simple cold extraction unless plant materials have been exhausted. The extracted material had been then concentrated utilizing a rotary evaporator at low pressure and temperature. A similar protocol had been conducted for aqueous extract by soaking the powder material in water. Tannins, alkaloids, terpenoids, flavonoids, and cardiac glycosides were found in methanolic and aqueous extracts. These components might be actively associated with silver and gold reduction. Ceylan et al. [60] took plants from *Sideritis* species (*S. argyrea*, *S. brevidens*, and *S. lycia*) for the synthesis of nanoparticles. The aerial part of plants had been washed thrice with deionized water, dried in the shade, and ground in a grinder mill. For every plant material, 5 g of powdered substance had been boiled (infusion) in 100 ml deionized water for 15–20 min. The water extracts have been filtered using Whatman filter paper No.1 and preserved at 4 °C before further usage. Chlorogenic acid was the major constituent present in plant extract which might responsible for the synthesis of silver nanoparticles. Rodríguez-Jiménez et al. [61] synthesized TiO_2 nanoparticles form *Syzygium aromaticum* extract. Polyphenolic compounds, flavonoids, hydroxybenzoic acid, hydroxycinnamic acid, hydroxyphenyl propane, and various terpenoids are all found in *S. aromaticum*. Eugenol is perhaps the most plentiful phenolic derivative (4-allyl-2-methoxyphenol), accounting for nearly 72–90% of the plant's phytochemicals. Eugenol predicated on its capacity to trap free radicals, might indeed act as a reducing agent due because of its powerful antioxidant and reducing activity. Furthermore, because eugenol does have the capacity to interact with metal ions, this could act as a natural capping agent to stabilize metal nanoparticles. Iqbal et al. [146] investigated *Capparis deciduas* leaf extract for the synthesis of CuO and NiO NPs. Phytochemicals found in *Capparis decidua* leaf extracts include phenols, ascorbic acid, flavonoids, polyphenolic, citric acid, alkaloids, and others. These bioactive compounds act as reducing and stabilizing agents, aiding in the reduction of metal ion substances. Hemlata et al. [25] describe a low-cost formulation of silver nanoparticles utilizing *Cucumis prophetarum* aqueous leaf extract. The dried leaves had been ground into a fine powder utilizing an electric lab blender. 250 mL of deionized water had been introduced to 25 g of leaf powder, and the combination had been heated at 80 °C for 3 h with constant stirring before being filtered through Whatman filter paper no. 1. The extract had been kept at 4 °C until it was needed. The plant extract contained tannins, alkaloids, triterpenoids, phenol, and saponins. The above phytochemicals might well be involved in reducing the amount of silver and serving as a capping agent to keep the NPs from aggregating and to provide them stability. Essien et al. [62] studied magnesium oxide nanoparticles (MgONPs) utilizing a leaf extract of *Chromolaena odorata*. The plant extract contains high amount of glycerin, 1-heptatriacotanol, β-sitosterol, 17-(1,5-dimethylhexyl)-2-hydroxyl -10,13-d, 7-oxabicyclo [4.1.0] heptan-2-ol, 1-naphthalenepropanol, 1,6-dideoxydulcitol, 2H-3,9a-methano-1-benzoxepin and Kauran-18-al. These all molecules obtained from plants are made up of the different types of carbon-based functional groups, namely, C-O, C = C, CC, C-O-C, and C = O, which are recognized to serve as reductants. Velsankar et al. [63] investigated phytochemicals of *Allium Sativum* for the synthesis of NPs. Triterpenoids (18.87%), terpenoids (14.36%), steroids (16.56%), tannin (18.28%), and glycoside (16.98%) are all present in high concentrations in this extractive value. Saponin (5.88%) and alkaloids (9.07%) are present in low concentrations. These phytochemicals are extremely toxic to the pathogens that cause severe diseases in humans. The biosynthesis of AgNPs was carried out in an aqueous medium using *Allium Sativum* fewer extract. Sasidharan et al. [64] reported *Myristica fragrans* fruit extract for the synthesis of copper oxide and silver nanoparticles. The plant extract contains isoeugeol (RT 5.8), caratol (RT 6.2), methoxy eugenol (RT 6.8), hydroxy benzofuranone (RT 8.6), lauric acid (RT 6.5), myristic acid (RT 8.04), palmitic acid (RT 9.8), oleic acid (RT 11.5), stearic acid (RT 11.6), stigmasterol (RT 20.1) and licarin a (RT 15.1). These phytochemicals were involved in the reduction and stabilization of nanoparticles. Jeevanandam et al. [65] studied *Aloe barbadensis* for the synthesis of nanoparticles. In a 250 ml conical flask, 20 g of latex powder was mixed with 100 ml of distilled water and 100 ml of ethanol. The mixture in the conical flask was constantly stirred (450 rpm) at room temperature for 24 h to completely elucidate the active materials and solubilize them in the liquid. The latex powder extract is then vacuum filtered, and the filtrate as well as the residues is collected. A liquid-liquid extract method is used to clean the filtrate (aqueous state) and the collected residues (solid-state). MgONPs were created using solid waste and aqueous filtrate, which included phytochemicals devoid of ethanol.

13.4 Factors influencing the synthesis of NMs

The biosynthesis and shape of NMs are influenced by a variety of factors that ultimately causes the variations in the size, efficiency and other properties of NMs. The regulation of crystalline structure, shape, size, and size distribution are the main issues in NMs' biosynthesis. The main factors affecting these features are pH, time, metal ion concentration, substrate concentration, etc. Researchers have connected these variations to the choice of adsorbent, i.e., plant and its constituents, and the catalyst used in the synthesis process.

13.4.1 pH

The size, shape, and synthesis of the plant-mediated NMs are significantly influenced by the pH of the reaction mixture [12]. Therefore, by adjusting the pH of the reaction media, NMs' size can be adjusted. Traiwatcharanon et al. [66] showed the impact of pH on the size and structure of the silver nanoparticles (AgNPs) synthesized from *Pistia stratiotes* extracts. Studies have revealed that the pH of a reaction solution significantly affects the synthesis process of NMs [67–70]. Changes in the pH of the reaction may have an impact on the shape and size of the NMs synthesized [71]. Larger particles are formed when comparing lower acidic pH values to higher acidic pH values [72, 73]. Akhtar et al. [74] reported that spherical particles of gold nanoparticles (AuNPs) with a diameter of 10–20 nm were formed at a pH rise of 7, while the decline in pH of the reaction to 4 causes the formation of nano-plates. In the process of AgNP synthesis, silver from $AgNO^3$ interacts with amino and sulfhydryl groups present in plant extracts at low pH levels, and these positive ions turn Ag^+ into Ag^0. Due to the presence of positively charged functional groups, ionic bonds and biomolecules – which were encouraged at lower pH levels – were mostly used to complete the reduction [11]. A huge number of biomolecules bind to the AgNPs, causing them to aggregate and grow in size. Small AgNPs were seen in the reaction mixture at pH 8 [75]. Another researcher showed that the size, reaction rate, and shape of PtNPs were all affected by the pH of the plant extract [76].

13.4.2 Temperature

Temperature is a significant element that affects the synthesis of NMs. The physical process requires a maximum temperature (>350 °C); however, the chemical method requires a lesser temperature. Generally, temperatures below 100 °C are needed for the green synthesis NMs. Asimuddin et al. [77] reported that an increase in temperature from 10 to 50 °C during AgNPs synthesis from *Azadirachta indica* leaves extract causes an increase in the rate of silver ions reduction. The researchers hypothesized that high temperatures may have had an impact on an essential enzyme involved in the production of NMs. But, the study's results revealed that temperatures above 60 °C favored the formation of larger-sized particles. This result is explained by the observation that quick Ag+ reduction at high temperatures, which promotes reduction and nucleation, occurs at the expense of secondary reduction on developing particle surfaces. Al-Radadi [76] showed that the average particle size of PtNPs was 3.4 nm at 20 °C and 2.6 nm at 30 °C. In the study, high temperatures were thought to change the interactions of plant macromolecules and prevent the coalescence of adjacent NPs.

13.4.3 Metal ion concentration

The amount, shape, and size of NMs are greatly affected by the presence of precursor metal ions. The majority of plant-mediated AgNP production processes used AgNO$_3$, and its concentration significantly affected the NP's particle sizes. For instance, the highest yield of larger-sized AgNPs was reported at concentrations of up to 1.25 mM of AgNO$_3$, while smaller-sized AgNPs were produced at higher concentrations of AgNO$_3$ [11, 51].

13.4.4 Substrate concentration (plant extract)

Both the reduction and the stability of the NMs depend on the biomolecules that are present in the extracts. It is well recognized that adding more biomolecules to the reaction can increase the synthesis of NMs, increase their size, and alter their structural properties for the best results. When the leaf extract concentration was raised, the absorbance spectra shrank, indicating that the NM's size had decreased. According to the Habeeb Rahuman et al. [11] study, the AgNPs synthesis reaction occurs promptly at 1.25 ml of leaf extract concentration with the highest yield.

13.4.5 Pressure

For NMs' production, pressure is crucial. The size and shape of the synthesized NMs are influenced by the pressure that is applied to the reaction media [78]. At ambient pressure circumstances, it has been discovered that biological agents significantly speed up the rate of reduction of metal ions [79].

13.4.6 Other factors

Secondary metabolites, which serve as reducing and stabilizing agents for the synthesis of NMs, are abundant in medicinal plants and the composition of these metabolites varies depending on the type of plant, plant part, and the procedure used for its extraction [80]. Additionally, the choice of purification technique can affect the amount and quality of NMs. Centrifugation is sometimes used to separate the NMs depending on the effects of gravity [81]. In other instances, chromatographic methods are employed to separate NPs based on variations in the coefficients of the mobile phase and stationary phase [82]. The extraction of NMs based on their solubility in two different miscible liquid phases (often water and organic solvents), followed by electrophoresis or chromatography, is one method used to efficiently separate the NMs [83, 84].

13.5 Applications of NMs

Nanotechnology is a significant area of modern research that deals with the synthesis, strategy, and manipulation of NMs and deal with its application in a wide range of fields, namely, health care, agriculture, food industries, drug-gene delivery, and the environment [3, 5]. NMs have a variety of applications such as antimicrobial activity, antioxidative capability, anticancer effect, antidiabetic properties, plant growth enhancement, dye degradation effects, and anti-larval properties [8, 11]. In this chapter we are discussing a few important applications of NMs synthesized by using medicinal plants as follows:

13.5.1 Agriculture

In agriculture sectors, pests and diseases are controlled by using chemicals like insecticides, fungicides, and herbicides that have adverse impacts on human health, pollinating insects and animals, and ecosystem-altering consequences. NM-based chemicals may be a suitable solution to all these problems [85]. Nanomaterials are employed as a pesticide or medication carrier and their pharmacokinetic parameters may shift to kinetic drug release features that lead to prolonged drug release, which eliminates the need for frequent drug administration [86, 87].

13.5.2 Food industry

Oxygen tends to act as one of the key factors that can lead to food spoilage and discoloration by oxidizing the food materials. The development of nano-based plastics for food packaging is one of the primary uses of nanotechnology in the food industries. NPs are spiked in the plastics and serve as a wall that inhibits the diffusion of oxygen in packaged foods. The nano-based plastics are also used to wrap fruits that prevent them from losing weight and shrinking [88, 89]. Ethylene, produced by ripening fruits, is absorbed by nanomaterials and extends the shelf life of foods [87, 90].

13.5.3 Drug delivery

Metal NMs are valuable in cancer treatment, gene therapy and diagnostics because their surfaces can be easily changed for additional targeting, enabling them to transport medications and genetic material to specific regions [91]. The combination of chemical components comprising AgNPs was successfully chosen for the acquisition of unique and effective drug delivery systems to target inflammation, highly infectious, and practical properties for nanoscale-derived healthcare settings [92, 93]. The nano-based technology used for drug administration also offers adjustable drug concentration and discharge properties that are appropriate. Additionally, it ought to maximize therapeutic effectiveness at lower doses than the primary components while minimizing the negative effects [94]. AgNPs have attracted a lot of attention and have been found to be effective antitumor drug delivery vehicles, acting as active or passive nano-carriers for anticancer medications [95, 96]. Several other NPs are also used in drug delivery such as AuNPs, iron oxide, silicon, and graphene [91, 96–98].

13.5.4 Environmental application

The pollutants from the textile industry, leather industry, steel industry, and fertilizer industry are the most important sources of environmental pollution. They produce undesirable turbidity in the water, soil, and air, which will reduce sunlight penetration, and this leads to the resistance of photochemical synthesis and biological attacks on aquatic and marine life as well as in a land ecosystem [99]. Metal NPs like ZnO, TiO2, SnO2, WO3, and CuO have been applied preferentially to remove the contamination from aquatic systems [8, 100]. Kolya et al. [101] reported the high efficiency of AgNPs synthesized by *Amaranthus gangeticus* for degradation of dye pollution in water bodies. AgNPs have also shown great potential as excellent antimicrobial agents for water purification [8].

13.5.5 Biomedical application

Due to their numerous uses in industries, electronics, the environment, energy, and more specifically in the biomedical fields, NPs are currently in high demand commercially. NPs, like the most well-known Ag and Au NPs, have been extensively researched in this field and hold great promise for biological applications [102, 103]. Currently, NPs are widely used in nanomedicine and human health protection as it has great potential for antimicrobial, antiparasitic, antiproliferative, pro-apoptotic, pro- or antioxidative, and anti-inflammatory activities [104–108].

13.5.5.1 Antibacterial activity

Wehling et al. [109] investigated the antibacterial activity of NPs and reported that they can form covalent connections with nearby proteins and various groups on cell walls and membranes of the bacterial cell surface with different reactive groups and facilitate the antibacterial activity. Combining with intracellular components may further inhibit important enzymes and proteins, disrupting bacterial metabolism, and ultimately leading to cell death [110]. Studies have shown that many NPs can prevent or overcome biofilm formation, including Au-based NPs [111–113] Ag-based NPs [29, 114], Mg-based NPs [115, 116], ZnO NPs [117, 118], CuO NPs [119, 120] and Fe_3O_4 NPs [116, 121].

13.5.5.2 Antifungal activity

The size distribution, shape, composition, crystallinity, agglomeration, and surface chemistry of the NPs are a few characteristics that affect their antifungal efficacy [122–124]. It is generally known that the synthesis routes can change and regulate the aforementioned parameters [125–127]. AgNPs currently play a crucial role in antifungal medications used to treat a variety of infections caused by fungi [11]. Numerous studies have been done on the use of metal NPs to control pathogenic fungi [128, 129]. Various NPs have been applied to combat pathogenic fungus thus far. For instance, outstanding antifungal capabilities have been demonstrated by Ag, Cu, Fe, Zn, Se, Ni, and Pd [122, 125, 130].

13.5.5.3 Antiviral property

The viral diseases that cause today's outbreaks are spreading fear and destroying the planet throughout humanity. Medicinal plants have great antiviral capabilities against viruses including dengue, herpes, HIV, influenza, Ebola, etc. [11]. AgNP is considered to be a powerful and powerful drug with antiviral activity [131, 132]. The interaction of NPs with virion surfaces and virus cores is involved in the viral replication mechanism. NPs bind to viral glycoproteins and then attach to the genome and cellular particles to block the viral replication process [133]. There are several articles that reported the antiviral properties of different types of NPs such as AgNPs [134, 135], AuNPs [136], graphene oxide [137, 138], and mesoporous silicon [139, 140].

13.5.5.4 Cancer therapy

NMs offer improved gene silencing, targeting, and drug delivery effectiveness in tumor cells; metal NPs have become an advantageous technique for cancer therapy. In addition to these therapeutic benefits, the metal NPs have also been used as a cancer cell detection tool. According to recent research, vascular endothelial growth factor (VEGF) antibodies with AuNPs and AgNPs attached can be used to treat B-chronic lymphocytic leukemia [31, 141]. The metal NP's tiny size makes it possible for them to deliver therapeutic agents to malignant cells [142]. Metal NPs can be used in a variety of formulations since they are very physiochemically stable and have no known technological limitations [143]. Varieties of NPs were used in cancer therapy like AgNPs, AuNPs, ZnO NPs, TiO_2 NPs, PtNPs, MoS_2 NPs, CeO_2 NPs, and CeO_2 NPs [142].

13.6 Future perspectives and conclusion

The use of NMs has completely changed how human body damage is found and treated. It is well known all over the world due to its numerous applications. Exciting possibilities are presented by the use of NMs in medicine, which also helps nanorobots repair at the cellular level. This minimizes the damage intended to the body's healthy cells, making it feasible to identify diseases earlier. In-depth work is being conducted on NMs as nanostructures for new and better biomedical uses. In summary, medicinal plants have been utilized for healing since ancient times. We think that certain plant-based chemicals have therapeutic benefits. This chapter provides a dataset of the latest reports that will help researchers in their future studies on the synthesis of plant-based NMs for use in biomedical applications. It summarizes the currently known potential biosynthesis of NMs from medicinal plant extracts. Because of their antibacterial, antifungal, and antiviral qualities, NMs are frequently used in the medical industry. The toxicity of NMs should be understood before use, and this suggests understanding the appropriate concentration so that it can function well without harming people or the environment. Future applications for NMs in medicine will concentrate on fluorescence imaging and cancer therapies, where NPs' enhanced stability, higher bioavailability, and sustained drug release in the target tissues make them ideal for targeted drug delivery. However, more research is necessary to clarify the mechanism of plant-mediated NMs used in biomedical applications.

It is advantageous to synthesize NMs from medicinal plants because it is affordable, energy-efficient, and cheap products. By eliminating waste and promoting healthy products, it can safeguard both human health and the environment. The use of plant-synthesized NPs may be superior to other biological entities because it can get around the time-consuming process of using microorganisms and maintaining their culture, which may lose its potential in the process of nanoparticle biosynthesis. Plant-synthesized NPs have important aspects of nanotechnology through unparalleled applications. As a result, this chapter provides a variety of recently reported literature to illustrate the significance of plant-mediated NM synthesis.

References

[1]   P. Katiyar, B. Yadu, J. Korram, M. L. Satnami, M. Kumar, S. Keshavkant. J Environ Sci. 2020, 92, 18–27.
[2]   B. Yadu, V. Chandrakar, J. Korram, M. L. Satnami, M. Kumar, S. Keshavkant. J Hazard Mater. 2018, 353, 44–52.
[3]   S. Ahmed, M. Ahmad, B. L. Swami, S. Ikram. J Adv Res. 2016, 7(1), 17–28.
[4]   H. Hiba, J. E. Thoppil. Bull Natl Res Cent. 2022, 46(1), 1–15.
[5]   H. Chopra, S. Bibi, I. Singh, M. M. Hasan, M. S. Khan, Q. Yousafi, A. A. Baig, M. M. Rahman, F. Islam, T. B. Emran, S. Cavalu. Front Bioeng Biotechnol. 2022, 10, 874742.

[6]   V. Gopinath, D. MubarakAli, S. Priyadarshini, N. M. Priyadharsshini, N. Thajuddin, P. Velusamy. Colloids Surf B Biointerfaces. 2012, 96, 69–74.

[7]   K. Selvam, C. Sudhakar, M. Govarthanan, P. Thiyagarajan, A. Sengottaiyan, B. Senthilkumar, T. Selvankumar. J Radiat Res Appl Sci. 2017, 10, 6–12.

[8]   N. S. Alharbi, N. S. Alsubhi, A. I. Felimban. J Radiat Res Appl Sci. 2022, 15, 109–124.

[9]   J. Singh, T. Dutta, K. H. Kim, M. Rawat, P. Samddar, P. Kumar. J Nanobiotechnol. 2018, 16(1), 1–24.

[10] R. K. Das, V. L. Pachapur, L. Lonappan, M. Naghdi, R. Pulicharla, S. Maiti, M. Cledon, L. M. A. Dalila, S. J. Sarma, S. K. Brar. Nanotechnol Environ Eng. 2017, 2, 1–21.

[11] H. B. Habeeb Rahuman, R. Dhandapani, S. Narayanan, V. Palanivel, R. Paramasivam, R. Subbarayalu, S. Thangavelu, S. Muthupandian. IET Nanobiotechnol. 2022, 16, 115–144.

[12] J. K. Patra, K. Baek. J Nanomater. 2014, (2014), 1–12.

[13] N. Khatoon, J. A. Mazumder, M. Sardar. J Nanosci: Curr Res. 2017, 02, 107.

[14] L. Chetia, D. Kalita, G. A. Ahmed. Sens Bio-Sens Res. 2017, 16, 55–61.

[15] W. A. El-Said, H. Y. Cho, C. H. Yea, J. W. Choi. Adv Mater. 2014, 26, 910–918.

[16] S. Ganguly, S. Mondal, P. Das, P. Bhawal, T. K. Das, M. Bose, S. Choudhary, S. Gangopadhyay, A. K. Das, N. C. Das. Nano-Struct Nano-Objects. 2018, 16, 86–95.

[17] S. Mukherjee, S. K. Nethi. Nanotechnology for Agriculture: Advances for Sustainable Agriculture. 2019.

[18] F. Niknejad, M. Nabili, R. Daie Ghazvini, M. Moazeni. Curr Med Mycol. 2015, 1, 17–24.

[19] S. Rajeshkumar, C. Kannan, G. Annadurai. Int J Pharm Biol Sci. 2012, 3, 502–510.

[20] S. F. Ahmed, M. Mofijur, N. Rafa, A. T. Chowdhury, S. Chowdhury, M. Nahrin, A. S. Islam, H. C. Ong. Environ Res. 2022, 204, 111967.

[21] P. K. Dikshit, J. Kumar, A. K. Das, S. Sadhu, S. Sharma, S. Singh, P. K. Gupta, B. S. Kim. Catalysts. 2021, 11(8), 902.

[22] K. S. Siddiqi, A. Husen. Biomater Res. 2020, 24(1), 1–15.

[23] J. A. George, S. Sundar, K. A. Paari. Nanosci Nanotechnol Asia. 2021, 11(5), 77–98.

[24] F. Moradi, S. Sedaghat, O. Moradi, S. Arab-Salmanabad. Mater Res Express. 2019, 6, 125008.

[25] M. P. R. Hemlata, A. P. Singh, K. K. Tejavath. ACS Omega. 2020, 5(10), 5520–5528.

[26] M. Asif, R. Yasmin, R. Asif, A. Ambreen, M. Mustafa, S. Umbreen. Dose-Response. 2022, 20(1), 15593258221088709.

[27] S. Rashid, M. Azeem, S. A. Khan, M. M. Shah, R. Ahmad. Colloids Surf. 2019, 179, 317–325.

[28] C. G. Yuan, C. Huo, S. Yu, B. Gui. Physica E Low Dimens Syst Nanostruct. 2017, 85, 19–26.

[29] P. Singh, I. Mijakovic. Front Microbiol. 2022, 13, 820048.

[30] M. Akhbari, R. Hajiaghaee, R. Ghafarzadegan, S. Hament, M. Yaghoobi. IET Nanobiotechnol. 2019, 13, 160–169.

[31] S. Jain, N. Saxena, M. K. Sharma, S. Chatterjee. Mater Today: Proc. 2020, 31, 662–673.

[32] A. Chakravarty, I. Ahmad, P. Singh, M. U. D. Sheikh, G. Aalam, S. Sagadevan, S. Ikram. Chem Phys Lett. 2022, 795, 139493.

[33] N. T. L. Chi, M. Narayanan, A. Chinnathambi, C. Govindasamy, B. Subramani, K. Brindhadevi, S. Pikulkaew. Environ Res. 2022, 208, 112684.

[34] K. R. Shubhashree, R. Reddy, A. K. Gangula, G. S. Nagananda, P. K. Badiya, S. S. Ramamurthy, N. Reddy. Mater Chem Phys. 2022, 280, 125795.

[35] K. Baruah, M. Haque, L. Langbang, S. Das, K. Aguan, A. S. Roy. Ocimum sanctum mediated green synthesis of silver nanoparticles: A biophysical study towards lysozyme binding and anti-bacterial activity. J Mol Liq. 2021, 337, 116422.

[36] R. Pachaiappan, S. Rajendran, G. Ramalingam, D. V. N. Vo, P. M. Priya, M. Soto-Moscoso. Chem Eng Technol. 2021, 44(3), 551–558.

[37] M. Devi, S. Devi, V. Sharma, N. Rana, R. Bhatia, A. K. Bhatt. J Tradit Complement Med. 2020, 10(2), 158–165.

[38] B. M. Satyanarayana, N. V. Reddy, J. V. Rao. Appl Sci. 2020, 2(11), 1–15.

[39] M. Kaur. Bio Nano Sci. 2020, 10(1), 1–10.

[40] R. Vinayagam, R. Selvaraj, P. Arivalagan, T. Varadavenkatesan. J Photochem Photobiol B Biol. 2020, 203, 111760.

[41] B. A. Fahimmunisha, R. Ishwarya, M. S. AlSalhi, S. Devanesan, M. Govindarajan, B. Vaseeharan. J Drug Delivery Sci Technol. 2020, 55, 101465.

[42] C. Chandra, F. Khan. J Radioanal Nucl Chem. 2020, 324(2), 589–597.

[43] N. Rabiee, M. Bagherzadeh, M. Kiani, A. Ghadiri. Adv Powder Technol. 2020, 31(4), 1402–1411.

[44] V. D. Doan, A. T. Thieu, T. D. Nguyen, V. C. Nguyen, X. T. Cao, T. L. H. Nguyen, V. T. Le. J Nanomater. 2020, 2020.

[45] P. Kharey, S. B. Dutta, A. Gorey. Chemis Select. 2020, 5(26), 7901–7908.

[46] D. Rueda, V. Arias, Y. Zhang, A. Cabot, A. C. Agudelo, D. Cadavid. Environ Nanotechnol Monit Manag. 2020, 13, 100285.

[47] T. Liang, X. Qiu, X. Ye, Y. Liu, Z. Li, B. Tian, D. Yan. Biotechnology. 2020, 10(1), 1–6.

[48] A. Aygun, F. Gülbagca, L. Y. Ozer, B. Ustaoglu, Y. C. Altunoglu, M. C. Baloglu, F. Sen. J Pharm Biomed Anal. 2020, 179, 112961.

[49] S. Sharma, K. Kumar, N. Thakur, S. Chauhan, M. S. Chauhan. J Environ Chem Eng. 2021, 9(4), 105395.

[50] J. F. Li, Y. Liu, M. Chokkalingam, E. J. Rupa, R. Mathiyalagan, J. Hurh, J. C. Ahn, J. K. Park, P. J. Yu, Y. Dc. Optik. 2020, 208, 164521.

[51] W. A. Shaikh, S. Chakraborty, R. U. Islam. Int J Environ Sci Technol. 2020, 17, 2059–2072.

[52] N. Aktepe, A. Baran. Medicine. 2022, 11(2), 794–799.

[53] M. Irfan, M. Moniruzzaman, T. Ahmad, M. F. R. Samsudin, F. Bashir, M. T. Butt, H. Ashraf. Inorg Nano-Metal Chem. 2022, 52(4), 519–532.

[54] E. R. Essien, V. N. Atasie, D. O. Nwude, E. Adekolurejo, F. T. Owoeye. South Afr J Sci. 2022, 118(1–2), 1–8.

[55] D. A. Lomelí-Rosales, A. Zamudio-Ojeda, O. K. Reyes-Maldonado, M. E. López-Reyes, G. C. Basulto-Padilla, E. J. Lopez-Naranjo, G. Velázquez-Juárez. Molecules. 2022, 27(5), 1692.

[56] V. Viswanath Konduri, N. Kumar Kalagatur, A. Nagaraj, V. Rao Kalagadda, U. Kiranmayi Mangamuri, C. Potla Durthi, S. Poda. Indian J Biochem Biophys. 2022, 59(5), 565–574.

[57] M. Pradeep, D. Kruszka, P. Kachlicki, D. Mondal, G. Franklin. ACS Sustain Chem Eng. 2021, 10(1), 562–571.

[58] J. A. Aboyewa, N. R. Sibuyi, M. Meyer, O. O. Oguntibeju. Nanomaterials. 2021, 11(1), 132.

[59] S. Bawazeer, A. Rauf, S. U. A. Shah, A. M. Shawky, Y. S. Al-Awthan, O. S. Bahattab, M. A. El-Esawi. Green Process Synth. 2021, 10(1), 85–94.

[60] R. Ceylan, A. Demirbas, I. Ocsoy, A. Aktumsek. Sustain Chem Pharm. 2021, 21, 100426.

[61] R. A. Rodríguez-Jiménez, Y. Panecatl-Bernal, J. Carrillo-López, M. Má, A. Romero-López, M. Pacio-Castillo, J. Alvarado. Chemis Select. 2021, 6(16), 3958–3968.

[62] E. R. Essien, V. N. Atasie, T. O. Oyebanji, D. O. Nwude. Chem Pap. 2020, 74(7), 2101–2109.

[63] K. Velsankar, R. Preethi, P. S. Ram, M. Ramesh, S. Sudhahar. Appl Nanosci. 2020, 10(9), 3675–3691.

[64] D. Sasidharan, T. R. Namitha, S. P. Johnson, V. Jose, P. Mathew. Sustain Chem Pharm. 2020, 16, 100255.

[65] J. Jeevanandam, Y. San Chan, Y. J. Wong, Y. S. Hii. IOP Conf Ser Mater Sci Eng IOP Publ. 2020, 943(1), 012030.

[66] P. Traiwatcharanon, K. Timsorn, C. Wongchoosuk. Advanced Materials Research. Trans Tech Publications Ltd. Vol. 1131, 2016, pp. 223–226.

[67] S. DV, K. T. RamyaDevi, N. Jayakumar, E. Sundaravadivel. AIP Adv. 2020, 10(9), 095119.

[68] I. Fernando, Y. Zhou. Chemosphere. 2019, 216, 297–305.

[69] N. Soni, S. Prakash. Am J Nanotechnol. 2011, 2, 112–121.

[70] S. Yazdani, A. Daneshkhah, A. Diwate, H. Patel, J. Smith, O. Reul, R. Cheng, A. Izadian, A. R. Hajrasouliha. ACS Omega. 2021, 6(26), 16847–16853.
[71] L. A. Gontijo, E. Raphael, D. P. Ferrari, J. L. Ferrari, J. P. Lyon, M. A. Schiavon. Matéria (Rio de Janeiro). 2020, 25.
[72] S. Mourdikoudis, R. M. Pallares, N. T. Thanh. Nanoscale. 2018, 10(27), 12871–12934.
[73] T. Stepan, L. Tété, L. Laundry-Mottiar, E. Romanovskaia, Y. S. Hedberg, H. Danninger, M. Auinger. Electrochimica Acta. 2022, 409, 139923.
[74] M. S. Akhtar, J. Panwan, Y. S. Yun. ACS Sustain Chem Eng. 2013, 1, 591–602.
[75] E. O. Nyakundi, M. N. Padmanabhan. Spectrochimica Acta Electronica Part A. 2015, 149, 978–984.
[76] N. S. Al-Radadi. Arab J Chem. 2019, 12, 330–349.
[77] M. Asimuddi, M. R. Shaik, S. F. Adil, M. R. H. Siddiqui, A. Alwarthan, K. Jamil, M. Khan. J King Saud Univ Sci. 2020, 32(1), 648–656.
[78] B. D. P. Abhilash, B. D. Pandey. IET Nanobiotechnol. 2012, 6, 144–148.
[79] Q. H. Tran, V. Q. Nguyen, A. T. Le. Adv Nat Sci. 2013, 4, 033001.
[80] Y. Park, Y. N. Hong, A. Weyers, Y. S. Kim, R. J. Linhardt. IET Nanobiotechnol. 2011, 69–78.
[81] S. Baker, D. Rakshith, K. S. Kavitha, P. Santosh, H. U. Kavitha, Y. Rao, S. Satish. Bio Impacts. 2013, 3, 111–117.
[82] V. L. Jimenez, M. C. Leopold, C. Mazzitelli, J. W. Jorgenson, R. W. Murray. Anal Chem. 2003, 75, 199–206.
[83] M. Hanauer, S. Pierrat, I. Zins, A. Lotz, C. Sönnichsen. Nano Lett. 2007, 7, 2881–2885.
[84] B. Kowalczyk, I. Lagzi, B. A. Grzybowski. Curr Opin Colloid Interface Sci. 2011, 16, 135–148.
[85] C. Pandit, A. Roy, S. Ghotekar, A. Khusro, M. N. Islam, T. B. Emran, S. E. Lam, M. U. Khandaker, D. A. Bradley. J King Saud Univ Sci. 2022, 34, 101869.
[86] I. Hussain, N. B. Singh, A. Singh, H. Singh, S. C. Singh. Biotechnol Lett. 2016, 38, 545–560.
[87] N. Khan, S. Ali, S. Latif, A. Mehmood. Nat Sci. 2022, 14, 226–234.
[88] N. Pantidos, L. E. Horsfall. J Nanomed Nanotechnol. 2014, 5, 233.
[89] P. Velusamy, G. V. Kumar, V. Jeyanthi, J. Das, R. Pachaiappan. Toxicol Res. 2016, 32, 95–102.
[90] I. Maliszewska, K. Szewczyk, K. Waszak. J Phys Conf Ser. 2009, 146, 012025.
[91] M. J. Mitchell, M. M. Billingsley, R. M. Haley, M. E. Wechsler, N. A. Peppas, R. Langer. Nat Rev Drug Discov. 2021, 20(2), 101–124.
[92] K. J. Prashob Peter. Int J Curr Pharm Rev Res. 2017, 9, 1–5.
[93] Y. Yao, Y. Zhou, L. Liu, Y. Xu, Q. Chen, Y. Wang, S. Wu, Y. Deng, J. Zhang, A. Shao. Front Mol Biosci. 2020, 7, 193.
[94] B. Kumar, K. Jalodia, P. Kumar, H. K. Gautam. J Drug Delivery Sci Technol. 2017, 41, 260–268.
[95] G. Bagherzade, M. M. Tavakoli, M. H. Namaei. Asian Pac J Trop Biomed. 2017, 7, 227–233.
[96] Y. Deng, X. Zhang, H. Shen, Q. He, Z. Wu, W. Liao, M. Yuan. Front Bioeng Biotechnol. 2020, 7, 489.
[97] K. W. Huang, F. F. Hsu, J. T. Qiu, G. J. Chern, Y. A. Lee, C. C. Chang, Y. T. Huang, Y. C. Sung, C. C. Chiang, R. L. Huang, C. C. Lin. Sci Adv. 2020, 6(3), 5032.
[98] M. Khafaji, M. Zamani, M. Golizadeh, O. Bavi. Biophys Rev. 2019, 11, 335–352.
[99] K. Jyoti, A. Singh. J Genet Eng Biotechnol. 2016, 14, 311–317.
[100] K. Thandapani, M. Kathiravan, E. Namasivayam. Environ Sci Pollut Res. 2017, 25, 1–12.
[101] H. Kolya, P. Maiti, A. Pandey, T. Tripathy. J Anal Sci Technol. 2015, 6, 1–7.
[102] R. Gul, H. Jan, G. Lalay, A. Andleeb, H. Usman, R. Zainab, Z. Qamar, C. Hano, B. H. Abbasi. Coatings. 2021, 11, 717.
[103] C. Hano, B. H. Abbasi. Biomolecules. 2021, 12, 31.
[104] A. Andleeb, A. Andleeb, S. Asghar, G. Zaman, M. Tariq, A. Mehmood, M. Nadeem, C. Hano, J. M. Lorenzo, B. H. Abbasi. Cancers. 2021, 13, 2818.
[105] S. Anjum, I. Anjum, C. Hano, S. Kousar. RSC Adv. 2019, 9, 40404–40423.

[106] S. Anjum, S. Ishaque, H. Fatima, W. Farooq, C. Hano, B. H. Abbasi, I. Anjum. Pharmaceuticals. 2021, 14, 707.

[107] M. Nadeem, R. Khan, K. Afridi, A. Nadhman, S. Ullah, S. Faisal, Z. U. I. Mabood, C. Hano, B. H. Abbasi. Int J Nanomed. 2020, 15, 5951–5961.

[108] K. Saleem, Z. Khursheed, C. Hano, I. Anjum, S. Anjum. Nanomaterials. 2019, 9, 1749.

[109] J. Wehling, R. Dringen, R. N. Zare, M. Maas, K. Rezwan. ACS Nano. 2014, 8, 6475–6483.

[110] L. Wang, C. Hu, L. Shao. Int J Nanomed. 2017, 12, 1227–1249.

[111] S. Sathiyaraj, G. Suriyakala, A. D. Dhanesh Gandhi, R. Babujanarthanam, K. S. Almaary, T. W. Chen, K. Kaviyarasu. J Infect Public Health. 2021, 14, 1842–1847.

[112] C. Su, K. Huang, H. H. Li, Y. G. Lu, D. L. Zheng. J Nanomater. 2020, (2020), 1–13.

[113] G. Suriyakala, S. Sathiyaraj, R. Babujanarthanam, K. M. Alarjani, D. S. Hussein, R. A. Rasheed, K. Kanimozhi. J King Saud Univ Sci. 2022, 34, 101830.

[114] A. B. A. Mohammed, A. Mohamed, E.-N. NEA, H. Mahrous, G. M. Nasr, A. Abdella, R. H. Ahmed, S. Irmak, M. S. A. Elsayed, S. Selim, A. Elkelish, D. H. M. Alkhalifah, W. N. Hozzein, A. S. Ali. J Nanomater. 2022, (2022), 1–15.

[115] T. Jin, Y. He. J Nanopart Res. 2011, 13, 6877–6885.

[116] N. T. T. Nguyen, L. M. Nguyen, T. T. Nguyen, T. T. Nguyen, D. T. C. Nguyen, T. V. Tran. Environmental Chemistry Letters. 2022, pp. 1–41.

[117] B. Alotaibi, W. A. Negm, E. Elekhnawy, T. A. El-Masry, M. E. Elharty, A. Saleh, D. H. Abdelkader, F. A. Mokhtar. Artif Cells Nanomed Biotechnol. 2022, 50, 96–106.

[118] N. Babayevska, P. Ł, I. Iatsunskyi, G. Nowaczyk, M. Jarek, E. Janiszewska, S. Jurga. Sci Rep. 2022, 12, 8148.

[119] H. T. Luong, C. X. Nguyen, T. T. Lam, T. H. Nguyen, Q. L. Dang, J. H. Lee, H. G. Hur, H. T. Nguyen, C. T. Ho. RSC Adv. 2022, 12, 4428–4436.

[120] K. Ssekatawa, D. K. Byarugaba, M. K. Angwe, E. M. Wampande, F. Ejobi, E. Nxumalo, M. Maaza, J. Sackey, J. B. Kirabira. Front Bioeng Biotechnol. 2022, 10, 820218.

[121] M. Caciandone, A. G. Niculescu, V. Grumezescu, A. C. Bîrcă, I. C. Ghica, V. Bş, O. Oprea, I. C. Nica, M. S. Stan, A. M. Holban, A. M. Grumezescu, I. Anghel, A. G. Anghel. Antibiotics. 2022, 11.

[122] A. R. Cruz-Luna, H. Cruz-Martínez, A. Vásquez-López, D. I. Medina. J Fungi. 2021, 7, 1033.

[123] R. C. Kasana, N. R. Panwar, R. K. Kaul, P. Kumar. Environ Chem Lett. 2017, 15, 233–240.

[124] J. R. Koduru, S. K. Kailasa, J. R. Bhamore, K. H. Kim, T. Dutta, K. Vellingiri. Adv Colloid Interface Sci. 2018, 256, 326–339.

[125] O. Burduniuc, A. C. Bostanaru, M. Mares, G. Biliuta, S. Coseri. Materials. 2021, 14, 7041.

[126] M. Rai, A. P. Ingle, P. Paralikar, N. Anasane, R. Gade, P. Ingle. Appl Microbiol Biotechnol. 2018, 102, 6827–6839.

[127] S. K. Srikar, D. D. Giri, D. B. Pal, P. K. Mishra, S. N. Upadhyay. Green Sustain Chem. 2016, 06, 34–56.

[128] M. D. Balakumaran, R. Ramachandran, P. Balashanmugam, S. Jagadeeswari, P. T. Kalaichelvan. Mater Technol. 2022, 37, 411–421.

[129] A. D. Mare, C. N. Ciurea, A. Man, M. Mareș, F. Toma, L. Berța, C. Tanase. Plants. 2021, 10, 2153.

[130] R. Sakthi Devi, A. Girigoswami, M. Siddharth, K. Girigoswami. Applied Biochemistry and Biotechnology. 2022, pp. 1–33.

[131] U. Ghosh, K. Sayef Ahammed, S. Mishra, A. Bhaumik. Chem Asian J. 2022, 17, e202101149.

[132] Q. He, J. Lu, N. Liu, W. Lu, Y. Li, C. Shang, X. Li, L. Hu, G. Jiang. Nanomaterials. 2022, 12, 990.

[133] N. Khandelwal, G. Kaur, K. K. Chaubey, P. Singh, S. Sharma, A. Tiwari, S. V. Singh, N. Kumar. Virus Res. 2014, 190, 1–7.

[134] O. Gherasim, A. M. Grumezescu, V. Grumezescu, F. Iordache, B. S. Vasile, A. M. Holban. Materials. 2020, 13, 1–15.

[135] K. M. S. D. B. Kulathunga, C. F. Yan, J. Bandara. Colloids Surf A Physicochem Eng Asp. 2020, 590, 124509.

[136] J. Kim, M. Yeom, T. Lee, H. O. Kim, W. Na, A. Kang, J. W. Lim, G. Park, C. Park, D. Song, S. Haam. J Nanobiotechnol. 2020, 18, 54.

[137] D. Iannazzo, A. Pistone, S. Ferro, L. De Luca, A. M. Monforte, R. Romeo, M. R. Buemi, C. Pannecouque. Bioconjug Chem. 2018, 29, 3084–3093.

[138] M. Yousaf, H. Huang, P. Li, C. Wang, Y. Yang. ACS Chem Neurosci. 2017, 8, 1368–1377.

[139] J. R. Compton, M. J. Mickey, X. Hu, J. J. Marugan, P. M. Legler. Biochemistry. 2017, 56, 6221–6230.

[140] A. E. LaBauve, T. E. Rinker, A. Noureddine, R. E. Serda, J. Y. Howe, M. B. Sherman, C. Amy Rasley, J. Brinker, D. Y. Sasaki, O. A. Negrete. Sci Rep. 2018, 8, 1–13.

[141] S. Gavas, S. Quazi, T. M. Karpiński. Nanoscale Res Lett. 2021, 16, 173.

[142] D. Mundekkad, W. C. Cho. Int J Mol Sci. 2022, 23, 1685.

[143] A. S. Nadukkandy, E. Ganjoo, A. Singh, L. D. Dinesh Kumar. Front Nanotechnol. 2022, 4, 39.

[144] D. Sivaselvi, N. Vijayakumar, R. Jayaprakash, V. Amalan, R. Rajeswari, M. R. Nagesh, G. Tailor, B. L. Yadav, J. Chaudhary, M. Joshi, C. Suvalka. Rasayan J. Chem. 2022, 15(2), 1166–1173.

[145] A. Iqbal, A. U. Haq, G. A. Cerrón-Calle, S. A. R. Naqvi, P. Westerhoff, S. Garcia-Segura. Catalysts. 2021a, 11(7), 806.

[146] J. Iqbal, B. A. Abbasi, T. Yaseen, S. A. Zahra, A. Shahbaz, S. A. Shah, P. Ahmad. Sci Rep. 2021b, 11(1), 1–13.

Shweta Singh Chauhan, Ravishankar Chauhan*,
Nagendra Kumar Chandrawanshi and Pramod Kumar Mahish*

Chapter 14
Bioactivity of nanoparticles synthesized from medicinal plants

Abstract: Progress in science is making life easy; nanoparticles are one of the modern achievements approaching social benefits in biomedical, agriculture, energy, industrial etc. Nano-drugs, nano-fertilizer, green synthesized nano-particles of antimicrobials, antioxidants and anticancerous agents are some examples. Conventionally, these are produced by chemicals; therefore, the products may be costly, limited, and nonenvironmentally friendly. Among the various alternatives nano particle-based molecules play an important role to overcome these problems. The natural content of the medicinal plants can be transferred as medicine by various methods like allopathic, Ayurvedic, homeopathic, and food. By the synthesis of nanoparticle natural content of medicinal plant combines with the metal ions of nanosize. The present chapter is focused on the bioactivity of nanoparticles synthesized from medicinal plants. The present chapter is definitely helpful in enriching the depth of knowledge among academician sand lay persons too.

14.1 Introduction

Nanoparticles belong to the small molecules of 1–100 nm but the term is also used for larger particles up to 1000 nm [1]. The nanoparticles may differ on the basis of optical characteristics, size, surface area, and some other chemical/physical properties. Obtaining high quality of nanoparticles with their renowned properties generally depends on the synthetic methods and reducing agents used during the preparation of the particles

*Corresponding author: Ravishankar Chauhan, Department of Botany, Pandit Ravishankar Tripathi Government College, Bhaiyathan 497231, Surajpur, Chhattisgarh, India; School of Studies in Biotechnology, Pt. Ravishankar Shukla University, Raipur 492010, Chhattisgarh, India,
e-mail: ravi18bt@gmail.com
*Corresponding author: Pramod Kumar Mahish, Department of Biotechnology and Research Centre, Government Digvijay (Autonomous) PG College, Rajnandgaon 491441, Chhattisgarh, India,
e-mail: drpramodkumarmahish@gmail.com
Shweta Singh Chauhan, Department of Biotechnology and Research Centre, Government Digvijay (Autonomous) PG College, Rajnandgaon 491441, Chhattisgarh, India
Nagendra Kumar Chandrawanshi, School of Studies in Biotechnology, Pt. Ravishankar Shukla University, Raipur 492010, Chhattisgarh, India

https://doi.org/10.1515/9783110791891-014

[2]. There are ample of methods applied for synthesizing nanoparticles such as chemical reduction, photochemical methods, microwave processing, thermal decomposition, laser ablation, gamma irradiation, and electron irradiation. Biological synthesis has advantages over the physical and chemical procedure in view of it being an alternate approach as well as being friendly to the environment [3]. The medicinal plants constitute rich natural compounds [4, 5], and these natural constituents support as reducing and stabilizing agents during biosynthesis of nanoparticle; therefore, further assistance is not required [6]. These natural compounds belong to the phytochemical class alkaloids, amines, peptides, amino acids, terpenes, steroids, saponin, flavonoids, tannins, phenylpropanoids, etc. [7]. The bio-production of nanoparticle with the aid of plant extract is becomes popular since few decades because the method of production is easy, requires less labor, and is cost-effective too [8]. Its application in the wide area is also a significant motive for considering plant extract–based nanoparticle among the scientific community [9]. There is enormous potential for the creation of NPs in biological entities. They are environmentally favorable and sustainable to reduce metal precursors via biogenesis to the appropriate NPs [10 and 11]. Green nanoparticles are also cost-effective [9], free of chemical contamination, and suitable for mass manufacturing [12]. The selection of plant extracts used to make NPs is dependent on the economic benefit of biological material [13]. Green synthesized nanoparticle from medicinal plant extract has been proven its application in various areas like medicine, agriculture, industrial, energy, waste treatment, and biosensor development, etc. Some of the major applications of nanoparticle synthesized from medicinal plants are described here.

14.2 Antibacterial activity

Plants are privileged to be used in the biosynthesis of AgNPs, which has a straightforward procedure, high efficiency, and quick response times [14]. The presence of many metabolites in plant extracts, such as ketones, aldehydes, phenols, amides, carboxylic acids, and proteins, etc., contributes to stabilize NPs significantly [15]. AgNPs antibacterial nature is still not fully understood and has not been fully elucidated. According to earlier research, it is possible that AgNPs commonly release silver ions (Ag+), which may be one of the processes underlying AgNPs antibacterial effect [16]. A diversity of plant and their parts were used to make AgNPs, which were then tested against a variety of pathogenic microorganisms, including multidrug-resistant bacteria. Antibiotic resistance is occurring across all microorganisms due to the forming of the layer over the surface of microorganisms. This results from a biofilm layer forming on top of the microorganisms. Understanding biofilm features can assist us in the treatment of infectious diseases brought on by microorganisms. Antibiotics cannot penetrate as deeply as nanoparticles, but by killing them, they aid in regulating microbial development. Many plant-based nanoparticles are very effective against different kind of bacteria. Silver nanoparticle synthesized from *Panaxginseng* active against *Staphylococcus aureus* and *Pseudomona-saeruginosa*; *Dioscoreabulbifera*, and *Droserabinate* AgNPs were active against *S. aureus* and *Plumbagozeylanica* active against *S .aureus, Acinetobacterbaumannii*, and *Escherichiacoli*. AgNP synthesized by *Carduuscrispus* exhibited inhibition on gram-negative bacteria *coli* [17]. Green particle of aqueous extract of *Berberis vulgaris, Brassica nigra, Capsella bursa-pastoris, Lavandula angustifolia, and Origanum vulgare* show antibacterial activity against *Listeriamonocytogenes, S. aureus, P. aeruginosa, E. coli*, and *Salmonella enterica* ser. *Typhimurium* [18]. AuNPs synthesized from *Punicagranatumf* ruits showed an excellent antibacterial activity against *Candida albicans, Aspergillus flavus, S. aureus, Salmonella typhi*, and *Vibrio cholerae* [19].

14.3 Antioxidant activity

Nuclear material, proteins, and enzymes can be damaged by oxidative stress, which is brought by endogenous and exogenous factors like smoking, ionizing radiation, pollution, organic solvents, pesticides, reactive oxygen species such as the hydroxyl radical, superoxide anion radical, superoxide anion radical, hydrogen peroxide, singlet oxygen, nitric oxide radical, and hypochlorite radical [20]. The extracts of plants are rich sources of biochemical that can be used as medicines likewise antioxidant as well as diet supplement. The components of natural product can serve as powerful reducing and capping agent, ensuring the stability of newly generated NPs. In addition to *Camellia sinensis* other plants such as *Ilex paraguariensis, Salvia officinalis, Tiliacorabarta, Levisticum officinale, Aegopodium podagraria, Urtica dioica*, and the marinealgae *Porphyrayezoensis* also exhibits antioxidant activity. However, *Camellia sinensis* exhibits highest antioxidant property [21]. Green nanoparticles have been found to have even more amazing antioxidant activity than nanoparticles made chemically, and they can be used in a wide range of biological applications [22].

14.4 Nano-fertilizer and pesticides

To achieve future food demand, it has been estimated that current crop production will need to grow by up to 70% [23]. Applying nitrogen nano-fertilizers to grasslands may be possible in new settings of escalating economic and environmental restrictions. By increasing the effectiveness of nitrogen availability to plants and lowering nitrogen losses to the environment, nitrogen nano-fertilizers are anticipated to boost NUE (nitrogen use efficiency) [24]. Several inorganic, organic, and composite nanomaterials have been studied on various plant species to determine how they might affect plant development, growth, and production [25]. In respect to the application of pesticides, nano-pesticides or nano plant-protection products represent a new technological advancement that may provide a number of advantages, such as improved efficacy, durability, and a reduction in the amounts of active components required [26].

14.5 Antidiabetic activity

It has been established that traditional medicines and green synthesis of AgNPs are reliable sources of treatments for diabetes mellitus [27]. Diabetes mellitus is a set of metabolic illnesses characterized by persistently elevated blood sugar levels. It has been established that inhibiting the carbohydrate-digesting enzymes (-glucosidase and -amylase) can treat hyperglycemia by blocking the conversion of carbohydrates into monosaccharides, which is a major contributor to elevated blood sugar levels [28].

14.6 Antiviral activity

Since viruses spread and proliferate quickly, treating viral infections is difficult. Numerous deadly viruses, like the coronavirus, Ebola virus, dengue virus, HIV, and influenza virus, are already doing terror worldwide. AgNPs are regarded as an influential and innovative pharmacological drug with strong antiviral efficacy against the feline coronavirus (FCoV) [29], HIV [30], adenovirus [31], herpes simplex virus [32], dengue virus [33], etc. One of the modes of action of AgNPs as viricidal in HIV-1 is described as AgNPs targets the gp120 and inhibits binding to host cell membrane. This leads to blocking entry, fusion, and infectivity [34].

14.7 Cytotoxic activity

Due to their nano size, nanoparticles differ from conventional medications in their biological effects, making them a possible therapeutic option for a number of disorders. Preventing the killing of noncancerous cells and destroying the tumor cells is the major task in the treatment of cancer. Only the malignant cells that are actively growing will be targeted for cytotoxicity by targeted medication therapies using nano-sized formulations. The use of nanostructure formulations in the treatment of cancer and other chronic diseases is truly remarkable [35]. Effective cancer treatment is being developed using plant-based AgNPs. The AgNPs of two sizes, 2 nm and 15 nm, have been tested for anticancer action against MCF-7 and T-47D cells lines, and it has been shown that they caused endoplasmic reticulum stress *via* Unfolded Protein Response (UPR), boosted Caspase-9 and Caspase-7 activation, and ultimately led to cell death [36]. Studies on AgNPs from *Cynarascolymus*(artichoke) for its antitumor effectiveness combined with photodynamic therapy, control mitochondrial apoptosis by regulating apoptotic proteins and causing demise of MCF-7 breast cancer cell lines [37]. Similarly, *Pinusroxburghii* against lung cancer and prostate cancer cell lines [38]. AgNP of *Nepeta deflersiana* plant extract is effective for the human cervical cancer (HeLa) cell lines [39]. *Juglans regia* (walnut fruits) AgNPs have been tested for cytotoxicity on the malignant cell line MCF-7 as well as on the normal cell line L-929 fibroblast cells [40], and it was revealed that the synthesized NPs were toxic to the targeted cancer cell lines and nontoxic to the normal cell lines.

A number of plant species have been utilized for the synthesis of nanoparticles with various applications in field of science and technology which is enlisted in Table 14.1. Medicinal plants are utilizing for treatment of various diseases from thousands of years ago and its development to the discovery drug and modern medicine is because of advanced extraction, purification, identification, and delivery methods [4, 41]. Plant extract-based nanoparticle is a focused research area because it may change traditional drug manufacturing and delivery system. The popularity of nano-medicine based on plant extract is not only due to the availability of plants, easy extraction, production method, and environmentally friendly approach, but also it may give rise to advantages like more effectiveness, less quantity, and more specific to the target.

14.8 Conclusions

The presence of proteins and metabolites in the plants, microbes, and animal cells acts as reducing agent for the conversion of inorganic metal ions to the metallic nanoparticles. This miracle leads to the delivery of plant extract from source to the target *via* nanoparticles. The chemical and physical methods are costly because we have to pay for both metal ion and reducing agent, while in the case of plant extract–based nanoparticles we only need to pay for the metal ion. Environmental concern is another issue while developing new resources in field of science. Plant extract–based nanoparticles overcome these problems too. Henceforth, in the future green synthesized nanoparticles will be a focused/targeted field among researchers and many applied products will be available for the society.

Table 14.1: Recent green synthesized nanoparticles (NPs) with their plant sources, characterization techniques, and significant bioactivities.

Medicinal plants	NPs	Characterization techniques	Bioactivities	References
Acca sellowiana	Ag	UV-vis	Antimicrobial activity against *Escherichia coli*, and *Staphylococcus aureus*; and antioxidant activity	[42]
Allium ampeloprasum	Ag	UV-vis, FT-IR, XRD, and SEM	Cytotoxic effect against HeLa cells; antioxidant, and antibacterial activity	[43]
Alpinia calcarata	Ag	UV-vis, FT-IR, XRD, TEM, DLS, and ICP-OES	Antibacterial against *E. coli* MTCC 44, *P. aeruginosa* MTCC 424, *S. aureus*, and antioxidant activity	[44]
Alternanthera bettzickiana	Au	UV-vis, XRD, FTIR, SEM, TEM, and ZP	Antimicrobial activity against *Bacillu subtilis*, *Staphylococcus aureus*, *Salmonella typhi*, *Pseudomonas aeroginosa*, *Micrococcus luteus*, and *Enterobacter aerogenes*; and Cytotoxic against A549 human lung cancer cell lines	[45]
Araucaria heterophylla	Ag	UV-vis, FT-IR, SEM, EDS, and DLS	Antimicrobial against *E. coli* and *Streptococcus* sp; anticancer against human breast cancer cell line [MCF- 7]; and for chromium removal	[46]
Azadirachta indica	Ag	UV-vis, SEM, TEM, and XRD	Antibacterial; antioxidant, and wound-healing property	[47]
Bauhinia acuminata	Ag	UV-vis, FTIR, XRD, and TEM	Larvicidal toxicity against malarial parasite, Zika virus and filariasis vectors- *Anopheles stephensi*, *Aedes aegypti*, and *Culex quinquefasciatus*	[48]
Berberis balochistanica	NiO	UV-vis, FT-IR, SEM, EDS, and XRD	Antioxidant, aytotoxic, antibacterial, antifungal activities, and enhanced seed germination via nano-fertilizer	[49]
Bergenia ciliata	Ag	UV-vis, FTIR, XRD, DLS, SEM, and EDX	Antioxidant activity, and cytotoxic against the human cervical cancer (HeLa] and human colon cancer [HT-29) cell lines	[50]
Brillantaisiapatula, *Crossopteryxfebrifuga*, and *Senna siamea*	Ag	UV-vis, FTIR, XRD, TEM and EDX	Antibacterial activity against *Staphylococcus aureus*, *Escherichia coli*, and *Pseudomonas aeruginosa*	[51]

Table 14.1 (continued)

Medicinal plants	NPs	Characterization techniques	Bioactivities	References
Calophyllumtomentosum	Ag	UV-vis, FTIR, XRD, and EDX	Antibacterial, antioxidant, antidiabetic, anti-inflammatory, and anti-tyrosinase activity	[52]
Canthiumdicoccum	ZnO	FT-IR, and DLS	Antibacterial activity against *Bacillus subtilis* and *Staphylococcus aureus*; antioxidant activity	[53]
Capsicum annuum	Ag	UV-vis, FTIR, SEM, and AFM	Antibacterial activity against *Pseudomonas aeruginosa*; and antioxidant activity	[54]
Cestrum nocturnum	Ag	UV-vis, FTIR, XRD, SEM, TEM, and XRD	Antibacterial activity against *Escherichia coli*, *Enterococcus faecalis*, and *Salmonella typhi*	[55]
Citrus limon	Ag	UV-vis, FT-IR, SEM, TEM, DLS, XRD, and ZP	Antioxidant; antimicrobial activity against *Escherichia coli*, *Staphylococcus aureus*, and *Candida albicans*	[56]
Corchorus olitorius	Au	UV-vis, FT-IR, XRD, and TEM	Cytotoxic activity in three human cancer cell lines, namely, colon carcinoma HCT-116, hepatocellular carcinoma HepG-2, and breast carcinoma MCF-7	[57]
Cratoxylum formosum, *Phoebe lanceolata*, *Scurrula parasitica*, *Ceratostigma minus*, *Mucuna birdwoodiana*, *Myrsineafricana*, and *Lindera strychnifolia*	Ag	UV-vis	Antioxidant, cytotoxic, apoptotic, and wound-healing activity	[58]
Crocus sativus	Au	UV-vis, TEM, and XRD	Antimicrobial against *Staphylococcus aureus*, *Bacillus cereus*, *Escherichia coli*, *Aspergillus* sp., and *Candida albicans*.	[59]
Croton caudatus	Au	UV-vis, FT-IR, XRD, SEM-EDAX, and TEM	Antimicrobial, antifungal, cytotoxic, and antioxidant activity	[60]
Curcuma pseudomontana	Au	SEM, HR-TEM, UV-vis, and FT-IR	Antimicrobial activity against *Pseudomonas aeruginosa*, *Staphylococcus aureus*, *Bacillus subtilis*, and *Escherichia coli*	[61]

Table 14.1 (continued)

Medicinal plants	NPs	Characterization techniques	Bioactivities	References
Dillenia indica	Ag	UV-vis, FT-IR, AFM, and HR-TEM	Antimicrobial activity against Enterococcus faecalis and E. coli	[62]
Geranium wallichianum	ZnO	UV-vis, FT-IR, TEM, EDS, XRD, and RS	Cytotoxicity, antioxidative, and antimicrobial activity	[63]
Glycyrrhiza glabra	TiO$_2$	UV-vis, FT-IR, SEM, and XRD	Antimicrobial and cytotoxic activity	[64]
Hypoxishemerocal lidea	Au	FTIR, HRTEM, DLS, and EDX	Immunomodulatory effects: hypoxoside on macrophage and natural killer cells	[65]
Hyssopus officinalis	ZnO	SEM	Antidiabetic, antioxidant, and anti-inflammatory activity	[66]
Lawsoniainermis	Ag	UV-vis, XRD, SEM, and TEM	Antibacterial activity against Escherichia Coli, Pseudomonas aeruginosa, Staphylococcus aureus, and Bacillus subtilis	[67]
Leucosidea sericea	Au	HRTEM, XRD, EDS, ZP, and DLS	Antidiabetic and antioxidant activity	[68]
Memecylon umbellatum	Ag	UV-vis, FTIR, HRTEM, ZP, and XRD	Antibacterial activity against Acinetobacter baumannii; Antitumor activity against the breast cancer cell line (MCF-7); and antioxidant activity	[69]
Mimusopselengi	Ag	UV-vis, XRD, and STEM	Antibiofilm; antibacterial against Enterococcus durans, Listeria innocua, Bacillus subtilis, Escherichia coli, Klebsiella pneumoniae, Salmonella enteritidis, Salmonella kentucky, Staphylococcus epidermidis, Staphylococcus aureus, Xanthomonas, and Proteus vulgaris; and anticancer activities against human colon (HT-29) and breast (MCF-7) cancer cell lines	[70]
Moringa oleifera	Ag	UV-vis, FT-IR, TEM, and DLS	Antioxidant, and cytotoxicity against growth of colon cancer cell lines	[71]
Moringa oleifera	Ag	UV-vis, FTIR, SEM, TEM, and XRD	Antibacterial activity against Staphylococcus aureus, Enterococcus faecalis, Escherichia coli, Pseudomonas aeruginosa, and Klebsiella pneumoniae	[72]

Table 14.1 (continued)

Medicinal plants	NPs	Characterization techniques	Bioactivities	References
Ocimumamericanum	ZnO	UV-vis, FT-IR, XRD, and SEM	Antimicrobial activity against gram-positive and gram-negative bacteria and antioxidant activity	[73]
Ocimumbasilicum	Ag	UV-vis, FTIR, and FESEM	Antimicrobial activity	[74]
Origanum majorana	CeO	UV-vis, FT-IR, TEM, XRD, and FESEM	Antioxidant, anti-inflammatory, and cytotoxic activity	[75]
Pinus roxburghii	Ag	UV-vis, EDS, TEM, XRD, SAED, and FESEM	Cytotoxicity toward A549 and prostatic small cell carcinomas [PC-3]	[38]
Piper longum	Ag	UV-vis, FTIR, TEM, XPS, GC-MS, and XRD.	Antimicrobial activity against *Bacillus cereus, Staphylococcus aureus, Escherichia coli, Proteus mirabilis, Klebsiella pneumoniae, Pseudomonas aeruginosa*, and *Salmonella typhi*	[76]
Prunus armeniaca	Au & Ag	UV-vis, FTIR, SEM, EDX, and XRD	Antimicrobial activity against *Staphylococcus aureus, Escherichia coli*, and *Pseudomonas aeruginosa*	[77]
Punica granatum	Ag	UV-vis, FE-SEM, EDS, and ZP	Antibacterial activity against *Pseudomonas aeruginosa*, and *Staphylococcus aureus*	[78]
Rhamnus virgata	CoO	UV-vis, FT-IR, SEM, DLS, XRD, and RS	Anticancer activity against human hepatoma HUH-7, and hepatocellular carcinoma HepG2	[79]
Rosa damascena	Au	UV-vis, EDS, TEM, XRD, and ZP	Cytotoxic against peripheral blood mononuclear lymphocytes (PBML], acute promyelocytic leukemia (HL60), and human lung adenocarcinoma [A549)	[80]
Rosa floribunda	MgO	UV-vis, FT-IR, XRD, SEM-EDX, and TEM	Antioxidant, antiaging, and antibiofilm activity; antimicrobial activity against *Staphylococcus epidermidis, Streptococcus pyogenes*, and *Pseudomonas aeruginosa*	[81]

Table 14.1 (continued)

Medicinal plants	NPs	Characterization techniques	Bioactivities	References
Rosmarinus officinalis	Pd	FT-IR, XRD, FESEM, TEM, and UV-vis	Antimicrobial against *Staphylococcus aureus, Staphylococcus epidermidis, E. coli*, and *Micrococcus lutens, Candida parapsilolis, Candida albicans, Candida glabrata, and Candida krusei*	[82]
Rubus fairholmianus	ZnO	UV-vis, FT-IR, SEM, and XRD	Antimicrobial activity against *Staphylococcus aureus*	[83]
Ruta graveolens	Ag	UV-vis, FT-IR, and SEM	Immune modulation, anticancer, and insecticidal activity against *Culex pipiens*	[84]
Salvia officinalis	Ag	UV-vis, FT-IR, XRD, SEM-EDX, and TEM	Cytotoxic effect against human cervix adenocarcinoma [HeLa] cells; antiplasmodial activity against *Plasmodium falciparum*	[85]
Sida acuta	CuO NP	FT-IR, FE-SEM, EDX, and TEM	Antimicrobial activity against *Escherichia coli, Proteus vulgaris*, and *Staphylococcus aureus*	[86]
Sida cordifolia	Ag	UV-vis, XRD, SEM, and TEM	Antibacterial activity against *Aeromonas hydrophila, Pseudomonas fluorescence, Flavobacteriumbranchiophilum, Edwardsiellatarda, Yersinia rukeri, Escherichia coli, Klebsiella pneumonia, Bacillus subtilis*, and *Staphylococcus aureus*	[87]
Spermacocehispida	Se	UV-vis, FT-IR, and SEM	Anti-inflammatory property, Antibacterial property, and Anticancer activity against human cervical cancer cell lines	[88]
Sutherlandia frutescens	ZnO	TEM, SEM, SAED, FTIR, EDS, and LCMS	Cytotoxic against human lung cancer cells [A549]; antimicrobial activity against *Escherichia coli, Pseudomonas aeruginosa, Staphylococcus aureus*, and *Enterobacter faecalis*	[89]
Syzygiumcumini	Ag	UV-vis, FT-IR, XRD, and SEM	Antibacterial activity against gram-positive and gram-negative bacteria; anti-inflammatory, and antioxidant activity	[90]

Table 14.1 (continued)

Medicinal plants	NPs	Characterization techniques	Bioactivities	References
Terminalia fagifolia	Ag	TEM	Antioxidant potential in microglial cells; Antimicrobial against gram-positive and gram-negative bacteria; Antifungal against yeasts and dermatophytes	[91]
Terminalia mantaly	Ag	UV-vis, FTIR, TEM, and DLS	Antimicrobial activity against *Streptococcus pneumoniae* and *Haemophilus influenzae*	[92]
Thymbra spicata	ZnO	UV-vis, FT-IR, XRD, and SEM	Antimicrobial activity against *Bacillus subtilis* ATCC 6633, *E. coli* ATCC 25952, *P. aeruginosa* ATCC 27853, and *Candida albicans* ATTC 90028; and Antioxidant activity	[93]
Thymus vulgaris	Au	XRD, EDS, FT-IR, UV-vis, TEM, and SEM	Effective for treating acute myeloid leukemia	[94]
Thymus vulgaris	Ag	UV-vis, FT-IR, XRD, SEM-EDX, and TEM	Cytotoxic against T47D human breast cancer; antioxidant activity	[95]
Tridax procumbens	Ag	XRD, ZP, TEM, and FT-IR	Antimicrobial activity against *Escherichia coli, Shigella* sp., *Aeromonas* sp., *Pseudomonas aeruginosa*, and *Candida tropicalis*; and anticancer activity	[96]
Withaniacoagulans Caesalpinia pulcherrima	Co_3O_4	UV-vis, and XRD	Antibacterial, and antifungal Antibacterial	[97]
Withaniasomnifera	Cu_2O	UV-vis, FTIR, and EDAX	Antibacterial activity against *Bacillus* sp., *Enterobacter* sp., *Escherichia coli*, and *Mycobacterium*	[98]

References

[1] N. Strambeanu, L. Demetrovici, D. Dragos, M. Lungu. M. Lungu, A. Neculae, M. Bunoiu, C. Biris eds. Springer, Cham, 2015, https://doi.org/10.1007/978-3-319-11728-7_1.

[2] P. G. Jamkhande, N. W. Ghule, A. H. Bamer, M. G. Kalaskar. J Drug Delivery Sci Technol, 2019, 53, 101174.

[3] M. Hasan, I. Ullah, H. Zulfiqar, K. Naeem, A. Iqbal, H. Gul, M. Ashfaq, N. Mahmood. Mater Today Chem, 2018, 8, 13-28.

[4] R. Chauhan, A. Quraishi, S. K. Jadhav, S. Keshavkant. Acta Physiologiae Plantarum, 2016, 38, 116.

[5] S. Chauhan, P. K. Mahish. Res J Pharm Tech. 2020, 13(11), 5647-5653.

[6] S. Iravani, H. Korbekandi, S. V. Mirmohammadi, B. Zolfaghari. Res Pharm Sci. 2014, 9(6), 385-406.

[7] J. N. Kabera, E. Semana, A. R. Mussa, X. He. J Pharm Pharmacol. 2014, 2(7), 377-92.

[8] R. G. Saratale, I. Karuppusamy, G. D. Saratale, A. Pugazhendhi, G. Kumar, Y. Park, G. S. Ghodake, R. N. Bharagava, J. R. Banu, H. S. Shin. Colloids Surf B Biointerfaces, 2018, 170, 20-35.

[9] A. K. Mittal, C. Yusuf, C. B. Uttam. Biotechnol Adv. 2013, 31(2), 346-356.

[10] C. Jayaseelana, A. A. Rahumana, A. V. Kirthi, S. Marimuthua, T. Santhoshkumara, A. Bagavana, et al. Spectrochimica Acta Part A. 2012, 90, 78–84.

[11] K. Gopinath, V. K. Shanmugam, S. Gowri, V. Senthilkumar, S. Kumaresan, A. Arumugam. J Nanostruct Chem, 2014, 4, 83.

[12] S. P. Chandran, M. Chaudhary, R. Pasricha, A. Ahmad, M. Sastry. Biotechnol Prog, 2006, 22, 577-583.

[13] I. Hussain, N. B. Singh, A. Singh, H. Singh, S. C. Singh. Biotechnol Lett, 2016, 38, 545–560.

[14] F. Moradi, S. Sedaghat, O. Moradi, S. Arab Salmanabadi. Inorg Nano-Metal Chem. 2021, 51(1), 133-42.

[15] S. MoradiF, Sedaghat, O. Moradi, S. Arab Salmanabadi. Inorg Nano-Metal Chem. 2020, 51(1), 133–142.

[16] K. P. Shejawal, D. S. Randive, S. D. Bhinge, M. A. Bhutkar, G. H. Wadkar, N. R. J. Jadhav. Genet Eng Biotechnol, 2020, 18, 43.

[17] E. Urnukhsaikhan, B. E. Bold, A. Gunbileg. Sukhbaatar N andMishig-Ochir T. 2021Sci Rep, 11(1), 21047.

[18] A. Salayová, Z. Bedlovičcová, N. Daneu, M. Baláž, Z. LukáčcováBuj ˇnáková, L. Balážová, C. Tká ̌. Nanomaterials. 2021, 11.

[19] S. Lokina, R. Suresh, K. Giribabu, A. Stephen, L. Lakshmi Sundaram, V. Narayanan. 2014, 129(2014), 484-490.

[20] L. MauJ, H. C. Lin, C. C. Chen. J Agric Food Chem, 2002, 50, 6072–6077.

[21] J. Flieger, W. Franus, R. Panek, C. M. Szyma, W. Flieger, M. Flieger, P. Kołodziej. Molecules, 2021, 26, 4986.

[22] E. Sreelekha, B. George, A. Shyam, et al. BioNanoSci. 2021, 11, 489–496.

[23] M. C. Hunter, R. G. Smith, M. E. Schipanski, L. W. Atwood, D. A. Mortensen. BioScience, 2017, 67, 386–391.

[24] J. H. Mejias, F. Salazar, L. Pérez Amaro, S. Hube, M. Rodriguez, M. Alfaro. Front Environ Sci, 2021, 9, 635114.

[25] R. Raliya, V. Saharan, C. Dimkpa, P. Biswas. J Agri Food Chem. 2018, 66(26), 6487-6503.

[26] R. S. Kookana, A. B. Boxall, P. T. Reeves, R. Ashauer, S. Beulke, Q. Chaudhry, G. Cornelis, T. F. Fernandes, J. Gan, M. Kah, I. Lynch. J Agri Food chem. 2014, 62(19), 4227-40.

[27] K. Anand, C. Tiloke, P. Naidoo, A. A. Chuturgoon. J Photochem Photobiol B Biol, 2017, 173, 626-639.

[28] U. Etxeberria, A. L. De La Garza, J. Campin, et al. Expert OpinTher Tar. 2012, 16, 269–297.

[29] Y. N. Chen, Y. H. Hsueh, C. T. Hsieh, D. Y. Tzou, P. L. Chang. Int J Environ Res. 2016, 13(4), 430.

[30] J. L. Elechiguerra, J. L. Burt, J. R. Morones, B. C. Bragado, X. Gao, H. H. Lara, M. J. Yacaman. Interaction of silver nanoparticles with HIV-1. J Nanobiotech. 2005, 3(6), 1–10.

[31] N. Chen, Y. Zheng, J. Yina, C. Lia, C. Zhenga. J Vir Met, 2013, 193, 470–477.

[32] R. L. Hu, S. R. Li, F. J. Kong, R. J. Hou, X. L. Guan, F. Guo. Inhibition efect of silver nanoparticles on herpes simplex virus 2. Gen Mol Res. 2014, 13(3), 7022–7028.

[33] V. Sujitha, K. Murugan, K. Paulpandi, C. Panneerselvam, U. Suresh, M. Roni, M. Nicoletti, et al. Parasitol Res. 2015, 114(9), 3315–3325.

[34] H. H. Lara, A. V. Ayala-Nuñez, L. Ixtepan-Turrent, C. Rodriguez-Padill. J Nanobiotech. 2010, 8(1), 1–10.

[35] S. Yesilot, C. Aydin. East J Med. 2019, 24(1), 111–116.

[36] J. C. Simard, I. Durocher, D. Girard. Apo, 2016, 21, 1279–1290.

[37] O. Erdogan, M. Abbak, G. M. Demirbolat, F. Birtekocak, M. Aksel, S. Pasa, et al. PLoS ONE. 2019, 14, 6.

[38] R. Kumari, A. K. Saini, A. Kumar, et al. J BiolInorg Chem. 2020, 25, 23–37.

[39] E. S. Al-Sheddi, N. N. Farshori, M. M. Al-Oqail, S. M. Al-Massarani, Q. Saquib, R. Wahab, J. Musarrat, A. A. Al-Khedhairy, M. A. Siddiqui. Bioinorganic Chemistry and Applications. 2018, 1–12.

[40] S. Khorrami, A. Zarrabi, M. Khaleghi, M. Danaei, M. R. Mozafari. *Int J Nanomed*, 2018, 13, 8013-8024.

[41] V. V. Makarov, A. J. Love, O. V. Sinitsyna, S. S. Makarova, I. V. Yaminsky, M. E. Taliansky, Kalinina. Acta Naturae. 2014, 6(1), 35-44.

[42] W. G. Sganzerla, M. Longo, J. L. de Oliveira, C. G. da Rosa, V. A. P. de Lima, R. S. de Aquino, M. R. Nunes. Physicochem Eng Aspect, 2020, 602, 125125.

[43] J. Fereshteh, C. Azam, S. Komail, F. Ali, S. Yalda. Adv Powder Technol. 2020, 31(3), 1323-1332.

[44] P. Khandel, S. K. Shahi, D. K. Soni, et al. Nano Convergence. 2018, 5, 37.

[45] M. Nagalingam, V. N. Kalpana, R. V. Devi, A. Panneerselvam. Biotechnol Rep, 2018, 19, e00268.

[46] A. V. Samrot, J. L. A. Angalene, S. M. Roshini, et al. J Clust Sci. 2019, 30, 1599–1610.

[47] G. Chinnasamy, S. Chandrasekharan, T. W. Koh, S. Bhatnagar. Front Microbiol, 2021, 12, 611560.

[48] S. A. Naiyf, G. Marimuthu, K. Shine, M. K. Jamal, N. A. Taghreed, A. A. Sami, N. A. Mohammed, G. Kasi, S. Arumugam. J Trace Elem Med Biol, 2018, 50, 146-153.

[49] S. Uddin, L. B. Safdar, S. Anwar, J. Iqbal, S. Laila, B. A. Abbasi, M. S. Saif, M. Ali, A. Rehman, A. Basit, Y. Wang, U. M. Quraishi. Molecules, 2021, 26, 1548.

[50] K. Dulta, A. G. Koşarsoy, P. Chauhan, et al. J Inorg Organomet Polym. 2021, 31, 180–190.

[51] K. K. Espoir, I. Christian, I. M. Blaise-Pascal, M. B. Alain, O. T. Daniel, I. L. Jean-Marie, W. M. K. Rui, B. M. Patrick. Heliyon. 2020, 6(8), 04493.

[52] M. Govindappa, B. Hemashekhar, M. Arthikala, V. R. Rai, Y. L. Ramachandra. Results Phys, 2018, 9, 400-408.

[53] C. Mahendra, N. C. Mohana, M. Murali, M. R. Abhilash, S. B. Singh, S. Satish, M. S. Sudarshana. Process Biochem, 2020, 89, 220-226.

[54] A. V. Samrot, N. Shobana, R. Jenna. BioNanoSci, 2018, 8, 632–646.

[55] K. K. Anand, S. Ragini, S. Payal, B. Y. Virendra, N. Gopal. J Ayurveda Integr Med. 2020, 11(1), 37-44.

[56] Y. Khane, K. Benouis, S. Albukhaty, G. M. Sulaiman, M. M. Abomughaid, A. A. Al, D. Aouf, F. Fenniche, S. Khane, W. Chaibi, A. Henni, H. D. Bouras, N. Dizge. Nanomaterials, 2022, 12, 2013.

[57] E. H. Ismail, A. M. A. Saqer, E. Assirey, A. Naqvi, R. M. Okasha. Cancer Cells Int J Mol Sci, 2018, 19, 2612.

[58] A. Eun-Young, J. Hang, P. Youmie. Mater Sci Eng C, 2019, 101, 204-216.

[59] O. Azizian-Shermeh, M. Valizadeh, M. Taherizadeh, et al. ApplNanosci. 2020, 10, 2907–2920.

[60] P. V. Kumar, S. M. Kala, K. S. Prakash. Mater Lett, 2019, 236, 19-22.

[61] N. Muniyappan, M. Pandeeswaran, A. Amalraj. Environ Chem Ecotoxicol, 2021, 3, 117-124.

[62] S. Nayak, M. P. Bhat, A. C. Udayashankar, T. R. Lakshmeesha, N. Geetha, S. Jogaiah. Appl Organomet Chem, 2020, 34, 5567.

[63] B. A. Abbasi, J. Iqbal, R. Ahmad, L. Zia, S. Kanwal, T. Mahmood, C. Wang, J. T. Chen. Biomolecules, 2020, 10, 38.

[64] M. Bavanilatha, L. Yoshitha, S. Nivedhitha, S. Sahithya. Agric Biotechnol, 2019, 19, 101131.

[65] A. M. Elbagory, A. A. Hussein, M. Meyer. Int J Nanomed. 2019, 19(14), 9007-9018.

[66] G. Rahimi, K. S. Mohammad, M. Zarei, et al. Biol Res. 2022, **55**, 24.

[67] H. Upadhyaya, S. Shome, R. Sarma, S. Tewari, M. Bhattacharya, S. Panda. Am J Plant Sci, 2018, 9, 1279-1291.

[68] U. M. Badeggi, E. Ismail, A. O. Adeloye, S. Botha, J. A. Badmus, J. L. Marnewick, C. N. Cupido, A. A. Hussein. Biomolecules, 2020, 10, 452.

[69] S. A. Mohamad, E. Kannan, J. A. R. Amirtham, P. Murali, S. Devanesan. Saudi J Biol Sci, 2019, 26(50, 970-978.

[70] K. Nesrin, C. Yusuf, H. Attia, K. Ahmet, S. B. Ali, N. A. Muhammad, Ç. Özge, Ş. Fatih. J Drug Delivery Sci Technol, 2020, 59, 101864.

[71] W. G. Shousha, W. M. Aboulthana, A. H. Salama, et al. Bull Natl Res Cent. 2019, 43, 212.

[72] S. M. Jerushka, B. Suresh, K. Naidu, P. S. Karen, G. Patrick. Adv Nat Sci Nanosci Nanotechnol. 2018, 9, 015011.

[73] H. K. Narendra Kumar, N. Chandra Mohana, B. R. Nuthan, K. P. Ramesha, D. Rakshith, N. Geetha, S. Satish. SN Appl Sci. 2019, 1(6), 1-9.

[74] P. Saba, G. Maryam, B. Saeid. Mater Sci Eng C, 2019, 98, 250-255.

[75] N. Aseyd, S. Es-haghi, A. M. H. Tabrizi. Appl Organomet Chem, 2020, 34, e5314.

[76] H. Huang, K. Shan, J. Liu, X. Tao, S. Periyasamy, S. Durairaj, J. A. Jacob. Bioorganic Chem, 2020, 103, 104230.

[77] U. I. Nazar, A. Raza, S. Muhammad, A. Muhammad. Arab J Chem. 2019, 12(8), 3977-3992.

[78] K. Sadegh, K. Fayezeh, Z. Ali. Biocatal Agric Biotechnol, 2020, 25, 101620.

[79] B. A. Abbasi, J. Iqbal, Z. Khan, et al. Microsc Res Tech. 2021, 84, 192-201.

[80] A. Kyzioł, S. Łukasiewicz, V. Sebastian, P. Kuśtrowski, D. Majda, A. Cierniak. J Inorg Biochem, 2021, 214, 111300.

[81] I. Y. Younis, S. S. El-Hawary, O. A. Eldahshan, et al. Sci Rep. 2021, 11, 16868.

[82] N. Rabiee, M. Bagherzadeh, M. Kiani, A. M. Ghadiri. Adv Powder Technol. 2020, 31(4), 1402-11.

[83] N. K. Rajendran, B. P. George, N. N. Houreld, H. Abrahamse. Molecules, 2021, 26, 3029.

[84] H. A. Ghramh, E. H. Ibrahim, M. Kilnay, Z. Ahmad, S. K. Alhag, K. A. Khan, R. Taha, F. M. Asiri. Bioinorg Chem Appl. 2020, 2020.

[85] K. Okaiyeto, H. Hoppe, A. I. Okoh. J Clust Sci, 2021, 32, 101–109.

[86] S. Selvam, V. Seerangaraj, B. Devaraj, S. Mythili, M. Elayaperumal, S. K. Smita, P. Arivalagan. J Photochem Photobiol B Biol, 2018, 188, 126-134.

[87] N. Panduranga, V. K. Pallela, U. Shameem, K. R. Lakshmi, S. V. N. Pammi, S. G. Yoon. Microb Pathogenesis, 2018, 124, 63-69.

[88] V. Krishnan, C. Loganathan, B. Rama, P. Thayumanavan. Adv. Nat Sci: Nanosci Nanotechnol. 2018, 9, 015005,

[89] L. M. Mahlaule-Glory, Z. Mbita, B. Ntsendwana, M. M. Mathipa, N. Mketo, N. C. Hintsho-Mbita. Mater Res Express. 2019, 6(8), 085006.

[90] A. Chakravarty, I. Ahmad, P. Singh, M. U. Sheikh, G. Aalam, S. Sagadevan, S. Ikram. Chem Phys Lett, 2022, 795, 139493.

[91] A. R. de Araujo, J. Ramos-Jesus, T. M. de Oliveira, A. M. A. de Carvalho, P. H. M. Nunes, T. C. Daboit, P. Eaton. Ind Crops Prod, 2019, 137, 52-65.

[92] M. S. Majoumouo, N. R. S. Sibuyi, M. B. Tincho, M. Mbekou, F. F. Boyom, M. Meyer. Int J Nanomed. 2019, 19(14), 9031-9046.

[93] G. Tuğba, M. Ismet, S. Hamdullah, B. Muhammed, S. Fatih. Environmental Research. Vol. 204, 2022, 111897.

[94] S. Hemmati, Z. Joshani, A. Zangeneh, M. M. Zangeneh (2020) Appl Organometal Chem. 34: e5267.

[95] Z. Heidari, A. Salehzadeh, S. A. Sadat Shandiz, et al. Biotech. 2018, 3(8), 177.

[96] P. Rohini, H. N. Shivraj, M. Nilesh, S. Ragini, P. S. Rana, S. K. Arun. Oxid Med Cell Longev. 2022, 9671594.

[97] M. Hasan, A. Zafar, I. Shahzadi, F. Luo, S. G. Hassan, T. Tariq, S. Zehra, T. Munawar, F. Iqbal, X. Shu. Molecules, 2020, 25, 3478.

[98] M. Dhara, K. Kisku, U. C. Naik (2022) ApplNanosci https://doi.org/10.1007/s13204-022-02452-3

Shushil Kumar Rai* and Ravishankar Chauhan*

Chapter 15
Phytochemicals in drug discovery

Abstract: Here, we summarize phytochemicals, their health benefits, pharmacokinetics and pharmacodynamic characteristics, toxicity, and their application in the development of lead compounds and related drug analogs. Natural products and their structural analogues have significant contributions to pharmacotherapy, especially for the treatment of cancer, malaria, diabetes, and infectious diseases. Drug discovery employing phytochemicals obtained from different plant sources is the basis of new drug discovery and lead compound development. The process of drug discovery involves the identification, isolation, and purification of phytochemicals from the selected plant species, biochemical characterization, and pharmacological investigation followed by preclinical and clinical trials. Drug discovery also requires bioinformatics computational tools and structure elucidation by cutting-edge analytical techniques such as high-performance liquid chromatography, gas chromatography, and mass spectrometry. The phytochemicals sometimes require modification in the native structure for the enhancement of biological activity and stability. Consequently, lead identification and new drug discovery is an important toolbox for tackling antimicrobial resistance. At present, phytochemicals are the main pillar in new drug discovery and the pharmaceutical industry revolution. Therefore, new opportunities would be expected in the new drug discovery using phytochemicals in near future.

15.1 Introduction

In nature, an astonishing range of structural and chemical diversity is found among various plant products. To fight emerging new ailments/diseases, phytochemicals-based drug discovery is a rational approach popular among scientists from the various disciplines of medicinal chemistry, pharmaceutics, and biotechnology. An in-depth knowledge of potential drug target, stability, efficiency, pharmacokinetics, and its formulation is required to select new targeted molecules of interest for the desired clinical applications [1, 2]. Novel drug discovery is a challenging task; despite that,

*Corresponding author: Shushil Kumar Rai,** Shri Rawatpura Sarkar Institute of Pharmacy, Kumhari, Durg 490042, Chhattisgarh, India; Center of Innovative and Applied Bioprocessing (CIAB), Sector – 81, Knowledge City, Mohali 140306, Punjab, India, e-mail: rainiper2411@gmail.com
*Corresponding author: Ravishankar Chauhan,** Department of Botany, Pandit Ravishankar Tripathi Government College, Bhaiyathan 497231, Surajpur, Chhattisgarh, India; School of Studies in Biotechnology, Pt. Ravishankar Shukla University, Raipur 492010, Chhattisgarh, India, e-mail: ravi18bt@gmail.com

https://doi.org/10.1515/9783110791891-015

natural products are believed to be excellent producers of new clinical drug candidates and medicinal compounds. Consequently, ample numbers of such compounds are currently at various stages of clinical trials, and among them, some are authorized for manufacturing, marketing, and sales in the global market [3, 2].

The recognition of new chemical entities (NCEs), possessing the necessary properties of druggability and chemistry, is essentially what new drug discovery entails. These NCEs can be obtained either chemically or by isolating them from natural sources. The first success stories in discovering novel drugs came from innovations in medicinal chemistry, which prompted the demand for the creation of more chemical libraries through combinatorial chemistry. However, it has been shown that this strategy has a lower overall success rate. Natural products have become another largest source of NCEs for prospective usage as therapeutic compounds. More than 80% of pharmacological compounds were entirely from natural sources before the development of high throughput screening, or they were modeled after molecules originating from natural products including semi-synthetic analogues [4].

Natural plant source provides useful primary metabolites, namely, carbohydrates, amino acids, and fatty acids responsible for growth and development. And, various secondary metabolites including glycosides, alkaloids, flavonoids, and volatile oils are responsible for different way of defense mechanisms, characteristics color, signaling, and regulation of its metabolic pathways. Some important intermediates used for the biological synthesis of plant secondary metabolites originated from shikimic acid, mevalonic acid, acetyl coenzyme A, and 1-deoxyxylulose-5-phosphate. These intermediates are the chief performer in the shikimate pathway, mevalonate pathway, acetate pathway, and deoxyxylulose phosphate pathway [5]. Galegine derived from *Galega officinalis* acts as a precursor for the production of metformin and few others biguanides-type antidiabetic drugs. About 200 years ago, opium made from chopped seed pods of *Papaver somniferum* was found to contain morphine. As a result of the discovery of penicillin, pharmaceutical research expanded after the Second World War and that includes extensive testing of microorganisms for novel antibiotics [6]. Penicillin, erythromycin, tetracycline, antimalarials (e.g., artemisinin and quinine), antiparasitics (e.g., avermectin), lipid-controlling medications (e.g., lovastatin and analogues), anticancer medications (e.g. irinotecan, and paclitaxel), and immunosuppressants for organ transplants (e.g., cyclosporine, rapamycin) are just a few examples [4].

Drug development from natural chemicals are thought to be critically influenced by several factors, including the effectiveness of plant's part, stability and the molecular size of natural compounds, pharmacodynamics, pharmacokinetics, toxicity, and also on its structural alteration. Finding novel, perhaps beneficial, chemicals is a necessary step in the drug discovery process. These are frequently isolated from natural resources or by chemical synthesis, or a combination of the two [7]. In the quest for new pharmaceuticals, the pharmaceutical industry is turning more and more to conventional treatments. Currently, isolation, purification, and screening of drug candidates are the primary phases in the development of drugs from natural materials. The utilization of large-scale extraction and bioprocessing is essential for the successful development of targeted bioactive molecule into clinically effective medications. Drug discovery is a time-consuming, tedious, laborious, and costly process. With it being more expensive than is realized in the form of new pharmaceuticals developed, it frequently leads to more failures than triumphs. A new drug's market introduction is expected to cost more than $1 billion in direct expenses [8].

15.2 Plants are the major source of phytochemicals

Plants are regarded as a traditional source for novel medication discoveries because they create poisonous substances that harm nearby species – a process known as allelopathy [9]. Only 6% and 15%, respectively, of the estimated up to 500,000 species of higher plants have undergone pharmacological and phytochemical research [10]. Some of the naturally occurring anticancer substances include vincristine and vinblastine from *Catharanthus roseus*, and podophyllotoxin from *Podophyllum peltatum*, which is used as the precursor for etoposide, a semi-synthetic anticancer drug. Indirubin, which is currently undergoing clinical development, was inspired by a chromane alkaloid of the Ayurvedic plant *Dysoxylum malabaricum* known as rohitukine, related to *Dysoxylum binectariferum*. This drug has been used to treat chronic myelogenous leukemia, an illness that affects chronic myelogenous leukemia patients. Like paclitaxel, which had *Taxus brevifolia* bark as its original source; CA-4-phosphate; and camptothecin, which was first isolated from *Camptotheca accuminata*; combretastatin, which was discovered in *Combretum*, an African plant species [11, 12]. The chief anticancerous drugs isolated from various species of plants are represented in Fig. 15.1.

15.3 What are phytochemicals?

Medicinal plants produce some bioactive compounds having nonnutritive, but desired disease preventive, properties. The intake of these nonessential phytochemicals promotes many health benefits. Nowadays, these bioactive compounds have a massive demand in the domestic and international markets due to their versatile applications. Bioresource is rich in medicinal plants engaged in the development of medicines, nutraceuticals, drug intermediates, fine chemicals, synthetic drugs, and food supplements [13]. Phytochemicals are present in many foods, whole grains, fruits, and vegetables. These phytochemicals have a high therapeutic potential in curing diseases. Foods rich in phytochemicals also possess nutraceutical properties and provide many health benefits to humans and prevent many serious health diseases such as diabetes, coronary heart disease, cancer, microbial infection, and hypertension.

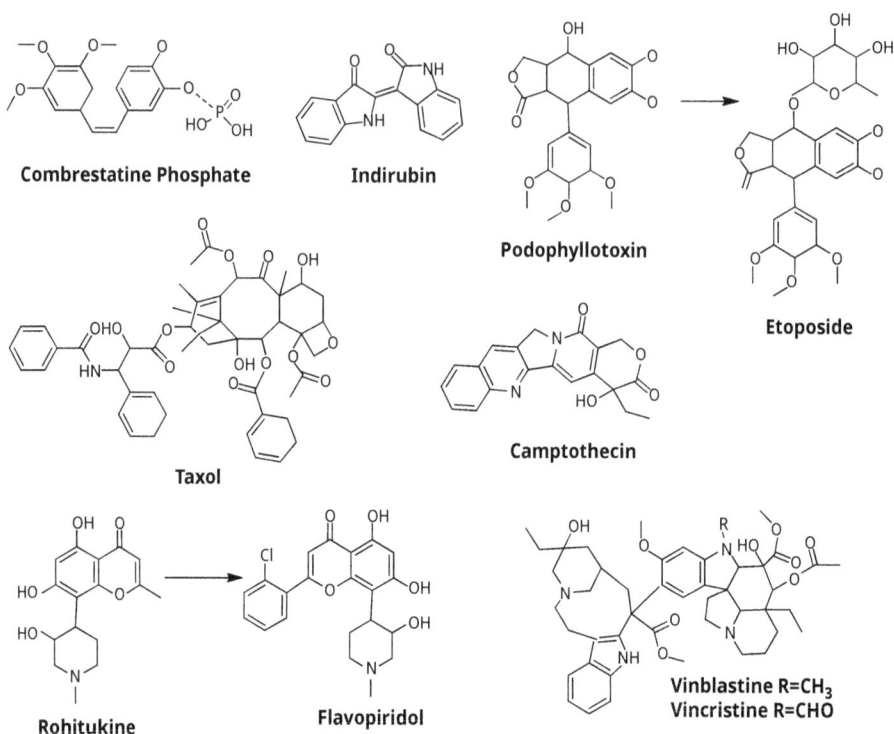

Fig. 15.1: Some anticancerous drugs isolated from different plants species.

15.3.1 Types of phytochemicals

Phytochemicals are thought to reduce the risk of oxidative cell damage. Antioxidants can be classified as *in vivo* and *in vitro* antioxidants. Free radical scavengers act as donors of protons and electrons, decompose peroxidase, quench singlet oxygen, inhibit some enzymes, and facilitate metal-chelation [14, 15]. Phytochemicals providing major health benefits are polyphenols, phytosterols, terpenoids, carotenoids, isoflavonoids, flavonoids, anthocyanidins, etc. (Table 15.1).

Tab. 15.1: The sources and health benefits of phytochemicals.

Phytochemicals	Sources	Health benefits	References
Limonoids	Citrus fruits	Anti-inflammatory	[16]
Carotenoids	Carrots, spinach, fenugreek	Antioxidant	[17]

Tab. 15.1 (continued)

Phytochemicals	Sources	Health benefits	References
Terpenoids	Mosses, liverworts, algae, mushrooms	Antimicrobial, antiallergic, antispasmodic	[18]
Glucosinolates	Cruciferous vegetables	Anticancer	[19]
Saponins	Safed musli roots, oats, leaves, flowers,	Aphrodisiac potential, antimicrobial, anti-inflammatory	[20]
Fibers	Fruits, vegetables	Reduce blood cholesterol and hypertension	[21]
Polyphenols	Fruits, vegetables, legumes	Antioxidant, free radical scavenging	[22]
Polysaccharides	Fruits, vegetables	Antimicrobial, anti-inflammatory	[23]
Phytoestrogen	Legumes, berries, whole grains, red wine	Protect bone loss and reduce risk of heart disease	[24]
Phytosterols	Vegetables, nuts, fruits, seeds	Anticancer	[25]

15.3.2 Health benefits of various phytochemicals

Nowadays, the demand for nutraceuticals has grown rapidly due to health awareness among people. Phytochemicals are considered very important due to its antioxidants, antiallergic, antispasmodic, antibacterial, anti-inflammatory, antifungal, antispasmodic, hepatoprotective, chemopreventive, neuroprotective, and many more pharmacological actions [26, 27].

For instance, galegine was synthesized from *Galega officinalis* and used for the development of metformin and several biguanides (an antidiabetic drug). Some other plant-derived drug intermediates include Khellin from *Ammi visnaga* used to develop sodium cromoglycate (a bronchodilator), and papaverine from *Papaver somniferum* used to form morphine, codeine, and verapamil (antihypertensive drug). Quinine is an antimalarial drug obtained from the *Cinchona* plant which is used as the building block for chloroquine and mefloquine. However, the emergence of resistance to these drugs leads to the further innovative discovery of artemisinin and its analogues [10, 28].

15.3.3 Pharmacokinetics and pharmacodynamic characteristics of phytochemicals

ADME meant for the absorption, distribution, metabolism, and excretion are the chief pharmacokinetic parameters studied during pre-clinical trials. Whereas the effect and mechanism of action of a drug is a pharmacodynamic characteristic studied to understand the effect of the drug to the human body. Unpredictably, pharmacokinetic behavior has been seen for some phytochemicals and, thus, required a detailed study of their efficacy for absorption, distribution, metabolism, excretion, and toxicity profile to establish their bioavailability. The ADME profile is directly related to their physicochemical properties which can be tuned via chemical modification in functional groups, structure-activity relationships, and pharmacophore based manipulation in molecular design on a natural template. For example, the chemotherapy drug cytarabine was a result of a small substitution of deoxyribose sugar with deoxycytidine used for treating myeloid leukemia and non-Hodgkin lymphoma. Phytochemicals have a range of pharmacological effects on human-being. For instance, the alkaloid class of phytochemicals; vinblastine obtained from *Catharanthus roseus* is an anticancer drug that binds with tubulin and interferes in microtubule formation. Similarly, harpagide and harpagoside isolated from *Harpagophytum procumbens* have antirheumatic and analgesic properties [29, 30].

15.3.4 Toxicity of phytochemicals

The toxicity of phytochemicals is an important reason for failure as a suitable drug candidate in a drug discovery process. The toxicity that arises in particular nonclinical trials is the safety-related aspect which mainly depends on the kind of phytochemicals tested in the laboratory. The mainly observed ones are liver and cardiac toxicity. It's a challenging task to predict with near accuracy the level of toxicity in a compound at an early stage of the drug discovery process, especially during lead optimization. This happens usually since some key features or information is missing at the time of lead optimization. The kinetics of a new compound is difficult to estimate or simulate at the time of *in vivo* drug profiling. As a result of this, many compounds were rejected under assay conditions due to safety standards [7].

15.4 Criteria of selection as candidate for lead screening

According to estimates, there are roughly 250,000 higher plants species, which include angiosperms or gymnosperms. Only 6% of them have reportedly undergone biological activity screening, and roughly 15% have undergone phytochemical activity screening. The initial enlisting of the targeted species for biological activity screening is a significant effort in and of itself. The following approaches are being used for the novel selection in drug discovery [5].

15.4.1 Random approach

For the objective of finding new drugs, two methods were used to screen the plants that were randomly chosen. The first one is screening for a particular class of chemicals, such as flavonoids and alkaloids: This technique is straightforward to follow, but its fault is that it does not indicate the biological effectiveness. However, this technique has a good possibility of producing unique structures. And the second one is screening certain bioassays on a sample of randomly chosen plants: This strategy was used around 30 years ago by the Central Drug Research Institute (CDRI), a leading R&D institute under the Council of Scientific and Industrial Research Laboratories of India. They examined the biological effects of about 2,000 plants. The screening, though, turned up no novel drugs. Approximately 35,000 higher plants were examined for anticancer activity by the National Cancer Institute (NCI) of the National Institutes of Health in the United States between 1960 and 1980. It led to the validation of two success stories: camptothecin and paclitaxel. Thus, the mentioned approach has been used for both targeted and general screening, with some degree of success in focused screening. Target-based bioassays, such as screening against PTP1B, would likely increase the likelihood of success. However, this method requires a sizable library of extracts, which relatively few companies have around the globe.

15.4.2 Ethnopharmacological approach

To find biologically active NCEs, ethnopharmacology relies heavily on actual experiences with the use of botanical medicines. This method, which is based on a variety of disciplines including anthropology, archaeology, history, and languages, comprises the observation, description, and experimental research of indigenous medications. This strategy based on the history of ethnomedicinal use has had some success. For instance, andrographolide isolated from *Andrographis paniculata* has been used to treat diarrhea. Similarly, morphine isolated from *Papaver somniferum*, picroside from *Picrorrhizakurroa*, and berberine from *Berberis aristata* haves been included in this ethnomedicinal strategy. Paclitaxel from *Taxus brevifolia* and L-Dopa from *Mucuna prurita* are two examples of plants with success stories that are not picked based on ethnomedical use.

15.5 Phytochemicals as lead compounds in drug discovery

Obviously, phytochemicals has active targeted compounds for novel drug discovery; their structural complexity, poor solubility, and stability issues restricted them and require certain modifications. The modified substance or semi-synthetic compounds based on SAR simulations can lead to the discovery of a new drug molecule [31].

The phytochemicals synthesized in plants have stereochemical specificity, chiral center, or cis/trans configuration which contributes to their biological activity. In short, modification is required to increase the activity, selectivity, metabolic stability, and physicochemical properties of compounds. For example, artemisinin is modified to synthesize dihydroartemisinin, artemether, and arteether from *Artemisia annua*. The complex molecules are simplified to reduce molecular weight by removing redundant functional groups to improve solubility, absorption, and metabolism of the natural compound [1, 3]. The development of novel drugs is possible because organic synthesis helps in the chemical modification of structures of a compound and improves their biological activity. The structural modification results in significant alteration of biological response for the receptor and leads to the design of an improved version of the therapeutically active drug molecule [32]. The disadvantage of traditional methods of structural modification can lead to enhanced bioactivity and reduced toxicity in long time study. Nowadays, new technologies such as computational drug design (CDD) and target prediction are available for modification in the native molecule of interest. For example, the natural compound tanshinone-I derived from a Chinese plant is used for cancer therapy and treatment of cardio- and cerebrovascular diseases. A modified version of tanshinone was synthesized later acting as antiproliferative agent, with structural similarity with β-lactam, and showed higher potency, improved solubility, metabolic stability, and pharmacokinetic parameters [1].

15.5.1 Discovery of anticancer analogues of combretastatins

The anticancer drug obtained from *Combretum latifolium* belonging to Combretaceae plant family functions as tubulin inhibitor against several cancer cell lines, antiangiogenensis agents, and vascular targeting agents. Compounds such as Combretastain A-4 (CA4) and Combretastain A-1 (CA1) are biologically active constituents obtained from *Combretum caffrum* to act as tubulin inhibitors. However, the poor solubility of the above compounds has been addressed and improved prodrugs CA4P and CA1P having potent activity were developed for cancer treatment. The structural characteristics of combretastatin are trimethoxy substituted A ring, and B ring consisting of C3, and C4 and an ethane bridge between the two rings which is responsible for the structural rigidity [33]. The fascinating drug development strategy involves an S_N2 reaction of 1-bromomethyl-3, 4, 5-trimethoxybenzene with tripheylphosphine to produce a phosphonium salt, and this intermediate was coupled with a benzaldehyde-derived B-ring owing to a substitution favoring Wittig olefination which results in many E and Z isomers of the compound. Normally, the cis-isomer of combretastatin has improved biological activity over the trans-isomer to inhibit the polymerization of tubulin and cytotoxicity. On the other side, a direct Perkin condensation reaction results in the synthesis of cis-combretastatin [34].

The drug development was initiated from colchicines and later various other tubulin inhibitors were discovered such as dolastatin, epothiline A, pacilitaxel, and vinblastine. The efficacy of analogues was improved by the desired modification in the functional groups. Anticancer drugs and its phosphate prodrugs are shown in Fig. 15.2.

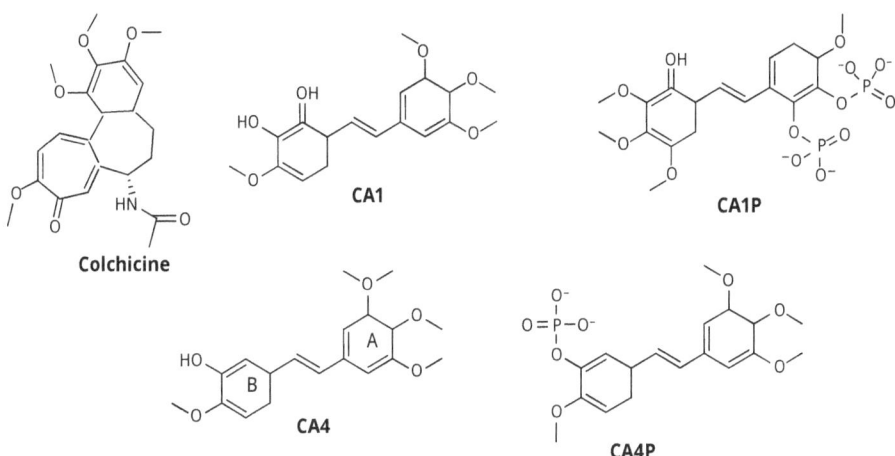

Fig. 15.2: Chemical structures of anticancer drug colchicine, combretastatin and its phosphate prodrugs.

15.5.2 Discovery of antiretroviral analogues of coumarin

Coumarins, an active ingredient with numerous uses in the pharmaceutical, food, cosmetic, and dye industries, are produced by plants like *Dipterya odorata*, *Anthoxanthum odoratum*, *Galium odoratum*, and others. Coumarins influence the operations of growth regulators, photosynthesis, and infection defense in plants. They also act as enzyme inhibitors and antioxidants. Coumarin draws interest from the scientific community for further study because of its wide range of pharmacological functions, including antimicrobial, antithrombotics, antibacterial, anticoagulants, antioxidant, antiinflammatory, antimalarial, hepatoprotective, anti-HIV, and anticancer activity [35].The powerful antioxidants and metal chelators found in coumarins with phenolic hydroxyl groups work to stop the production of free radicals. Coumarins have a wide range of biological activity and structural variation due to the possibility of structural replacement. The synthesis of coumarins has been carried out using several named reactions, including the Perkin reaction, the Wittig reaction, the Pechmann reaction, the Knoevenagel condensation, the Reformatsky reaction, etc. A quantitative structure activity relationship and docking study was conducted to examine the antiretroviral action of coumarins, and various coumarin analogues were created and synthesized (Fig. 15.3) that exhibit exceptional antiretroviral activity [36].

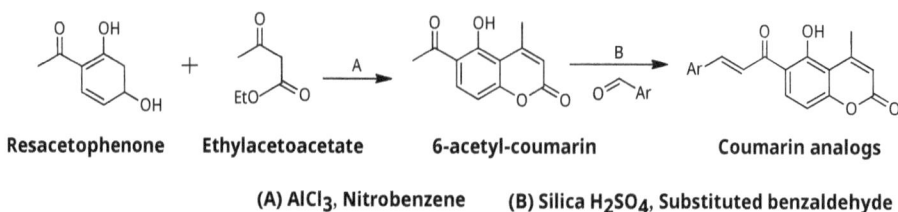

| Resacetophenone | Ethylacetoacetate | 6-acetyl-coumarin | Coumarin analogs |

(A) AlCl$_3$, Nitrobenzene   (B) Silica H$_2$SO$_4$, Substituted benzaldehyde

Fig. 15.3: Synthesis of coumarin analogs from resacetophenone.

15.5.3 Development of oseltamivir phosphate as an anti-influenza drug

In 1999, the FDA initially approved oseltamivir phosphate (Tamiflu) as the sole orally accessible medication for treatment of H5N1 avian flu/human influenza and prophylaxis. The H5N1 avian flu and human influenza have killed many people since 2003, according to WHO records, and the virulence is steadily rising, posing a threat to human being in the near future. To shield individuals from an outbreak of H5N1 avian influenza/human influenza, oseltamivir phosphate should be produced and kept on hand in every nation. Scientists from Gilead Sciences, Inc. and F. Hoffmane La Roche Ltd. created a useful synthetic pathway that is now applied to the production of oseltamivir phosphate. The recent approach, however, has numerous shortcomings, including a lengthy synthetic process and a relatively low total yield. As a result, a unique eight-step synthesis beginning with (-)-shikimic acid was reported as a more effective way to make oseltamivir phosphate [37]. N-acetyl aziridine 7 was created by reacting compound 6 with 2 equivalents of acetic anhydride and 3 equivalents of triethylamine in ethyl acetate. After that, compound 7 was treated with 1.5 equiv of boron trifluoride etherate in 3-pentanol to generate, in an 86% yield, ring-opening product 8. The (R)-configuration of C-5 is inverted to the (S)-configuration by treating compound 8 with 4 equiv of sodium azide in aqueous ethanol (EtOH:H$_2$O::5:1) under refluxing for 8 h at a temperature of roughly 75 °C. Finally, using the described method, azide 9 was changed into the title compound 1 (Fig. 15.4). This particular example is given here only because shikimic acid, an extracted secondary metabolite from the plant *Illicium verum*, is the initial material for the production of oseltamivir phosphate. It is like a blessing that (-)-shikimic acid is present in significant amounts in Chinese star anise (1.1 kg of (-)-shikimic acid from 30 kg of the dried *Illicium verum*) and is widely sold there. It is the best example of a secondary metabolite from a species of plant that was used in the production of a medicine. This drug, oseltamivir phosphate, helped humans fight H5N1 avian flu/human influenza until they were completely eradicated from the world [38].

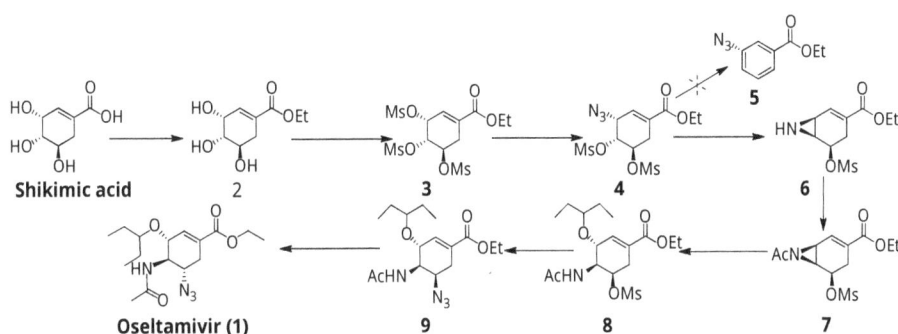

Fig. 15.4: Synthesis of anti-influenza drug oseltamivir from shikimic acid.

15.5.4 Role of phytochemicals in antimalarial drug discovery

Mefloquine and chloroquine, which were shown to be more effective than quinine for treating malaria, were synthesized using the antimalarial medication quinine from the *Cinchona* plant. However, the development of resistance against these medications made room for the finding of artemisinin, an antimalarial drug produced from plants. Many analogues were created to increase the activity and utility of artemisinin. Dimeric analogues have been proven to treat mice with malaria more effectively than artemisinin after a single dosage. Further investigation could benefit from the considerable anticancer activity that artemisinin and similar compounds have demonstrated *in vitro*.

The two flavonoids 30-formyl-20,40-dihydroxy-60-methoxychalcone and 8-formyl-7-hydroxy-5-methoxyflavanone, which were extracted from the leaves of *Friesodielsia discolour*, both demonstrated antiplasmodial activity, as was noted in numerous investigations [39]. Miliusacunines A and B are phytoconstituents isolated from acetonic extract of *Miliusa cuneatas* leaves and twigs. As opposed to Miliusacunines B, which inhibited the K1 malarial strain, Miliusacunines A showed antimalarial efficacy by inhibiting the TM4 malarial strain at IC_{50} 10.8 mM. An Australian tree species called *Mitrephora diversifolia* yielded the alkaloid 5-hydroxy-6-methoxy onychia, which has antiplasmodial action with IC_{50} values of 11.4 and 9.9 mM against the Dd2 and 3D7 clones of *Plasmodium falciparum*, respectively [40]. Seven phytocompounds isolated from *Rhaphidophora decursiva* leaves and stem methanolic extract exhibited antibacterial action against W2 and D6 clones. Moreover, rhaphidecurperoxin and polysyphorin both exhibited antimalarial efficacy against the W2 and D6 *Plasmodium* strains. Rhaphidecursinols A and B, grandisin, epigrandisin, and decursivine are some additional phytoconstituents with antiplasmodial action that were likewise active against *Plasmodium falciparum*. Through an antimalarial bioassay-directed separation process, a novel steroidal glycoside from *Gongronema nepalense* was discovered and named gongroneside

A. The study detailed this plant's antimalarial properties. Luteolin 7-O-glucoside and Apigenin 7-O-glucoside, two flavonoid glycosides isolated from *Achillea millefolium*, exhibited antimalarial activity against W2 and D10 strains. Another bioactive substance found in *Carpesium divaricatum*, 2-isopropenyl-6-acetyl-8-methoxy-1,3 benzo-dioxin-4-one, has antimalarial action that inhibits D10 strains. For the treatment of malaria, plants have proved to be a precious source of active components. The herbs mentioned above have been discovered to have malaria-curing potential [41, 42]. Some structures of antimalarial drugs and their analogues are shown in Fig. 15.5.

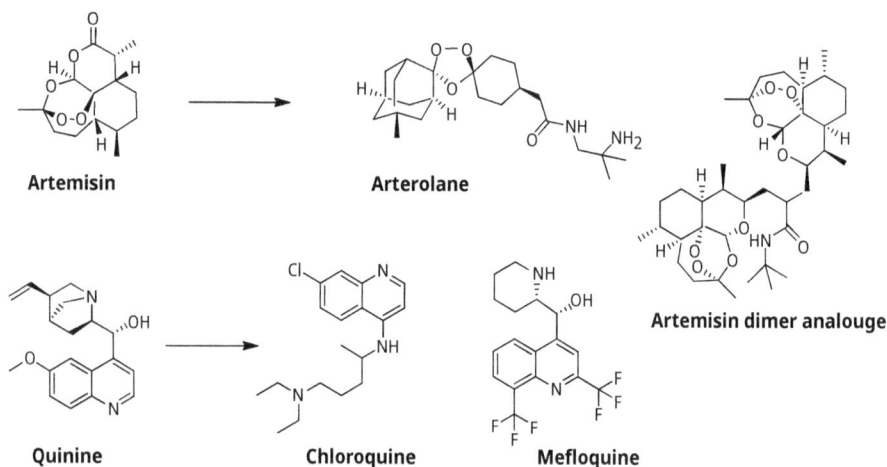

Fig. 15.5: Structures of antimalarial drugs and their analogues.

15.5.5 Role of antidiabetic lead compounds for drug development

Clinical and pharmaceutical studies on medicinally important plants have revealed antidiabetic properties as well as the recovery of the islet of Langerhans beta cells. Finding lead compounds has increasingly focused on the Kingdom Plantae. Finding a unique compound from the myriad of other chemicals in the genuine extract is necessary for the discovery of lead compounds from natural sources [43]. Numerous lignans and neolignans serve as lead molecules in the development of novel medications. *Pongamia pinnata* fruits are the source of the lead chemicals karanjin and pongamol (Fig. 15.6), which have antihyperglycemic properties. These substances are capable of suppressing protein tyrosine phosphatases (PTPase), which are crucial for signal transduction pathways and responsible for the dephosphorylation of tyrosine residues. Because they block the PTPase-mediated negative insulin signaling pathway, PTPase inhibitors increase insulin sensitivity and may be a promising treatment for T2DM and its consequences [44]. Caffeic acid with 3-(3, 4-dihydroxyphenyl)-lactic acid forms the acid ester rosmarinic acid which is also known as danshensu. Herbs from the Lamiaceae family have been identified to contain this rosmarinic acid. They consist of *Thymus vulgaris, Melissa officinalis, Salvia officinalis, Origanum vulgare,* and *Rosmarinus officinalis.* These plant species are exceptional for both their therapeutic purposes and abilities to taste food. It has the potential to reduce plasma lipid levels while also enhancing insulin sensitivity. Rosmarinic acid is a bioactive compound that can be found in many different herbs used as folk remedies. The protective effects of rosmarinic acid are b-cell dysfunction and glucolipotoxicity-mediated oxidative stress, as well as its inhibitory activity against sugar-digesting enzymes. Previous research has shown how effective rosmarinic acid is at regulating plasma glucose levels and improving insulin sensitivity in hyperglycemia [45].

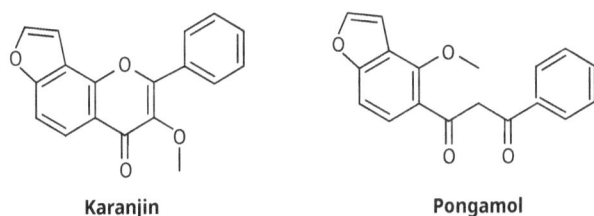

Karanjin   Pongamol

Fig. 15.6: Chemical structure of antidiabetic lead compounds karanjin and pongamol.

15.6 Conclusion

In conclusion, natural products are a promising tool for the discovery of new drugs and their structural analogues. Plant phytochemicals are mainly concerned due to their high beneficial pharmacological properties such as anticancer, antidiabetic, and antimicrobial. The diversity in the structure of phytochemicals is the reason for new drug discovery and lead optimization. Drug development from the bioactive phytochemicals faced critical challenges in accessibility, supply chain, and geographical constraints. However, new technological advancements supported a new drug discovery approach through high-throughput computational screening and combinatorial structural modifications in the development of the lead bioactive compound. In this chapter, some important plant phytochemicals are discussed in detail for specific structural modifications that result in enhanced pharmacological effects and the discovery of new drug lead compounds.

References

[1]   Z. Guo. Acta Pharm Sin B. 2017, 7(2), 119–136.
[2]   S. Mathur, C. Hoskins. Biomedical Rep. 2017, 6(6), 612–614.
[3]   H. F. Ji, X. J. Li, H. Y. Zhang. EMBO Reports. 2009, 10(3), 194–200.
[4]   A. L. Harvey. Drug Discovery Today. 92008, 13:894–901
[5]   C. Katiyar, A. Gupta, S. Kanjilal, S. Katiyar. Ayurveda. 2012, 33(1), 10–19.
[6]   F. E. Koehn, G. T. Carter. Nat Rev Drug Discovery. 2005, 4, 206–220
[7]   E. A. Blomme, Y. Will. Chem Res Toxicol. 2016, 29(4), 473–504.
[8]   M. Hay, D. W. Thomas, J. L. Craighead, C. Economides, J. Rosenthal Nat Biotechnol. 2014, 32(1),
      40–45.
[9]   H. Bais, R. Vepachedu, S. Gilroy, R. Callaway, J. Vivanco. Science. 2003, 301(5638), 1377–1380.
[10] G. M. Cragg, D. J. Newman. Biochimica Et Biophysica Acta – Gener Subj. 2013, 1830, 3670–3695.
[11] D. Newman, G. Cragg. J Nat Prod. 2007, 70(3), 461–477.
[12] D. Newman, G. Cragg, K. Snader. J Nat Prod. 2003, 66(7), 1022–1037.
[13] N. S. Ncube, A. J. Afolayan, A. I. Okoh. Afr J Biotechnol. 2008, 7(12), 1797–1806.
[14] E. Cieslik, A. Greda, W. Adamus. Food Chem. 2006, 94, 135–142.
[15] A. Scalbert, C. Manach, C. Morand, C. Remesy. Crit Rev Food Sci Nutr. 2005, 45, 287–306.
[16] G. Sudhakaran, P. Prathap, A. Guru, R. Rajesh, S. Sathish, T. Madhavan, M. V. Arasu, N. A. Al-Dhabi,
      K. C. Choi, P. Gopinath, J. Arockiaraj. Cell Biol Int. 2022, 46(5), 771–791.
[17] P. Swapnil, M. Meena, S. K. Singh, U. P. Dhuldhaj, Harish, A. Marwal. Curr Plant Biol. 2021, 26,
      100203.
[18] B. Biswas, M. Golder, A. M. Abid, K. Mazumder, S. K. Sadhu. Heliyon. 2021, 7, 75–80.
[19] D. B. Nandini, R. S. Rao, B. S. Deepak, P. B. Reddy. J Oral Maxillofacial Pathol. 2020, 24(2), 405.
[20] R. Chauhan, S. Keshavkant, A. Quraishi. Ind Crops Prod. 2018, 113, 234–239.
[21] K. Marcinek, Z. Krejpcio. RocznikiPanstwowegoZakladuHigieny. 2017, 68(2), 123–129.
[22] S. Upadhyay, M. Dixit. Oxid Med Cell Longev. (2015), 2015, 504253.
[23] D. T. Dave, G. B. Shah. J Pharmacol. 2015, 4(6), 307–310.
[24] C. Gupta, D. Prakash. J Complement Integr Med. 2014, 11(3), 151–169.
[25] P. G. Bradford, A. B. Awad. Mol Nutr Food Res. 2007, 51(2), 161–170.
[26] D. Prakash, R. Dhakarey, A. Mishra. Indian J Agric Biochem. 2004, 17, 1–8.
[27] L. Packer, S. U. Weber. The Role of Vitamin E in the Emerging Field of Nutraceuticals. Marcel Dekker,
      New York, 2001, pp. 27–43.
[28] D. S. Fabricant, N. R. Farnsworth. Environ Health Perspect. 2001, 109, 69–75.
[29] B. E. Van-Wyk, M. Wink. Medicinal Plants of the World. Timber Press Inc, Oregon, 2004.
[30] Z. Xiao, S. L. Morris-Natschke, K. H. Lee. Med Res Rev. 2016, 36(1), 32–91.
[31] H. Yao, J. Liu, S. Xu, Z. Zhu, J. Xu. Expert Opin Drug Discovery. 2017, 12(2), 121–140.
[32] C. Ding, Q. Tian, J. Li, M. Jiao, S. Song, Y. Wang, Z. Miao, A. Zhang. J Med Chem. 2018, 61(3), 760–776.
[33] R. Singh, H. Kaur. Synthesis. 2009, 2009, 2471–2491.
[34] M. Ma, L. Sun, H. Lou, M. Ji. Chem Cent J. 2013, 7, 179.
[35] E. Jameel, T. Umar, J. Kumar, N. Hoda. Chem Biol Drug Des. 2016, 87, 21–38.
[36] V. Srivastav, M. Tiwari, X. Zhang, X. Yao. Indian J Pharm Sci. 2018, 80, 108–117.
[37] L. D. Nie, X. X. Shi, K. H. Ko, W. D. Lu. J Org Chem. 2009, 74, 3970–3973.
[38] L. D. Nie, X. X. Shi. Tetrahedron. 2009, 20, 124–129.
[39] U. Prawat, D. Phupornprasert, A. Butsuri, A. Salae, S. Boonsri, P. Tuntiwachwuttikul. Phytochem Lett.
      2012, 5(4), 809–813.
[40] D. Mueller, R. Davis, S. Duffy, V. Avery, D. Camp, R. Quinn. J Nat Prod. 2009, 72(8), 1538–1540.
[41] I. Chung, S. Seo, E. Kang, W. Park, S. Park, H. Moon. Phytother Res. 2010, 24(3), 451–453.

[42]   S. Vitalini, G. Beretta, M. Iriti, S. Orsenigo, N. Basilico, S. Dall'Acqua, M. Iorizzi, G. Fico. Acta Biochimica Polonica. 2011, 58(2), 203–209.

[43]   S. Apers, A. Vlietinck, L. Pieters. Phytochem Rev. 2003, 2(3), 201–217.

[44]   A. K. Tamrakar, P. P. Yadav, P. Tiwari, R. Maurya, A. K. Srivastava. J Ethnopharmacol. 2008, 118(3), 435–439.

[45]   W. S. Waring. Medicine. 2016, 44(3), 138–140.

[46]   N. Choudhary, G. L. Khatik, A. Suttee. Curr Diabetes Rev. 2021, 17(2), 107–121.

[47]   H. Zhang, P. Tamez, Z. Aydogmus, G. Tan, Y. Saikawa, K. Hashimoto, M. Nakata, N. Hung, N. Cuong, D. Soejarto, J. Pezzuto. Planta Medica. 2002, 68(12), 1088–1091.

Anton Soria-Lopez, Nuno Muñoz-Seijas, Rosa Perez-Gregorio,
Jesus Simal-Gandara and Paz Otero*

Chapter 16
Extraction and production of drugs from plant

Abstract: Plants are producers of many secondary metabolites with biological activities like alkaloids, terpenes, lignans, tannins, and phenolic compounds which can be used in the development of pharmaceutical drugs. Because of the complex chemical structure, these compounds are difficult to synthesize at an industrial level. Thus, using appropriate extraction methodologies to obtain extract rich in these phytochemicals is crucial. In this context, infusion, maceration, percolation, and soxhlet are currently used traditional extraction methodologies to produce drug from medicinal plants. However, to improve the extraction performance, other innovative techniques like ultrasound-assisted extraction, microwave-assisted extraction, pressurized liquid extraction, and supercritical fluid extraction have been developed in the recent years. The parameters and procedures of these techniques play an important role in the extraction performance and yield of these metabolites. This chapter aims to present and compare extraction procedures of phytochemicals from medicinal plants as well as describe some trends in the natural compounds biology to improve the production of the secondary metabolites from medicinal plants.

Acknowledgment: The authors would like to thank the FPU grant for A. Soria-Lopez (FPU2020/06140).

*Corresponding author: Paz Otero**, Nutrition and Bromatology Group, Department of Analytical Chemistry and Food Science, Faculty of Science, Universidade de Vigo – Ourense Campus, E-32004 Ourense, Spain, e-mail: paz.otero@uvigo.es
Anton Soria-Lopez, Jesus Simal-Gandara, Nutrition and Bromatology Group, Department of Analytical Chemistry and Food Science, Faculty of Science, Universidade de Vigo – Ourense Campus, E-32004 Ourense, Spain
Nuno Muñoz-Seijas, Faculty of Science, Universidade de Vigo – Ourense Campus, E-32004 Ourense, Spain
Rosa Perez-Gregorio, Associated Laboratory for Green Chemistry of the Network of Chemistry and Technology (LAQV-REQUIMTE), Departamento de Química e Bioquímica, Facultade de Ciencias da Universidade de Porto, Porto, Portugal; Nutrition and Bromatology Group, Department of Analytical Chemistry and Food Science, Faculty of Science, Universidade de Vigo – Ourense Campus, E-32004 Ourense, Spain

https://doi.org/10.1515/9783110791891-016

16.1 Introduction

Plants contain essential compounds for their physiological functions and their development. Because of their secondary metabolism, they are producers of many bioactive compounds susceptible to be used as a starting point in the development of pharmaceutical drugs [1, 2]. One of the challenges for drug discovery is gaining access to medicinal plants, despite their metabolites are generally found in small quantities [3]. According to their biosynthetic route, secondary metabolites are classified into three main groups: phenolic compounds (phenolic acids, coumarins, stilbenes, flavonoids, lignans, tannins, and lignin) which are synthesized in the shikimate pathway; terpenes (like terpenoids, carotenoids, and sterols) synthesized in the mevalonic pathway; and nitrogen-containing compounds (like alkaloids and glucosinolates) synthesized in the tricarboxylic acid cycle pathway [4]. Among them, the major plant secondary metabolites are alkaloids, terpenes, and phenolic compounds [5].

Alkaloids are secondary metabolites of plants characterized by a chemical structure containing a heterocyclic ring [6]. These are mostly soluble in aqueous solutions and are easily extracted in water upon protonation of nitrogen. Depending on their biosynthetic pathway, they are classified as true or proper alkaloids, protoalkaloids (both originated from amino acids), and pseudoalkaloids which are not originated from amino acids [7]. However, the most established classification is that based on their ring structure: pyridine, tropane, isoquinoline, phenanthrene, phenylethylamine, indole, purine, imidazole, or terpenoid group [2]. Some alkaloids obtained from plants with medicinal effects are vinblastine, nicotine, quinine, cinchonine, nibidine, strychnine, rawelfinine, glycyrrhizin, and punarnavine [8]. Caffeine, nicotine, cocaine, theobromine, and morphine present stimulant activity on the central nervous system and exert anxiolytic, analgesic, and hallucinogenic effects [9]. In plants, alkaloids are present in the form of salts of organic acids mainly malate, acetate, and citrate, or combined with other molecules, such as tannins [10]. In general, free alkaloids are soluble in organic solvents, and alkaloid salts are generally soluble in water, less in alcohol, and mostly nearly insoluble or sparingly soluble in organic solvents. There are some exceptions like caffeine and ephedrine, both soluble in water. The alkaloid can be recovered from the plant matrix in the free base form by basifying the aqueous extract by mixing the pulverized and degreased plant material with an alkaline solution that displaces the alkaloids from their saline combinations. The released bases are immediately solubilized in an organic solvent of medium polarity. However, some procedures like the method of purifying the morphine does not appear to be sustainable and efficient, since considerable quantities of alcohol required, owing to the slight solubility [11].

Terpenes, also called isoprenoids, are the largest family of compounds found in essential oils. They are hydrocarbons made up of isoprene units [12]. Based on their number in the carbon skeleton, terpenes are classified into hemi-, mono-, sesqui-, di-, sester-, tri-, sesquar-, tetra-, and polyterpenes [12]. The terpenes containing additional oxygen-containing functional groups are called terpenoids [13]. Terpenes and terpenoids have been reported to exhibit diverse biological activities including anticancer, antitubercular, anxiolytic, and mutagenic active molecules [14–17]. Some current terpenes in therapeutic use are limonene for their anticarcinogenic properties and linalool due to their antibacterial effects [15, 16]. Other example is thymol, a natural monoterpene derivative of cymene, obtained at the pharmaceutical industry level from *Thymus vulgaris* L. and *Thymus zygis* L., which contains antifungal, antiviral, antileishmanial, anticancer, and antibiofilm properties. One technique to isolate terpenes is based on solvent-free extractions (distillation and hydrodistillation) in which heat and pressure are used. The advantage of these methods is the fact that there is no danger of external chemical residues caused by solvents. However, they require high temperatures and long extraction times, which can alter and even destroy many plant compounds, including terpenes. A method that offers better yields is the use of gases such as CO_2 at a lower boiling point. These methods incorporate vacuum pumps to reduce the initial heat, keeping temperatures low enough to avoid degrading any bioactive compounds [18, 19].

Phenolic compounds contain at least one phenol group in their structure and several other functional groups such as esters, methyl esters, and glycosides [20]. These compounds mainly contain antiviral, antimicrobial, antihypertensive, antioxidant, and lipid-lowering effects [21, 22]. Some phenolic compounds of interest because of their antiviral activity are punicalin, punicalagin, and furin [22]. Furthermore, resveratrol has been shown to have antithrombotic and anti-inflammatory effects [23]. Some of the most traditional used techniques for the extraction of phenolic compounds are maceration, infusion, percolation, and soxhlet [24, 25]. The polarity of the solvent used for the extraction of phenolic compounds from plants determines the yield of the extraction. The use of ethanol, methanol, acetone, and their mixtures with water in different proportions has been reported, but there is no defined method and solvent, as this will depend on the chemical nature of the compounds to be extracted, their amount and position of its hydroxyl groups, molecular size, as well as factors such as the temperature, the extraction time, and the particle size [26, 27]. Extraction with organic solvents is efficient and simple, but expensive since large amounts of solvents are needed. It can also be harmful for human use due to traces of organic solvent that may remain in the extract. In the last years, other techniques have been developed to minimize the use of solvent and to increase the extraction yield. They include supercritical fluid extraction (SFE), ultrasonic-assisted extraction (UAE), pressurized liquid extraction (PLE), and microwave extraction [18, 28].

Considering the numerous extraction procedures to obtain secondary metabolites of interest from medicinal plants (Fig. 16.1) and the need to obtain secondary metabolites more affordable, this chapter aims to show and compare the different extraction methods of phytochemicals from medicinal plants, and to describe some trends in the natural compounds biology, such as strategies in situ or cell culture systems, to enhance the production of the secondary metabolites from medicinal plants.

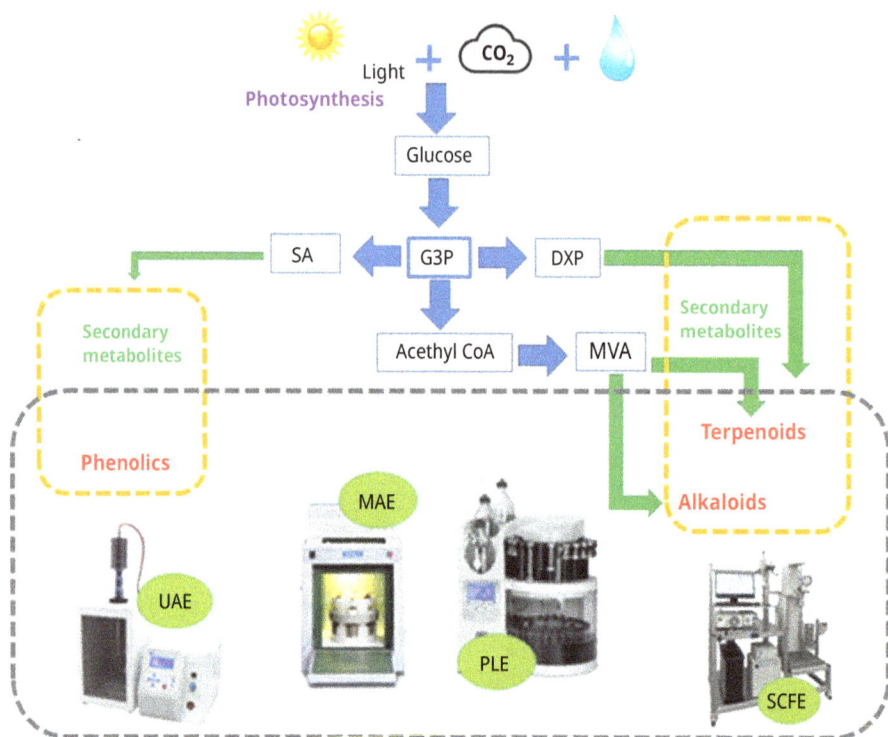

Figure 16.1: Biosynthetic routes of main secondary metabolites from plants and the main innovate extraction techniques for drug production from plants. SA: shikimic acid, G3P: glyceraldehyde-3-phosphate, DXP: 1-deoxyxylulose-5-phosphate, MVA: mevalonic acid. UAE: ultrasound assisted extraction, MAE: microwaves assisted extraction, PLE: pressurized liquid extraction, SCFE: supercritical fluid extraction.

16.2 Traditional extraction technologies of drugs from plants

16.2.1 Infusions

Infusion is a traditional extraction method, which consists of immersing the plant containing the drug of interest in water [29]. The process can be with hot or cold water, according to the nature of phytochemicals to be extracted. If they can be degraded by temperature, cold infusion must be performed [29]. In most cases, the plant material is first brought into cold water for a short period of time, and then in boiling water for longer time [30]. The process can be with or without agitation, and finally a filtration process is carried out to eliminate the present residues. Currently, the preparation of infusions usually consists in making a concentrate of phytochemicals and then, from this solution, a 10 times dilution is made with water [31]. Generally, infu-sions are realized for immediate use because they do not have preservatives for their conservation. Those infusions that require some preservation, approximately 25% of alcohol can be added to preserve phytochemicals [29]. Table 16.1 shows some applica-tions of infusions to obtain bioactive compounds from plants, like the extraction of vanillic acid from aracá (*Psidium cattleianum*) [32] and phenolic compounds from *Hel-ichrysum plicatum* and *Maytenus boaria* [33]. In this last case, Soto-Maldonado and colleagues used this technique for the extraction of phenolic compounds from the leaves of *Maytenus boaria* submerging the entire leaf in water for 5 min and obtaining a recovery of 2.36 ± 0.36 g GAE/ kg leaves [34].

16.2.2 Maceration

Maceration is a solid–liquid extraction, in which the drugs are mixed with a liquid or extractant for 24–72 h with frequent stirring. The agitation is of vital importance be-cause the extraction process works by molecular diffusion, and the agitation favors the correct dispersion of the solution as well as increases the diffusion speed [31]. After that, the drug of interest is dissolved and filtered [31]. The ratio between the solvent and the drug widely varies in relation to the type of drug to be extracted. Usu-ally, maceration employs a solid:liquid ratio between 1:3 and 1:20. In organized drugs, with a well-defined cellular structure, minimal volume of solvents per amount of drug is needed [31].

There are different types of maceration. Based on the temperature, they can be classified as cold or hot maceration. The main advantages of cold maceration are that the equipment does not need a high energy requirement, and cold maceration can be used for thermolabile compounds [35]. However, the time required to obtain the drug is higher than in hot maceration, so when the compounds are thermostable, maceration with heat is chosen [35]. Maceration with heat has the advantage of shorter extraction times that are required since the temperature applied favors the speed of the extrac-tion. Their main disadvantage is that it does not completely extract the drug of interest because certain compounds of the drug can be altered by being in contact with high temperatures [29]. The use of steam currents to shorten the extraction times is often used to solve this issue.

Table 16.1 shows several examples of drug extraction using maceration. For exam-ple, rosmarinic acid, which is an anti-inflammatory and antioxidant compound, was extracted from plants of the Lamiaceae family using 70% ethanol at 25 °C for a period of time higher than 30 min, obtaining an extraction yield of 28.9 ± 0.1 mg/g [36]. Paral-lelly, Jiang and colleagues extracted curcumin, a compound presenting a broad anti-inflammatory capacity, from *Curcuma longa* L. It was obtained using acetone in a ratio 1:8 (solid:solvent), at 30 °C, for 3 h, obtaining yields of 69.67% [37].

16.2.3 Percolation

Percolation is a traditional extraction process in which the raw material is in contact with a solvent that is renewed continuously in a container called percolator, allowing the filtration of the solvent when it is saturated [38]. Since the dry plant material suffers a strong swelling in contact with the solvent, it is important not to introduce the dry drug into the percolator, which can affect negatively the flow of the solvent and can result in loss of small particles generated during the percolation process [39]. This issue can be solved by uniformly moistening the plant material before starting the process. After the drug is introduced into the percolator, enough liquid to saturate the plant material is added and left for several hours at temperature below 25 °C. After that, the percolation process begins properly by opening the key and allowing the flow of solvent in a uniform mode being in contact with the plant material constantly. When 85% of the liquid (approx.) has passed through the system, the container is changed and an extract with a very high concentration of the drug is obtained [39]. Sometimes a countercurrent flow to bring the plant material in contact with the new solvent is used. It increases the concentration gradient between them accelerating the extraction. Table 16.1 shows some examples of drug extraction by percolation. Daud and coworkers have obtained several flavonoids from residues of *Artocarpus heterophyllus* with 70% ethanol, at 25 °C and 1 h extraction, obtaining 79.5 ± 0.1 mg GAE/g dw [40]. Pyrrolizidine alkaloids from *Symphytum* spp. were obtained using methanol at room temperature for 2 h [41]. This is really one of the most used methods as it does not require much manipulation or time. It is important to have in mind that in water: alcohol mixtures, the alcohol can be evaporated, generating a practically aqueous extract which may cause drug precipitation.

16.2.4 Soxhlet

This is a traditional extraction method, which uses a particular type of condenser known as the soxhlet apparatus [42]. The extractor is composed of a thimble-holder, where the drug is placed inside the thimble, and a distillation flask where the fresh solvent is added. The process consists of placing the plant material on a filter paper by inserting it in the central tube of the extractor, placing in the flask the solvent. When the solvent reaches the boiling point, it vaporizes and enters the matrix, dissolving the drug. After that, the solvent hits the cooling tubes of the condenser and condenses back into the initial flask with the drug. The process is repeated until the drug is completely extracted [42, 43]. Some applications of soxhlet with different solvents are in Tab. 16.1. Flavonoids from *Morus nigra* are obtained with petroleum ether at 50 °C for 3 h [44] and the andrographolide from *Andrographis paniculata* using methanol for 6 h [45].

Tab. 16.1: Traditional methodology employed for the extraction of phytochemicals from plants.

Phytochemical	Plants	Dissolvent/temperature/time	Recovery	Ref.
Infusions				
Total phenolic content	*Maytenus boaria mol.*	S: H_2O, T = boiling, E.t = 5 min, 1 g/200 mL (m/v)	2.36 ± 0.36 g GAE/ kg leaves	[34]
Phenolic compounds	*Helichrysum plicatum* DC. subsp. *plicatum*	S: H_2O, T = cold, E.t = 30 min	27.66%	[33]
Vanillic acid (PC)	*Psidium cattleianum* Sabine	S: H_2O, T = 80 °C, E.t = 10 min, 1:50 (m/v)	28.20 ± 0.65 µg/g	[32]
Maceration				
Benzyl cyanide (A)	*Lepidium sativum* L.	S: 96% EtOH, T = 50 °C, E.t = 24 h	17.14 ± 0.86 µg/g	[87]
Rosmarinic acid (PC)	Lemon balm	S: EtOH:H_2O:HCl (70:29:1), T = 25 °C, E.t = 30 min	28.9 ± 0.1 mg/g	[36]
Curcumin (PC)	*Curcuma longa* L.	S: Ac, T = 30 °C, E.t = 3 h, 1:8 (m/m)	69.67%	[37]
Total phenolic contents	*Papaver rhoeas* L.	S: MeOH, T = room, E.t = 48 h	95.4 ± 2.42 mg GAE/g	[25]
Polyphenols	*Thymus serpyllum* L.	S: 50% EtOH, T = room, E.t = 1 h, 1:30 (m/m)	26.6 mg GAE/L	[88]
Percolation				
Total phenolic content	*Artocarpus heterophyllus*	S: 70% EtOH, T = 25 °C, E.t = 1 h, 1:10 (m/v), flow rate = 1 mL/min	79.5 ± 0.1 mg GAE/g dw	[40]
Isoquercetin	*Moringa oleifera* L.	S: 70% EtOH, T = room; flow rate = 1 mL/min	0.0713 ± 0.011%	[89]
Crypto-chlorogenic acid	*Moringa oleifera* L.	S: 70% EtOH, T = room; flow rate = 1 mL/min	0.0956 ± 0.011%	[89]
Pyrolizidine alkaloids (A)	Symphytum	S: MeOH, T = room; E.t = 2 h	640 ppm	[41]
Soxhlet				
Flavonoids	*Morus nigra*	S: petroleum ether, T = 50 °C; E.t = 3 h	58.94%	[44]
Total alkaloid content (A)	*Annona muricata*	S: EtOH, T = 80 °C; E.t = 7 h	16.58 ± 0.43 mg AE/g	[47]

Tab. 16.1 (continued)

Phytochemical	Plants	Dissolvent/temperature/time	Recovery	Ref.
Aristolene (TE)	*Renealmia petasites* Gagnep	S: petroleum ether, $T = 80\ °C$; E.t = 1 h	1.07 ± 0.16	[17]
Andrographolide (TE)	*Andrographis paniculata*	S: MeOH, T = nd.; E.t = 6 h	0.452%	[45]

Ac, acetone; A, alkaloid; E.t, extraction time; EtOH, ethanol; MetOH, methanol; nd., not determined; PC, phenolic compounds; S, solvent; T, temperature; TE, terpene; H_2O, water.

The use of soxhlet for the extraction of drugs has several advantages. The temperature is maintained throughout the process and does not require drug filtration, being perfectly separated from the original plant material. In addition, it allows the extraction of several samples simultaneously at easy operational process and low cost [43, 46]. However, it is not useful with thermolabile drugs, since the temperature used is high [42]. Another disadvantage is the large amount of organic solvents used and the long periods of extraction time [43]. To solve these issues, in the last years, soxhlet has evolved to more automating processes, aiming to shorten the extraction times. In addition, it is being coupled to innovative technologies like supercritical fluid-soxhlet extraction, ultrasounds, or microwaves, resulting in a more efficient technique than the traditional soxhlet [43, 46]. Table 16.1 shows some examples. Lee extracted alkaloids from *Annona muricata*, using more complex methods such as the ultrasound-mechanical stirrer technique. Ethanol is used as a solvent at a fairly high temperature, close to 80 °C for approximately 7 h, obtaining 16.58 ± 0.43% [47]. Recently, dos Santos and colleagues extracted aristolene from *Renealmia petasites* Gagnep, using petroleum ether at 80 °C for 1 h and obtained a recovery of 1.07 ± 0.16% [17].

16.3 Green extraction technologies of drugs from plants

16.3.1 Ultrasound-assisted extraction (UAE)

UAE is an innovative technique in which solvent volumes, extraction times, and economic and environmental impacts are lower than conventional techniques to extract drug from plants. Ultrasound equipment is considered a "clean technology" because an environmentally friendly extraction is used [48]. This equipment employs sound waves with frequency ranges between 20 and 100 kHz for the extraction and process applications [48, 49]. Table 16.2 shows studies in which several phenolic compounds, alkaloids, and terpenes are extracted from medicinal plants by UAE, obtaining good yields. The technology is based on the principle of acoustic cavitation. This physico-chemical phenomenon, induced by ultrasonic waves, encompasses all formation processes of bubble, as its growth (expansion and compression cycles) and collapse produced in liquid medium. The bubble collapse leads to high temperatures at a localized level (>5,000 K) and an important release of free radicals [48]. There are a series of parameters affecting extraction efficiency [50]. The physical parameters can be associated with ultrasonic waves such as power (P), frequency (F), and ultrasonic intensity (UI), or associated with the equipment such as extraction time (E.t) and the shape and size of ultrasonic reactors [48].

Tab. 16.2: Novel analytical methodology employed for the extraction of phytochemicals from plants.

Phytochemical	Plant	Parameters	Recovery/yield	Ref.
Ultrasonic-assisted extraction (UAE)				
Total flavonoids (PC)	*Myrtus communis*	S: 50% EtOH, E.t = 15 min, F = 20 kHz, 50 mL/g (v/m)	3.88 ± 0.45 mg QE/g	[28]
Condensed tannins (PC)			23.32 ± 0.01 mg/g	
Caffeine (A)	*Coffea arabica*	S: H_2O, E.t = 5 min, F = 35 kHz, T = 80 °C	27.65 mg/g	[55]
Chlorogenic acid (PC)		S: H_2O, E.t = 10 min, F = 35 kHz, T = 80 °C	16.66 mg/g	
Curcumin (PC)	*Curcuma longa*	S: Ac, E.t = 7 min, P = 150 W	71.42%	[52]
Shikimic acid (PC)	*Eichhornia crassipes*	S: H_2O, E.t = 30 min, F = 26 kHz, P = 200 W	3.1% (stem) 0.56% (leaves) 0.33% (roots)	[53]
Caffeic acid (PC)	*Physalis angulata* L.	S: 57% H_2O + 35% EtOH + 8% MetOH, E.t = 10 min, F = 50/60 Hz, P = 90 W	25.0–32.5 µg/g	[54]
Rutin (PC)			75.6–88.2 µg/g	
Chlorogenic acid (PC)	*Lonicera japonica* Thunb	S: ionic liquid, E.t: 40 min, P = 200 W, T = 60 °C, pH = 1.2	94.6–103.2%	[56]
Microwave-assisted extraction (MAE)				
Total flavonoids (PC)	*Myrtus communis*	S: 42% EtOH, E.t = 1.04 min, P = 500 W, 32 mL/g (v/m)	5.02 ± 0.05 mg QE/g	[28]
Condensed tannins (PC)			32.65 ± 0.01 mg /g	
Curcumin (PC)	*Curcuma longa*	S: Ac, E.t = 5 min, P = 140 W	68.57%	[52]

Tab. 16.2 (continued)

Phytochemical	Plant	Parameters	Recovery/yield	Ref.
Theobromine (A)	*Theobroma cacao* L.	S: 30% H_2O/70% DES mixture, E.t = 10 min, P: 600–800 W, T = 60 °C	5.004 mg/g	[61]
Caffeine (A)			1.599 mg/g	
Andrographolide (TE)	*Andrographis paniculata* Nees	S: 60% MetOH, E.t = 6 min, P = 80 W, flow rate = 1 mL/min	97.7%	[62]
Dehydroandro-grapholide (TE)			98.7%	
Neferine (A)	*Nelumbo nucifera* Gaertn.	S: [C4MIM][BF4], E.t: 90 s, P = 280 W, 1:10 (v/m)	15.76 mg/g 104.9% (yield)	[64]
Berberine (A)	*Berberis jaeschkeana*	S: 100% MetOH, E.t: 2 min, P = 598 W	46.38 mg/g	[63]
Palmatine (A)			20.54 mg/g	
Pressurized liquid extraction (PLE)				
Caffeic acid (PC)	*Ziziphus jujuba* Mill.	S: 70% EtOH, E.t: 105 min, P: 5–10 bar	70.00 ± 1.23 µg/g	[90]
Luteolin (PC)			5.59 ± 0.14 µg/g	
Salvianolic acid (PC)	*Symphytum officinale* L.	S: 85% EtOH, E.t: 20 min, P: 1,500 psi, T = 63 °C	–	[68]
Vanillic acid (PC)	*Psidium cattleianum* Sabine	S: H_2O, E.t: 60 min, P = 1,500 psi, T = 50 °C	2326.92 ± 3.13 µg/g	[32]
Catechin (PC)		S: H_2O, E.t: 60 min, P = 1500 psi, T = 50 °C	1999.78 ± 13.36 µg/g	
Rosmarinic acid (PC)	*Melissa officinalis*	S:EtOH, E.t: 20 min, T = 150 °C	90.53 ± 4.74 µg/mg	[69]
Thymol (TE)	*Thyme vulgaris*	S: EtOH, E.t: 10 min, P = 10 MPa, T = 130 °C	10.98 ± 1.49 mg/g	[18]
Supercritical fluid extraction (SFE)				
β-Carboline (A)	*Peganum harmala* L.	S: CO_2 + EtOH (3% w/w), E.t: 35 min, P = 176 atm, CO_2 flow rate: 0.3 mL/min, T = 40 °C	3.6%	[72]
Vinblastine (A)	*Catharanthus roseus* L.	S: CO_2 + EtOH (5% w/w), E.t = 130 min, P = 300 bar, flow rate = 1,000 g/h, T = 40 °C	301.7 µg/g	[74]

Tab. 16.2 (continued)

Phytochemical	Plant	Parameters	Recovery/yield	Ref.
Colchicine (A)	*Colchicum speciosum* L.	S: CO_2 + MetOH (5% w/w), E.t: 55 min, P = 247 bar, CO_2 flow rate = 1.5 mL/min, modifier flow rate = 0.5 mL/min, T = 35 °C	1.44%	[77]
Pyrrolidine (A)	*Piper amalago* L.	S: CO_2 + MetOH (5% w/w), E.t = 60 min, P = 200 bar, CO_2 flow rate = 3 mL/min, T = 40 °C	3.8 ± 0.8 mg/g	[73]
Thymol (TE)	*Thyme vulgaris*	S: CO_2 + EtOH (10% w/w), E.t = 240 min, P = 15 mPa, CO_2/thyme ratio = 35 kg/kg, CO_2 flow rate: 60–70 g/min, T = 40 °C	10.27 mg/g	[18]
Camphor (TE)	*Artemisia sieberi* L.	S: CO_2, E.t: 40 min, P = 30.4 MPa, CO_2 flow rate = 0.30 ± 0.05 mL/min, T = ~ 55 °C	77.43%	[19]

Ac, acetone; A, alkaloid; DESs, deep eutectic solvents; E.t, extraction time; EtOH, ethanol; F, frequency; MAE, microwave-assisted extraction; MetOH, methanol; PLE, pressurized liquid extraction; UAE, ultrasonic-assisted extraction; P, power or pressure; R, ratio; S, solvent; SFE, supercritical fluid extraction; T, temperature; TE, terpene.

Concerning P, the higher P, the higher UAE efficiency in terms of yield and extract composition because of the generation of strong shear forces [51]. In fact, the most used powers are those that correspond to the intermediate range (100–300 W). Several studies employed these ranges, and good extraction results were obtained [52–54] (Tab. 16.2). Regarding F, this parameter plays a relevant role in UAE since it modulates the type of effect that predominates during the extraction process, and also determines the duration of the expansion/compression cycles of bubble before its collapse [48]. In several extractions of drugs from plants, frequencies less than 50 kHz are used, resulting in excellent extractions [28, 53, 55] (Tab. 16.2).

UI is another physical parameter that plays a remarkable role in UAE, since a minimum intensity is required for the cavitation process, being higher this minimum at higher power. Extraction time is a parameter associated with the equipment and depends on the properties of the plant material. As a general rule, yields are higher when extraction times are increased [48]. However, excessive extraction times can lead to undesirable changes in the drugs. For this reason, the most ultrasonic equipment contains extraction times below 1 h [49]. In Tab. 16.2, all drug extractions using ultrasounds are less than 45 min [28, 52–56]. Finally, the shape and size of ultrasonic reactors (ultrasonic bath and probe) can also affect extraction yields. Depending on the form and thickness of these reactors, the extraction yield and the energy transference can present different values [48]. The solvent nature and its properties, extraction temperature, and the presence of gases are considered as medium parameters [48]. In the case of the first parameter, the solvent polarity depends on the target compound. Nowadays, there are green solvents (e.g., ionic liquid or deep eutectic solvents (DES)) that can be good substitutes for conventional solvents. For instance, Zhang and colleagues used ultrasonics to extract chlorogenic acid from *Lonicera japonica* using ionic liquid as a solvent [56]. Moreover, their properties such as viscosity or vapor pressure are also important since they can affect the cavitation process.

In addition, *T* is another parameter that requires an adequate optimization to improve the extraction properties of solvent and to protect the structure of target components. High extraction temperatures lead to the destruction of interest compounds and can prevent the noninertial cavitation processes [51]. Intermediate temperatures are normally used (60–80 °C). Duangjai and coworkers employed an extraction temperature of 80 °C to extract caffeine and chlorogenic acid from *Coffea arabica* [55] (Tab. 16.2). Lastly, the presence of gases in the extraction medium is another important parameter. It is necessary for the formation of cavitation bubbles. In addition, most ultrasonic equipments use external pressures close to atmospheric pressure, since less UI is required to induce cavitation in this pressure [48].

The type of drug, their structure, pretreatment, particle size, and solid–liquid ratio are examples of matrix parameters [48]. The effectiveness of extraction depends on the structure of matrix, and the size of particle, showing good results for small sizes [50]. Moreover, many studies showed that the application of pretreatments is an important factor that affects the extraction efficiency. Finally, the solid–liquid ratio parameter may affect the extraction yield and depends on the matrix and the target compounds. In the study performed by Dahmoune in 2015, a solid–liquid ratio of 50 mL/g is used to obtain good yields of flavonoids and tannins [28] (Tab. 16.2).

16.3.2 Microwave-assisted extraction (MAE)

Microwave-assisted extraction (MAE) is another nonconventional technique to extract drugs from medicinal plants [57]. Microwave technology is mainly applied to destroy cell matrices using microwave energy (0.3–300 GHz), heating solvents by ionic conduction and dipole rotation mechanisms [58], and releasing compounds that are both diffused out and dissolved in extraction solvents. The main mechanism consists of introducing the electromagnetic waves in the medium to cause vibration of water molecules and being responsible for the temperature increase in the intracellular contents. Consequently, the water evaporated exerts pressure on the cell walls causing cell wall to break down and release the intracellular content in the medium or extraction solvent [57, 59, 60]. Table 16.2 shows different studies in which alkaloids and terpenes are extracted using MAE. The employment of this technique has several advantages

with respect to conventional techniques such as the possibility to analyze simultaneously various samples, the use of low solvent volumes, and elevated temperatures
during the extraction [59]. The extraction time in MAE is usually very short in relation
to conventional techniques. Normally, 10 min is enough to obtain good recoveries, for
instance, the extraction of theobromine and caffeine alkaloids [61]. Other studies obtained excellent yields using extraction times below 10 min [52, 62, 63], and even some
authors used duration of about 1 min [28]. However, MAE can present some disadvantages to produce drugs at an industrial scale. Among others, high operating costs, the
requirement of a cleanup step before the final analysis, and the possibility to obtain a
high content of impurities in the extract obtained [57].

There is a series of parameters influencing on the drug extraction process like the
type of solvent and volume, temperature, pressure, power, and particle size [57]. Concerning solvent composition, the microwave-absorbing properties of solvent, the interaction of the solvent with the matrix, and the analyte solubility in the solvent are
aspects to be considered when an extraction solvent is chosen [58]. In fact, the solvent
should present a high selectivity toward the drug of interest. Those solvents that absorb microwave energy there will be heating and therefore effective extraction. Polar
molecules (water, ethanol, methanol, or acetone) and ionic solutions (generally acids)
have the ability to absorb this energy [57]. Methanol or ethanol is used to extract several flavonoids, tannins, alkaloids, and terpenes, obtaining good results [28, 62, 63]. In
addition, acetone is used to extract phenolic compounds [52]. Moreover, the use of different polarity solvent mixtures is also common. For instance, Pavlović and colleagues
used a mixture of H_2O and DES to extract theobromine and caffeine from *Theobromine cacao*, obtaining good yields [61]. Ionic liquids are recent solvents that have
gained the attention nowadays [64]. Solvent volume is another parameter that can affect the extraction process. Generally, the amount of solvent required for a sample is
approximately between 10 and 30 mL [57, 61]. The use of low solvent volumes in MAE
leads to higher drug extraction yields.

Temperature is another parameter that widely affects the extraction yields. In
MAE, temperatures are usually above the solvent boiling point. The desorption of analytes from its active sites in the matrix is higher when temperature increases. Moreover, at high temperatures, the solvent presents lower surface tension and viscosity,
and a higher capacity to solubilize analytes [57]. Nonetheless, if the drug is thermolabile, high temperatures may lead to its degradation, resulting in lower recoveries.
Therefore, depending on the compound to analyze, the optimum temperatures can
considerably vary [59]. For example, theobromine and caffeine were extracted at different temperatures (30, 60, or 90 °C), showing the best results at 60 °C [61]. Nevaden-
sin was extracted using a range of temperatures between 40 and 80 °C, showing that
the best yields are obtained at 80 °C [65].

Finally, other parameters that can also affect the concentration of the drug obtained are pressure, power, and particle size. MAE usually employs elevated pressures
to keep the solvents at temperatures higher than their boiling point at atmospheric

pressures. The power value used depends on the total volume of solvent and will minimize the time required to reach the set temperature [57, 59]. Good yields of berberine and palmatine are obtained using a power of 598 W in 2 min [63], and a high amount of neferine at 280 W for 90 s is obtained [64]. The particle size can also be an important factor affecting extraction yields [65].

16.3.3 Pressurized liquid extraction (PLE)

PLE is another innovative extraction technique that is widely employed for the extraction of secondary metabolites from plants. Table 16.2 collects extraction conditions and yields for phenolic compounds and terpenes extracted from medicinal plants by PLE. This technique is considered as an alternative to traditional extraction methods such as soxhlet, maceration, or percolation [58]. Some advantages with respect to traditional methods are low extraction times and solvent volumes required. Likewise, due to the possibility of choosing different solvents for the extraction, PLE could become a versatile technique. However, the extraction could present a low selectivity [66]. The extraction efficiency depends on the matrix nature, mass transfer, and solubility, which at the same time depends on the selection of temperature, pressure, flow rate, and extraction time [67].

The temperature plays an important role in the PLE performance. The higher the temperature, the higher the extraction yield, unless thermolabile compounds are extracted since the use of high temperatures could produce their degradation. The temperature optimization is a very important consideration. The mass transfer is improved at high temperatures, since a decrease in the viscosity and the solvent surface tension results in a weak solid–matrix interaction and increases diffusivity. Moreover, the solvent polarizability is widely modified when altering the temperature [67]. In fact, the polarity of pure water at 200 °C and 1.5 MPa is like methanol at room temperature. For this reason, PLE could be a low selective technique. The temperature in PLE is typically between 50 and 200 °C [18, 32, 68, 69] (Tab. 16.2).

The high pressure employed in PLE is mainly to keep the solvent in a liquid state and wet the sample matrix [67]. The typical values of pressure varied between 500 and 3,000 psi, being 1,500 psi, a common pressure used (Tab. 16.2). Parallelly, the extraction time should be as minimal as possible, but enough to allow the drug to be extracted. There are other parameters such as the solvent to sample ratio, size of particle, dispersants, and moisture content that may affect the extraction performance, being the solvent to sample ratio as the most critical parameter [67].

16.3.4 Supercritical fluid extraction (SFE)

Besides the mentioned extraction technologies, SFE is another novel technology to extract drug from plants by supercritical fluid solvents. The supercritical condition occurs when the temperature and pressure of a certain substance are above its critical point [70]. These fluids behave both as a gas and as a liquid. In fact, they can diffuse through solids like a gas and dissolve materials like a liquid [71]. Solvents employed in SFE are CO_2, water, nitrous oxide, ethylene, propylene, propane, n-heptane, or ethanol. CO_2 is the most used solvent because it is cheap, nontoxic, nonflammable, low cost, easily removed, and with a low critical temperature and pressure. This technique is widely employed to extract vitamin E and polycyclic aromatic hydrocarbons, also to remove fat, pesticides, and alcohol from vine or beer [70]. Recently, SFE has been used to extract secondary metabolites (alkaloids and terpenes) from the medicinal plants [19, 72, 73], using CO_2 as a solvent (Tab. 16.2).

Recent years, researches have received much attention on supercritical fluids as a solvent because of their extraordinary properties [70]. There are two main advantages of SFE in relation to other mentioned techniques: (I) Solvents in critical conditions have a better capacity to dissolve solutes. According to this, the solubility can decrease by the reduction of pressure or density. The expansion is a key process in SFE, since it leads to a reduction of solvent power. For this reason, a rapid expansion of supercritical solvents leads to precipitation of solute, resulting in solvent separation. (II) Supercritical fluids have a high density and high diffusivity, and low surface tension. Due to these properties, these fluids can penetrate easily the solid matrix and dissolve solutes, resulting in good extractions of compounds of interest [18, 19, 71]. There are various important parameters that can affect the drug extraction yield by SFE with CO_2 as a solvent, namely temperature and pressure, extraction time, flow rate, and cosolvents or modifiers. They are explained further.

Temperature and pressures above CO_2 critical point (T_c = 304 K and P_c = 73 atm or 74 bar) are needed to become critical conditions. Since the CO_2 solubility increases when increasing pressure, the use of those pressure values in which its solubility reaches the maximum is preferable. Most studies used temperatures around 313 K and pressures between 150 and 300 atm [18, 19, 73, 74] (Tab. 16.2).

The main objective of any extraction is to extract the maximum compound of interest amounts in the shortest possible time. Although Norsyamimi and coworkers concluded that the extraction time is considered as a crucial parameter in addition to temperature and pressure [75], it has been observed that this parameter can vary considerably in different studies, from 10 to 240 min [18, 76]. Therefore, this parameter depends on various factors, and more studies are needed.

The flow rate must be carefully optimized, since the extraction is completely diffusion limited when the values are very high, and the very low values lead to drug extractions with limited solubility. Therefore, the optimization of this parameter is very important, since it is preferable to use the flow rate value in which both solubility and diffusion are significant. Flow rates go from 0.3 to 3 mL/min [72, 73, 77] (Tab. 16.2).

Since CO_2 is a nonpolar solvent, it is not adequate to solubilize strong polar compounds with this gas. However, the use of modifiers or cosolvents, generally of a polar nature, may also be needed together with CO_2 to better solubilize the polar compounds [78]. Table 16.2 shows that the mostly used cosolvents to improve solubility are ethanol and methanol [18, 72–74, 77]. In fact, CO_2 alone as a solvent is very seldom used. However, a study carried out by Ghasemi et al. used CO_2 alone as a solvent to extract camphor, obtaining a good yield of 77.43% [19].

16.4 New trends in the production of drugs from plants

While growing the pharmaceutically interesting plant species may be a solution, sometimes it is not possible to culture the organism outside its natural habitat and use the biochemical synthesis pathway of metabolites at an industrial level [79]. Secondary metabolites are complicated structures to synthesize in a laboratory, and most plants accumulate them in small amounts. In this sense, strategies can be applied in situ through development of improved forms of controlled use of naturally growing plants [80]. For example, a traditional approach to enhance the alkaloid content in plant species is modifying nutritional additives of the crop, that is, quinoline alkaloids can be increased by the plant culture with supplementation of mannitol and sucrose [81].

Another method to increase secondary metabolites are cell culture systems. This approach is based on culturing tissues, cells, or protoplasts on a synthetic nutrient media under sterile and controlled environmental conditions independent of climatic changes or soil conditions [82]. The major advantages of this method over the conventional cultivation of whole plants are that it can provide a continuous and reliable source of useful natural products, the cells of any tropical or alpine plant are easily grown to release their specific metabolites, cells are free of microorganisms and insects, and automated control of cell growth can reduce costs to improve productivity [83]. Using plant tissue culture technology resulted in the production of a wide variety of pharmaceuticals like phenolics, alkaloids, terpenoids, steroids, saponins, flavonoids, and amino acids. However, the cell culture systems need to optimize culture conditions and select the high metabolite producing strains and growth regulators [81]. In addition, some other limitations concern the inability to produce certain metabolites due to a lack of specialized cells [84, 85]. Moreover, reducing flux through competitive pathways, reducing catabolism, and overexpression of regulatory genes are other strategies that can be employed for enhanced secondary metabolites production through metabolic engineering [83, 86].

16.5 Conclusions

Plants, concretely medicinal plants, are producers of valuable secondary metabolites. In fact, many of these compounds can be used to develop pharmaceutical drugs due to the presence of bioactive compounds. Phenolic compounds, terpenes, and nitrogen-containing compounds are the main groups of secondary metabolites of medicinal plants, being phenolic compounds and alkaloids the most abundant ones. These metabolites are mainly extracted by conventional extraction methods such as infusion, maceration, percolation, and soxhlet. This chapter has collected and reviewed various studies using these techniques obtaining high extraction yields. However, as previously mentioned, these extraction methods contain a series of disadvantages such as high solvent volumes, long extraction times, and low selectivity. For this reason, other novel extraction techniques like UAE, MAE, PLE, and SFE (known as "green techniques") are the focus of this study in the natural products field. The chapter has also collected the current uses of these techniques as alternatives to extract drug from medicinal plants. These techniques present some advantages as they use lower solvent volumes and minor extraction time than conventional methods. Numerous studies that employed these extraction methods proved excellent recoveries. Finally, researches began with the search for strategies to improve the production of secondary metabolites from medicinal plants. Strategies in situ or cell culture systems are described as possible trends in the production of drugs from plants. However, more research is required in this area.

References

[1]   Naushad, M.; Durairajan, S.S.K.; Bera, A.K.; Senapati, S.; Li, M. Natural compounds with anti-BACE1 activity as promising therapeutic drugs for treating Alzheimer's disease. *Planta Med.* **2019**, *85*, 1316–1325, doi:10.1055/a-1019-9819.
[2]   Swamy, M.K. Plant-derived bioactives: Chemistry and Mode of Action; 2020; ISBN 9789811523618.
[3]   Brugnerotto, P.; Seraglio, S.K.T.; Schulz, M.; Gonzaga, L.V.; Fett, R.; Costa, A.C.O. Pyrrolizidine alkaloids and beehive products: A review. Food Chem. 2021, 342, doi:10.1016/j. foodchem.2020.128384.
[4]   Jan, R.; Asaf, S.; Numan, M.; Kim, K. Plant secondary metabolite biosynthesis and transcriptional regulation in response to biotic and abiotic stress conditions. Agronomy. **2021**, 1–31.
[5]   Yeshi, K.; Cragn, D.; Ritmejergte, E.; Wangchok, P. Plant secondary metabolites produced in response to abiotic stresses has potential application in pharmaceutical product development. Molecules. 2022, 27, 313.
[6]   Kohnen-Johannsen, K.L.; Kayser, O. Tropane alkaloids: Chemistry, pharmacology, biosynthesis and production. Molecules 2019, 24, 1–23, doi:10.3390/molecules24040796.
[7]   Rios, J.L.; Simeon, S.; Villar, A. Pharmacological activity of aporphinoid alkaloids. A review. Fitoterapia 1989, 60, 387–412, doi:10.63019/ajb.v1i2.467.

[8]   Muthaura, C.N.; Rukunga, G.M.; Chhabra, S.C.; Mungai, G.M.; Njagi, E.N.M. Traditional phytotherapy of some remedies used in treatment of malaria in Meru district of Kenya. South African Journal of Botany 2007, 73, 402–411, doi:10.1016/j.sajb.2007.03.004.

[9]   Griffiths, M.R.; Strobel, B.W.; Hama, J.R.; Cedergreen, N. Toxicity and risk of plant-produced alkaloids to Daphnia magna. Environ. Sci. Eur. 2021, 33, 10.

[10]   Kopp, T.; Abdel-Tawab, M.; Mizaikoff, B. Extracting and analyzing pyrrolizidine alkaloids in medicinal plants: A review. Toxins (Basel). 2020, 12, 7–10, doi:10.3390/toxins12050320.

[11]   Dittbrenner, A.; Mock, H.P.; Börner, A.; Lohwasser, U. Variability of alkaloid content in Papaver somniferum L. J. Appl. Bot. Food Qual. 2009, 82, 103–107.

[12]   Taejoon, K.; Bokyeong, S.; Cho, K.S.; Im-Soon, L. Therapeutic potential of volatile terpenes and terpenoids from forests for inflammatory diseases. Int. J. Mol. Sci. Rev. 2020.

[13]   Ji, W.; Ji, X. Comparative analysis of volatile terpenes and terpenoids in the leaves of Pinus species – A potentially abundant renewable resource. Molecules 2021, 26, doi:10.3390/molecules26175244.

[14]   García-Risco, M.R.; Mouhid, L.; Salas-Pérez, L.; López-Padilla, A.; Santoyo, S.; Jaime, L.; Ramírez de Molina, A.; Reglero, G.; Fornari, T. Biological activities of Asteraceae (*Achillea millefolium* and *Calendula officinalis*) and Lamiaceae (*Melissa officinalis* and *Origanum majorana*) plant extracts. Plant Foods Hum. Nutr. 2017, 72, 96–102, doi:10.1007/s11130-016-0596-8.

[15]   Espina, L.; Gelaw, T.K.; de Lamo-Castellví, S.; Pagán, R.; García-Gonzalo, D. Mechanism of bacterial inactivation by (+)-limonene and its potential use in food preservation combined processes. PLoS One 2013, 8, 1–10, doi:10.1371/journal.pone.0056769.

[16]   Taniguchi, S.; Hosokawa-Shinonaga, Y.; Tamaoki, D.; Yamada, S.; Akimitsu, K.; Gomi, K. Jasmonate induction of the monoterpene linalool confers resistance to rice bacterial blight and its biosynthesis is regulated by JAZ protein in rice. Plant, Cell Environ. 2014, 37, 451–461, doi:10.1111/pce.12169.

[17]   Dos Santos, L.C.; Álvarez-Rivera, G.; Sánchez-Martínez, J.D.; Johner, J.C.F.; Barrales, F.M.; de Oliveira, A.L.; Cifuentes, A.; Ibáñez, E.; Martínez, J. Comparison of different extraction methods of Brazilian "pacová" (Renealmia petasites Gagnep.) oilseeds for the determination of lipid and terpene composition, antioxidant capacity, and inhibitory effect on neurodegenerative enzymes. Food Chem. X 2021, 12, doi:10.1016/j.fochx.2021.100140.

[18]   Bermejo, D.V.; Angelov, I.; Vicente, G.; Roumiana, P.; García-risco, M.R.; Reglero, G.; Fornari, T. Extraction of thymol from different varieties of thyme plants using green solvents. J Sci Food Agric 2014, 95, 2901–2907, doi:10.1002/jsfa.7031.

[19]   Ghasemi, E.; Yamini, Y.; Bahramifar, N.; Sefidkon, F. Comparative analysis of the oil and supercritical CO2 extract of Artemisia sieberi. J. Food Eng. 2007, 79, 306–311, doi:10.1016/j.jfoodeng.2006.01.059.

[20]   Olennikov, D.N.; Chirikova, N.K.; Kashchenko, N.S.; Vikolaev, V.M.; Kim, S-W.; Vennos, C. Bioactive phenolics of the genus Artemisia (Asteraceae): HPLC-DAD-ESI-TQ-MS/MS profile of the Siberian species and their inhibitory potential against α -amylase and α- glucosidase. Frontiers in Pharmacology **2018**, *9*, doi:10.3389/fphar.2018.00756.

[21]   Das, A.; Pandita`, D.; Kumar, G.; Agarwal, P.; Singh, A.; Khar, R.K.; Lather, V. Chemico-biological interactions role of phytoconstituents in the management of COVID-19. Chem. Biol. Interact. 2021, 341, 109449, doi:10.1016/j.cbi.2021.109449.

[22]   Suručić, R.; Tubić, B.; Stojiljković, M.P.; Djuric, D.M.; Travar, M.; Grabež, M.; Savikin, K.; Skrbic, R. Computational study of pomegranate peel extract polyphenols as potential inhibitors of SARS-CoV-2 virus internalization. Molecular and cellular Biochemistry 2020.

[23]   Giordo, R.; Zinellu, A.; Eid, A.H. Therapeutic potential of resveratrol in COVID-19-associated hemostatic disorders. Molecules. 2021, 26, 856.

[24]   Alara, O.R.; Abdurahman, N.H.; Ukaegbu, C.I. Extraction of phenolic compounds: A review. Curr. Res. Food Sci. 2021, 4, 200–214, doi:10.1016/j.crfs.2021.03.011.

[25] Marsoul, A.; Ijjaali, M.; Oumous, I.; Bennani, B.; Boukir, A. Determination of polyphenol contents in Papaver rhoeas L. Flowers extracts (Soxhlet, maceration), antioxidant and antibacterial evaluation. In Proceedings of the Materials Today: Proceedings; Elsevier Ltd, 2020; Vol. 31, pp.S183–S189.

[26] Alu'datt, M.H.; Alli, I.; Ereifej, K.; Alhamad, M.N.; Alsaad, A.; Rababeh, T. Optimisation and characterisation of various extraction conditions of phenolic compounds and antioxidant activity in olive seeds. Nat. Prod. Res. 2011, 25, 876–889, doi:10.1080/14786419.2010.489048.

[27] Horta, A.; Pinteus, S.; Alves, C.; Fino, N.; Silva, J.; Fernandez, S.; Rodrigues, A.; Pedrosa, R. Antioxidant and antimicrobial potential of the Bifurcaria bifurcata epiphytic bacteria. Mar. Drugs 2014, 12, 1676–1689, doi:10.3390/md12031676.

[28] Dahmoune, F.; Nayak, B.; Moussi, K.; Remini, H.; Madani, K. Optimization of microwave-assisted extraction of polyphenols from Myrtus communis L. leaves. Food Chem. 2015, 166, 585–595, doi:10.1016/j.foodchem.2014.06.066.

[29] Yalavarthi, C.; Thiruvengadarajan, U.S. A review on identification strategy of phyto constituents present in herbal plants. Int. J. Res. Pharm. Sci. 2013, 4, 2, 123–140.

[30] Handa, S.S. An Overview of Extraction Techniques for Medicinal and Aromatic Plants. In Extraction Technologies for Medicinal and Aromatic Plants; Harda, S.S.; Khanuja, S.P.S.; Longo, G.; Rakesh, D.D., Eds.; 2008.

[31] Thakur, R.; Jain, N.; Pathak, R.; Sandhu, S.S. Practices in wound healing studies of plants. Evidence-based complement. Altern. Med. 2011, 2011.

[32] Zandoná, G.P.; Bagatini, L.; Woloszyn, N.; de Souza Cardoso, J.; Hoffmann, J.F.; Moroni, L.S.; Stefanello, F.M.; Junges, A.; Rombaldi, C.V. Extraction and characterization of phytochemical compounds from Araçazeiro (Psidium cattleianum) leaf: Putative antioxidant and antimicrobial properties. Food Res. Int. 2020, 137, 109573, doi:10.1016/j.foodres.2020.109573.

[33] Ergen, N.; Hoşbaş, S.; Delıorman Orhan, D.; Aslan, M.; Sezık, E.; Atalay, A. Evaluation of the lifespan extension effects of several Turkish medicinal plants in Caenorhabditis elegans. Turkish J. Biol. 2018, 42, 163–173, doi:10.3906/biy-1711-5.

[34] Soto-Maldonado, C.; Fernández-Araya, B.; Saavedra-Sánchez, V.; Santis-Bernal, J.; Alcaíno-Fuentes, L.; Arancibia-Díaz, A.; Zúñiga-Hansen, M.E. Antioxidant and antimicrobial capacity of Maytenus boaria leaves, recovery by infusion and solvent extraction. Electron. J. Biotechnol. 2022, 56, 47–53, doi:10.1016/j.ejbt.2022.02.002.

[35] Catania, C.; Avagnina, S. 21. La maceración. Curso superior de Degurtación de vinos. 2007, 1, 1–12.

[36] Sik, B.; Hanczné, E.L.; Kapcsándi, V.; Ajtony, Z. Conventional and nonconventional extraction techniques for optimal extraction processes of rosmarinic acid from six Lamiaceae plants as determined by HPLC-DAD measurement. J. Pharm. Biomed. Anal. 2020, 184, doi:10.1016/j.jpba.2020.113173.

[37] Popuri, Ashok K. & Papala, B. Extraction of curcumin from turmeric roots. Int. J. Innov. Res. Stud. 2013, 2, 290–299.

[38] Benítez-Benítez, R.; Sarria-Villa, R.A.; Gallo-Corredor, J.A.; Pérez Pacheco, N.O.; Álvarez Sandoval, J.H.; Giraldo Aristizabal, C.I. Obtención y rendimiento del extracto etanólico de dos plantas medicinales. Rev. Fac. Ciencias Básicas 2020, 15, 31–40, doi:10.18359/rfcb.3597.

[39] Singh, J. Maceration, Peridation and Infusion Techniques for the Extraction of Medicinal Plants. In Extraction Technologies for Medicinal and Aromatic Plants; Harda, S.S.; Khanuja, S.P.S.; Longo, G.; Rakesh, D.D., Eds.; 2008.

[40] Daud, M.N.H.; Fatanah, D.N.; Abdullah, N.; Ahmad, R. Evaluation of antioxidant potential of Artocarpus heterophyllus L. J33 variety fruit waste from different extraction methods and identification of phenolic constituents by LCMS. Food Chem. 2017, 232, 621–632, doi:10.1016/j.foodchem.2017.04.018.

[41] Mroczek, T.; Widelski, J.; Gowniak, K. Optimization of extraction of pyrrolizidine alkaloids from plant material. Chem. Anal. 2006, 51, 567–580.

[42] Ashraf, R.; Ghufran, S.; Akram, S.; Mushtaq, M.; Sultana, B. Cold Pressed Coriander (Coriandrum Sativum L.) Seed Oil; Elsevier Inc., 2020; ISBN 9780128181881.

[43] Garcia-Vaquero, M.; Rajauria, G.; Tiwari, B. Conventional extraction techniques : Solvent extraction. In Sustainable Seaweed Technologies; Torres, M.D., Kraan, S., Dominguez, H., Eds.; Elsevier Inc., 2020; pp. 171–189, ISBN 9780128179437.

[44] Feng, R.; Wang, Q.; Tong, W. et al. Extraction and antioxidant activity of flavonoids of Morosnig. International Journal of Clinical and Experimental Medicine. 2015, 8(12), 22328–22336. ISSN: 19405901.

[45] Kandanur, S.G.S.; Tamang, N.; Golakoti, N.R.; Nanduri, S. Andrographolide: A natural product template for the generation of structurally and biologically diverse diterpenes. Eur. J. Med. Chem. 2019, 176, 513–533.

[46] López-Bascón-Bascon, M.A.; Luque de Castro, M.D. Soxhlet extraction. Liq. Extr. 2019, 327–354, doi:10.1016/B978-0-12-816911-7.00011-6.

[47] Lee, C.H.; Lee, T.H.; Ong, P.Y.; Wong, S.L.; Hamdan, N.; Elgharbawy, A.A.M.; Azmi, N.A. Integrated ultrasound-mechanical stirrer technique for extraction of total alkaloid content from Annona muricata. Process Biochem. 2021, 109, 104–116, doi:10.1016/j.procbio.2021.07.006.

[48] Carreira-Casais, A.; Otero, P.; Garcia-Perez, P.; Garcia-Oliveira, P.; Pereira, A.G.; Carpena, M.; Soria-Lopez, A.; Simal-Gandara, J.; Prieto, M.A. Benefits and drawbacks of ultrasound-assisted extraction for the recovery of bioactive compounds from marine algae. Int. J. Environ. Res. Public Health 2021, 18, doi:10.3390/ijerph18179153.

[49] Tiwari, B.K. Ultrasound: A clean, green extraction technology. TrAC – Trends Anal. Chem. 2015, 71, 100–109, doi:10.1016/j.trac.2015.04.013.

[50] Chemat, F.; Rombaut, N.; Sicaire, A.G.; Meullemiestre, A.; Fabiano-Tixier, A.S.; Abert-Vian, M. Ultrasound assisted extraction of food and natural products. Mechanisms, techniques, combinations, protocols and applications. A review. Ultrason. Sonochem. 2017, 34, 540–560, doi:10.1016/j.ultsonch.2016.06.035.

[51] Sandra Kentish and Muthupandian Ashok Kumar. The physical and chemical effects of ultrasound. In Ultrasound Technologies for Food Bioprocessing; Feng, Hao; Barbosa-Cánovas, Gustavo; Weiss, J., Ed.; New York, 2011; pp. 1–12 ISBN 9788578110796.

[52] Wakte, P.S.; Sachin, B.S.; Patil, A.A.; Mohato, D.M.; Band, T.H.; Shinde, D.B. Optimization of microwave, ultra-sonic and supercritical carbon dioxide assisted extraction techniques for curcumin from Curcuma longa. Sep. Purif. Technol. 2011, 79, 50–55, doi:10.1016/j.seppur.2011.03.010.

[53] Ganorkar, P. V.; Jadeja, G.C.; Desai, M.A. Extraction of shikimic acid from water hyacinth (Eichhornia crassipes) using sonication: An approach towards waste valorization. J. Environ. Manage. 2022, 305, 114419, doi:10.1016/j.jenvman.2021.114419.

[54] Moreira, G.C.; de Souza Dias, F. Mixture design and Doehlert matrix for optimization of the ultrasonic assisted extraction of caffeic acid, rutin, catechin and trans-cinnamic acid in Physalis angulata L. and determination by HPLC DAD. Microchem. J. 2018, 141, 247–252, doi:10.1016/j.microc.2018.04.035.

[55] Duangjai, A.; Saokaew, S.; Goh, B.H.; Phisalprapa, P. Shifting of physicochemical and biological characteristics of coffee roasting under ultrasound-assisted extraction. Front. Nutr. 2021, 8, 1–8, doi:10.3389/fnut.2021.724591.

[56] Zhang, L.; Liu, J.; Zhang, P.; Yan, S.; He, X.; Chen, F. Ionic liquid-based ultrasound-assisted extraction of chlorogenic acid from Lonicera japonica Thunb. Chromatographia 2011, 73, 129–133, doi:10.1007/s10337-010-1828-y.

[57] Sparr Eskilsson, C.B. Analytical-scale microwave-assisted extraction. J. Chromatogr. A 2000, 902, 227–250, doi:10.1109/SMC.2016.7844685.

[58] Christen, P.; Kaufmann, B. Recent extraction techniques for natural products : microwave-assisted extraction and pressurised solvent extraction. Phytochem. Anal. 2002, 113, 105–113.

[59] Delazar, A.; Nahar, L.; Hamedeyazdan, S.; Sarker, S.D. Microwave-assisted extraction in natural products isolation. Methods Mol. Biol. 2012, 864, 89–115, doi:10.1007/978-1-61779-624-1_5.

[60] Fierascu, R.C.; Fierascu, I.; Ortan, A.; Georgiev, M.I.; Sieniawska, E. Innovative approaches for recovery of phytoconstituents from medicinal/aromatic plants and biotechnological production. Molecules 2020, 25, doi:10.3390/molecules25020309.

[61] Pavlović, N.; Jokić, S.; Jakovljević, M.; Blažić, M.; Molnar, M. Green extraction methods for active compounds from food waste – Cocoa bean shell. Foods 2020, 9, 1–14, doi:10.3390/foods9020140.

[62] Chen, L.; Jin, H.; Ding, L.; Zhang, H.; Wang, X.; Wang, Z.; Li, J.; Qu, C.; Wang, Y.; Zhang, H. On-line coupling of dynamic microwave-assisted extraction with high-performance liquid chromatography for determination of andrographolide and dehydroandrographolide in Andrographis paniculata Nees. J. Chromatogr. A 2007, 1140, 71–77, doi:10.1016/j.chroma.2006.11.070.

[63] Belwal, T.; Pandey, A.; Bhatt, I.D.; Rawal, R.S. Optimized microwave assisted extraction (MAE) of alkaloids and polyphenols from Berberis roots using multiple-component analysis. Sci. Rep. 2020, 10, 1–10, doi:10.1038/s41598-020-57585-8.

[64] Lu, Y.; Ma, W.; Hu, R.; Dai, X.; Pan, Y. Ionic liquid-based microwave-assisted extraction of phenolic alkaloids from the medicinal plant Nelumbo nucifera Gaertn. J. Chromatogr. A 2008, 1208, 42–46, doi:10.1016/j.chroma.2008.08.070.

[65] Zhou, T.; Xiao, X.; Li, G.; Cai, Z. W. Study of polyethylene glycol as a green solvent in the microwave-assisted extraction of flavone and coumarin compounds from medicinal plants. J. Chromatogr. A 2011, 1218, 3608–3615, doi:10.1016/j.chroma.2011.04.031.

[66] Barros, F.; Dykes, L.; Awika, J.M.; Rooney, L.W. Accelerated solvent extraction of phenolic compounds from sorghum brans. J. Cereal Sci. 2013, 58, 305–312, doi:10.1016/j.jcs.2013.05.011.

[67] Alvarez-Rivera, G.; Bueno, M.; Ballesteros-Vivas, D.; Mendiola, J.A.; Ibañez, E. Pressurized liquid extraction. In Liquid-Phase Extraction; 2020; pp. 375–398 ISBN 9780128169117.

[68] Nataša Nastic, Isabel Borrás-Linares, Jesús Lozano-Sánchez, J.Š.-G. and A.S.-C. Comparative assessment of phytochemical profiles of Comfrey (Symphytum officinale L.) Root Extracts. Molecules 2020, 25, 1–17.

[69] Miron, T.L.; Herrero, M.; Ibáñez, E. Enrichment of antioxidant compounds from lemon balm (Melissa officinalis) by pressurized liquid extraction and enzyme-assisted extraction. J. Chromatogr. A 2013, 1288, 1–9, doi:10.1016/j.chroma.2013.02.075.

[70] Tony, C. Fundamentals of Supercritical Fluids; Oxford Uni.; 1999; ISBN 978-0198501374.

[71] Sapakale, G. N.; Patil, S. M.; and Bhatbhage, P.K. Supercritical fluid extraction. Int. J. Chem. Sci. 2010, 8, 729–743, doi:10.1016/B978-0-12-384947-2.00675-9.

[72] Salehi, H.; Karimi, M.; Rezaie, N.; Raofie, F. Extraction of β-carboline alkaloids and preparation of extract nanoparticles from Peganum harmala L. capsules using supercritical fluid technique. J. Drug Deliv. Sci. Technol. 2020, 56, doi:10.1016/j.jddst.2020.101515.

[73] Carrara, V.S.; Filho, L.C.; Garcia, V.A.S.; Faiões, V.S.; Cunha-Júnior, E.F.; Torres-Santos, E.C.; Cortez, D.A.G. Supercritical fluid extraction of pyrrolidine alkaloid from leaves of Piper amalago L. Evidence-based Complement. Altern. Med. 2017, 2017, doi:10.1155/2017/7401748.

[74] Falcão, M.A.; Scopel, R.; Almeida, R.N.; Do Espirito Santo, A.T.; Franceschini, G.; Garcez, J.J.; Vargas, R.M.F.; Cassel, E. Supercritical fluid extraction of vinblastine from Catharanthus roseus. J. Supercrit. Fluids **2017**, 129, 9–15, doi:10.1016/j.supflu.2017.03.018.

[75] Norsyamimi, Hassim; Masturah, Markom; Masli Irwan, Rosli; Shuhaida, H. Effect of static extraction time on supercritical fluid extraction of bioactive compounds from Phyllanthus niruri. J. Comput. Theor. Nanosci. 2020, 17, 918–924, doi:https://doi.org/10.1166/jctn.2020.8742.

[76] Gede, L.; Yuan, L.C.; At, Y.; Daim, H. Effect of static extraction time on extraction efficiencies using on-line supercritical fluid extraction – high performance liquid chromatograph y for lipoquinone analysis in activated sludge. In Proceedings of the 5th Annual International Conference Syiah Kuala University; 2015; pp. 130–136.

[77] Bayrak, S.; Sökmen, M.; Aytaç, E.; Sökmen, A. Conventional and supercritical fluid extraction (SFE) of colchicine from Colchicum speciosum. Ind. Crops Prod. 2019, 128, 80–84, doi:10.1016/j.indcrop.2018.10.060.

[78] Brondz, I.; Sedunov, B.; Sivaraman, N. Influence of modifiers on supercritical fluid chromatography (SFC) and supercritical fluid extraction (SFE), Part I. Int. J. Anal. Mass Spectrom. Chromatogr. 2017, 05, 17–39, doi:10.4236/ijamsc.2017.52002.

[79] Hussain, M.S.; Fareed, S.; Ansari, S.; Rahman, M.A.; Ahmad, I.Z.; Saeed, M. Current approaches toward production of secondary plant metabolites. J. Pharm. Bioallied Sci. 2012, 4, 10–20, doi:10.4103/0975-7406.92725.

[80] Wiersum, K.F.; Dold, A.P.; Husselman, M.; Cocks, M. Cultivation of medicinal plants as a tool for biodiversity conservation and poverty alleviation in the Amatola Region, South Africa. Med. Aromat. Plants 2007, 43–57, doi:10.1007/1-4020-5449-1_3.

[81] Ahmad, S.; Garg, M.; Tamboli, E.; Abdin, M.; Ansari, S. In vitro production of alkaloids: Factors, approaches, challenges and prospects. Pharmacogn. Rev. 2013, 7, 27–33.

[82] Bhatia, S. History and Scope of Plant Biotechnology. Mod. Appl. Plant Biotechnol. Pharm. Sci. 2015, 1–30, doi:10.1016/B978-0-12-802221-4.00001-7.

[83] Vanisree, M.; Lee, C.Y.; Lo, S.F.; Nalawade, S.M.; Lin, C.Y.; Tsay, H.S. Studies on the production of some important secondary metabolites from medicinal plants by plant tissue cultures. Bot. Bull. Acad. Sin. 2004, 45, 1–22.

[84] St-Pierre, B.; Vazquez-Flota, F.A.; De Luca, V. Multicellular compartmentation of Catharanthus roseus alkaloid biosynthesis predicts intercellular translocation of a pathway intermediate. Plant Cell 1999, 11, 887–900, doi:10.1105/tpc.11.5.887.

[85] Ratnadewi, D. Alkaloids in plant cell cultures. In Alkaloids – Alternatives in Synthesis, Modification and Application; IntechOpen, 2017 ISBN 978-953-51-3392-6.

[86] Gaosheng, H.; Jingming, J. Production of Useful Secondary Metabolites Through Regulation of Biosynthetic Pathway in Cell and Tissue Suspension Culture of Medicinal Plants. Recent Adv. Plant Vitr. Cult. 2012, doi:10.5772/53038.

[87] Rafińska, K.; Pomastowski, P.; Rudnicka, J.; Krakowska, A.; Maruśka, A.; Narkute, M.; Buszewski, B. Effect of solvent and extraction technique on composition and biological activity of Lepidium sativum extracts. Food Chem. 2019, 289, 16–25, doi:10.1016/j.foodchem.2019.03.025.

[88] Jovanović, A.A.; Đorđević, V.B.; Zdunić, G.M.; Pljevljakušić, D.S.; Šavikin, K.P.; Gođevac, D.M.; Bugarski, B.M. Optimization of the extraction process of polyphenols from Thymus serpyllum L. herb using maceration, heat- and ultrasound-assisted techniques. Sep. Purif. Technol. **2017**, *179*, 369–380, doi:10.1016/j.seppur.2017.01.055.

[89] Vongsak, B.; Sithisarn, P.; Mangmool, S.; Thongpraditchote, S.; Wongkrajang, Y.; Gritsanapan, W. Maximizing total phenolics, total flavonoids contents and antioxidant activity of Moringa oleifera leaf extract by the appropriate extraction method. Ind. Crops Prod. 2013, 44, 566–571, doi:10.1016/j.indcrop.2012.09.021.

[90] Trifonova, D.; Stoilova, I.; Marchev, A.; Denev, P.; Angelova, G.; Lante, A.; Krastanov, A. Phytochemical constituents of pressurized liquid extract from Ziziphus jujuba mill. (Rhamnaceae) fruits and in vitro inhibitory activity on α-glucosidase, pancreatic α-amylase and lipase. Bulg. J. Agric. Sci. 2021, 27, 391–402.

Index

https://doi.org/10.1515/9783110791891-017